# Communications
# in Computer and Information Science　　1357

More information about this series at http://www.springer.com/series/7899

Wil M. P. van der Aalst ·
Vladimir Batagelj · Alexey Buzmakov ·
Dmitry I. Ignatov · Anna Kalenkova ·
Michael Khachay et al. (Eds.)

# Recent Trends in Analysis of Images, Social Networks and Texts

9th International Conference, AIST 2020
Skolkovo, Moscow, Russia, October 15–16, 2020
Revised Supplementary Proceedings

 Springer

For the full list of editors *see next page*

ISSN 1865-0929          ISSN 1865-0937 (electronic)
Communications in Computer and Information Science
ISBN 978-3-030-71213-6          ISBN 978-3-030-71214-3 (eBook)
https://doi.org/10.1007/978-3-030-71214-3

This Springer imprint is published by the registered company Springer Nature Switzerland AG
The registered company address is: Gewerbestrasse 11, 6330 Cham, Switzerland

*Editors*

Wil M. P. van der Aalst (ID)
RWTH Aachen University
Aachen, Germany

Alexey Buzmakov (ID)
National Research University Higher
School of Economics
Perm, Russia

Anna Kalenkova (ID)
University of Melbourne
Melbourne, VIC, Australia

Olessia Koltsova (ID)
National Research University Higher
School of Economics
Saint-Petersburg, Russia

Sergei O. Kuznetsov (ID)
National Research University Higher
School of Economics
Moscow, Russia

Natalia Loukachevitch (ID)
Lomonosov Moscow State University
Moscow, Russia

Amedeo Napoli (ID)
LORIA
Vandœuvre-lès-Nancy, France

Panos M. Pardalos (ID)
University of Florida
Gainesville, FL, USA

Andrey V. Savchenko (ID)
National Research University Higher
School of Economics
Nizhny Novgorod, Russia

Vladimir Batagelj (ID)
University of Ljubljana
Ljubljana, Slovenia

Dmitry I. Ignatov (ID)
National Research University Higher
School of Economics
Moscow, Russia

Michael Khachay (ID)
Krasovskii Institute of Mathematics
and Mechanics of RAS
Ekaterinburg, Russia

Andrey Kutuzov (ID)
University of Oslo
Oslo, Norway

Irina A. Lomazova (ID)
National Research University Higher
School of Economics
Moscow, Russia

Ilya Makarov (ID)
National Research University Higher
School of Economics
Moscow, Russia

Alexander Panchenko (ID)
Skolkovo Institute of Science
and Technology
Moscow, Russia

Marcello Pelillo (ID)
Università Ca' Foscari Venezia
Venezia, Italy

Elena Tutubalina (ID)
Kazan Federal University
Kazan, Russia

# Preface

This volume contains the refereed proceedings of the 9th International Conference on Analysis of Images, Social Networks, and Texts (AIST 2020)[1]. The previous conferences (during 2012–2019) attracted a significant number of data scientists – students, researchers, academics, and engineers – working on interdisciplinary data analysis of images, texts, and social networks.

The broad scope of AIST make it an event where researchers from different domains, such as image and text processing, exploiting various data analysis techniques, can meet and exchange ideas. As the test of time has shown, this leads to the cross-fertilisation of ideas between researchers relying on modern data analysis machinery.

Therefore, AIST 2020 brought together all kinds of applications of data mining and machine learning techniques. The conference allowed specialists from different fields to meet each other, present their work, and discuss both theoretical and practical aspects of their data analysis problems. Another important aim of the conference was to stimulate scientists and people from industry to benefit from knowledge exchange and identify possible grounds for fruitful collaboration.

The conference was held during October 15–16, 2020. The conference was organised in the Skolkovo Innovation Center, Russia, on the campus of the Skolkovo Institute of Science and Technology[2], but held entirely online due to the COVID-19 pandemic.

This year, the key topics of AIST were grouped into six tracks:

1. Data Analysis and Machine Learning chaired by Sergei O. Kuznetsov (HSE University, Russia) and Amedeo Napoli (Loria, France)
2. Natural Language Processing chaired by Natalia Loukachevitch (Lomonosov Moscow State University, Russia), Andrey Kutuzov (University of Oslo, Norway)
3. Social Network Analysis chaired by Vladimir Batagelj (University of Ljubljana, Slovenia) and Olessia Koltsova (HSE University, Russia)
4. Computer Vision chaired by Marcello Pelillo (University of Venice, Italy) and Andrey V. Savchenko (HSE University, Russia)
5. Theoretical Machine Learning and Optimization chaired by Panos M. Pardalos (University of Florida, USA) and Michael Khachay (IMM UB RAS and Ural Federal University, Russia)
6. Process Mining chaired by Wil M. P. van der Aalst (RWTH Aachen University, Germany) and Irina A. Lomazova (HSE University, Russia)

---

[1]  https://aistconf.org.

[2]  https://www.skoltech.ru/en/.

To facilitate easy communication and negotiation of the area chairs and our authors via only digital channels, due to the virtual character of the event, we invited additional area and program co-chairs for certain tracks, respectively: Alexei Buzmakov (HSE University, Perm, Russia), Ilya Makarov (HSE University, Moscow, Russia), Anna Kalenkova (University of Melbourne, Australia), and Elena Tutubalina (Kazan Federal University, Russia).

The Programme Committee and the reviewers of the conference included 134 well-known experts in data mining and machine learning, natural language processing, image processing, social network analysis, and related areas from leading institutions of many countries including Australia, Austria, Czech Republic, France, Germany, Greece, India, Iran, Ireland, Italy, Japan, Lithuania, Norway, Qatar, Romania, Russia, Slovenia, Spain, Taiwan, Ukraine, United Kingdom, and the USA. This year, we received 115 submissions: mostly from Russia but also from Algeria, Brazil, Finland, Germany, India, Norway, Pakistan, Serbia, Spain, Ukraine, United Kingdom, and the USA.

Out of 115 submissions (not taking into account seven automatically rejected papers), only 27 full papers and four short papers were accepted as regular oral papers in the main volume. Invited talks were also included in the main volume. In order to encourage young practitioners and researchers, we included 14 full and nine short papers in this companion volume after their short presentation at the conference and four non-indexed poster abstracts as well. Thus, the acceptance rate of this volume was around 30%. Each submission was reviewed by at least three reviewers, experts in their fields, in order to supply detailed and helpful comments.

The conference featured several invited talks dedicated to current trends and challenges in the respective areas.

The invited talks from academia were on Computer Vision and Natural Language Processing, respectively:

- Marcello Pelillo (Ca' Foscari University of Venice, Italy): "Graph-Theoretic Methods in Computer Vision: Recent Advances"
- Miguel Couceiro (LORIA, Université de Lorraine, France): "Making Models Fairer Through Explanations"
- Leonard Kwuida (Bern University of Applied Sciences): "On Interpretability and Similarity in Concept Based Machine Learning"
- Santo Fortunato (Indiana University Network Science Institute, USA): "Consensus Clustering in Networks"

The invited industry speakers gave the following talks:

- Nikita Semenov (MTS AI, Russia): "Text and Speech Processing Projects at MTS AI"
- Ivan Smurov (ABBYY; Moscow Institute of Physics and Technology, Russia): "When CoNLL-2003 is not Enough: are Academic NER and RE Corpora Well-Suited to Represent Real-World Scenarios?"

An extended part of Ivan's industry talk was included in the main volume under the title "RuREBus: a Case Study of Joint Named Entity Recognition and Relation Extraction from e-Government Domain."

We would like to thank the authors for submitting their papers and the members of the Programme Committee for their efforts in providing exhaustive reviews.

According to the programme chairs, and taking into account the reviews and presentation quality, the Best Paper Awards were granted to the following papers:

- Track 1. Data Analysis and Machine Learning: "Gradient-Based Adversarial Attacks on Categorical Sequence Models via Traversing an Embedded World" by Ivan Fursov, Alexey Zaytsev, Nikita Klutchnikov, Andrey Kravchenko, and Evgeny Burnaev;
- Track 2. Natural Language Processing: "Do Topics Make a Metaphor? Topic Modeling for Metaphor Identification and Analysis in Russian" by Yulia Badryzlova, Anastasia Nikiforova, and Olga Lyashevskaya;
- Track 3. Social Network Analysis: "Detecting Automatically Managed Accounts in Online Social Networks: Graph Embedding Approach" by Ilia Karpov and Ekaterina Glazkova;
- Track 4. Computer Vision: "Deep Learning on Point Clouds for False Positive Reduction at Nodule Detection in Chest CT Scans" by Ivan Drokin and Elena Ericheva;
- Track 5. Theoretical Machine Learning and Optimization: "Fast Approximation Algorithms for Stabbing Special Families of Line Segments with Equal Disks" by Konstantin Kobylkin;
- Track 6. Process Mining: "Checking Conformance between Colored Petri Nets and Event Logs" by Julio Cesar Carrasquel, Khalil Mecheraoui, and Irina Lomazova.

We would also like to express our special gratitude to all the invited speakers and industry representatives.

We deeply thank all the partners and sponsors, especially the hosting university and our main sponsor and the co-organiser this year, the Skolkovo Institute of Science and Technology, as well as the National Research University Higher School of Economics (including its subdivisions). Our special thanks go to Springer for their help, starting from the first conference call to the final version of the proceedings. Last but not least, we are grateful to Evgeny Burnaev and all the organisers, especially our secretary Irina Nikishina, and the volunteers, whose endless energy saved us at the most critical stages of the conference preparation.

Here, we would like to mention that the Russian word "aist" is more than just a simple abbreviation (in Cyrillic) – it means "a stork". Since it is a wonderful free bird, a symbol of happiness and peace, this stork gave us the inspiration to organise the AIST conference series. So we believe that this conference will still likewise bring inspiration to data scientists around the world!

October 2020

Wil M. P. van der Aalst
Vladimir Batagelj
Aleksey Buzmakov
Dmitry Ignatov
Anna Kalenkova
Michael Khachay
Olessia Koltsova
Andrey Kutuzov
Sergei O. Kuznetsov
Irina Lomazova
Natalia Loukachevitch
Ilya Makarov
Amedeo Napoli
Alexander Panchenko
Panos M. Pardalos
Marcello Pelillo
Andrey Savchenko
Elena Tutubalina

# Organisation

The conference was organised by a joint team from Skolkovo Institute of Science and Technology (Skoltech), the divisions of the National Research University Higher School of Economics (HSE University), and Krasovskii Institute of Mathematics and Mechanics of the Russian Academy of Sciences.

## Organising Institutions

- Skolkovo Institute of Science and Technology (Moscow, Russia)
- Krasovskii Institute of Mathematics and Mechanics, Ural Branch of the Russian Academy of Sciences (Yekaterinburg, Russia)
- School of Data Analysis and Artificial Intelligence, HSE University (Moscow, Russia)
- Laboratory of Algorithms and Technologies for Networks Analysis, HSE University (Nizhny Novgorod, Russia)
- International Laboratory for Applied Network Research, HSE University (Moscow, Russia)
- Laboratory for Models and Methods of Computational Pragmatics, HSE University (Moscow, Russia)
- Laboratory for Social and Cognitive Informatics, HSE University (St. Petersburg, Russia)
- Laboratory of Process-Aware Information Systems, HSE University (Moscow, Russia)
- Research Group "Machine Learning on Graphs", HSE University (Moscow, Russia)

## Program Committee Chairs

| | |
|---|---|
| Wil van der Aalst | RWTH Aachen University, Germany |
| Vladimir Batagelj | University of Ljubljana, Slovenia |
| Michael Khachay | Krasovskii Institute of Mathematics and Mechanics of Russian Academy of Sciences, Russia & Ural Federal University, Yekaterinburg, Russia |
| Olessia Koltsova | HSE University, Russia |
| Andrey Kutuzov | University of Oslo, Norway |
| Sergei Kuznetsov | HSE University, Moscow, Russia |
| Amedeo Napoli | LORIA – CNRS, University of Lorraine, and Inria, Nancy, France |
| Irina Lomazova | HSE University, Moscow, Russia |
| Natalia Loukachevitch | Computing Centre of Lomonosov Moscow State University, Russia |

| Panos Pardalos | University of Florida, US |
| Marcello Pelillo | University of Venice, Italy |
| Andrey Savchenko | HSE University, Nizhny Novgorod, Russia |
| Elena Tutubalina | Kazan Federal University, Russia |

## Additional Area Chairs

| Aleksey Buzmakov | HSE University, Perm, Russia |
| Anna Kalenkova | University of Melbourne, Australia |
| Ilya Makarov | HSE University, Moscow, Russia |

## Proceedings Chair

| Dmitry I. Ignatov | HSE University, Moscow, Russia |

## Steering Committee

| Dmitry I. Ignatov | HSE University, Moscow, Russia |
| Michael Khachay | Krasovskii Institute of Mathematics and Mechanics of Russian Academy of Sciences, Russia & Ural Federal University, Yekaterinburg, Russia |
| Alexander Panchenko | Skolkovo Institute of Science and Technology, Moscow, Russia |
| Andrey Savchenko | HSE University, Nizhny Novgorod, Russia |
| Rostislav Yavorskiy | Tomsk Polytechnic University, Russia |

## Program Committee

| Anton Alekseev | St. Petersburg Department of V.A.Steklov Institute of Mathematics of the Russian Academy of Sciences, Russia |
| Ilseyar Alimova | Kazan Federal University, Russia |
| Vladimir Arlazarov | Smart Engines Ltd. and Federal Research Centre "Computer Science and Control" of the Russian Academy of Sciences, Russia |
| Aleksey Artamonov | Neuromation, Russia |
| Ekaterina Artemova | HSE University, Moscow, Russia |
| Jaume Baixeries | Universitat Politècnica de Catalunya, Spain |
| Amir Bakarov | HSE University, Moscow, Russia |
| Nikita Basov | St. Petersburg State University, Russia |
| Vladimir Batagelj | University of Ljubljana, Slovenia |
| Tatiana Batura | Ershov Institute of Informatics Systems, Siberian Branch of the Russian Academy of Sciences and Novosibirsk State University, Russia |
| Malay Bhattacharyya | Indian Statistical Institute, India |
| Michael Bogatyrev | Tula State University, Russia |

| | |
|---|---|
| Elena Bolshakova | Moscow State Lomonosov University, Russia |
| Ivan Bondarenko | Novosibirsk State University, Russia |
| Evgeny Burnaev | Skolkovo Institute of Science and Technology, Russia |
| Aleksey Buzmakov | HSE University, Perm, Russia |
| Dmitry Chaly | Demidov Yaroslavl State University, Russia |
| Mikhail Chernoskutov | Krasovskii Institute of Mathematics and Mechanics of Russian Academy of Sciences and Ural Federal University, Russia |
| Alexey Chernyavskiy | Philips Innovation Labs, Russia |
| Massimiliano de Leoni | University of Padua, Italy |
| Oksana Dereza | National University of Ireland, Ireland |
| Boris Dobrov | Moscow State University, Russia |
| Ivan Drokin | BrainGarden.ai, Russia |
| Aleksandr Drozd | Tokyo Institure of Technology, Japan |
| Shiv Ram Dubey | Indian Institute of Information Technology, Sri City, India |
| Olga Gerasimova | HSE University, Moscow, Russia |
| Dmitry Granovsky | Yandex, Russia |
| Vera Ignatenko | HSE University, St. Petersburg, Russia |
| Dmitry Ignatov | HSE University, Moscow, Russia |
| Dmitry Ilvovsky | HSE University, Moscow, Russia |
| Max Ionov | Goethe University Frankfurt, Germany and Moscow State University, Russia |
| Vladimir Ivanov | Innopolis University, Kazan, Russia |
| Anna Kalenkova | University of Melbourne, Australia |
| Ilia Karpov | HSE University, Moscow, Russia |
| Egor Kashkin | Vinogradov Russian Language Institute of the Russian Academy of Sciences, Russia |
| Yury Kashnitsky | Elsevier, The Netherlands |
| Alexander Kazakov | Matrosov Institute for System Dynamics and Control Theory, Siberian Branch of the Russian Academy of Science, Russia |
| Michael Khachay | Krasovskii Institute of Mathematics and Mechanics of Russian Academy of Sciences, Russia |
| Vladimir Khandeev | Sobolev Institute of Mathematics, Siberian Branch of the Russian Academy of Sciences, Russia |
| Javad Khodadoust | Payame Noor University, Iran |
| Gregory Khvatsky | HSE University, Moscow, Russia |
| Donghyun Kim | Georgia State University, USA |
| Denis Kirjanov | HSE University, Moscow, Russia |
| Dmitrii Kiselev | HSE University, Moscow, Russia |
| Sergei Koltcov | HSE University, St. Petersburg, Russia |
| Olessia Koltsova | HSE University, St. Petersburg, Russia |
| Evgeny Komotskiy | Ural Federal University, Russia |
| Jan Konečný | Palacký University Olomouc, Czech Republic |
| Anton Konushin | Moscow State University and Samsung, Russia |

Andrey Kopylov — Tula State University, Russia
Evgeny Kotelnikov — Vyatka State University, Russia
Ekaterina Krekhovets — HSE University, Nizhny Novgorod, Russia
Tomas Krilavičius — Vytautas Magnus University, Lithuania
Sofya Kulikova — HSE University, Perm, Russia
Maria Kunilovskaya — University of Wolverhampton, UK
Anvar Kurmukov — Kharkevich Institute for Information Transmission Problems of the Russian Academy of Sciences, Russia
Andrey Kutuzov — University of Oslo, Norway
Elizaveta Kuzmenko — University of Trento, Italy
Andrey Kuznetsov — Samara National Research University, Russia
Sergei O. Kuznetsov — HSE University, Moscow, Russia
Stepan Kuznetsov — Steklov Mathematical Institute of the Russian Academy of Sciences, Russia
Florence Le Ber — Université de Strasbourg, France
Alexander Lepskiy — HSE University, Moscow, Russia
Bertrand M. T. Lin — National Chiao Tung University, Taiwan
Irina Lomazova — HSE University, Moscow, Russia
Konstantin Lopukhin — Scrapinghub Inc., Ireland
Natalia Loukachevitch — Research Computing Center of Moscow State University, Russia
Ilya Makarov — HSE University, Moscow, Russia
Tatiana Makhalova — HSE University, Russia and Loria, Inria, France
Alexey Malafeev — HSE University, Nizhny Novgorod, Russia
Yury Malkov — Institute of Applied Physics of the Russian Academy of Sciences, Russia
Valentin Malykh — Institute for Systems Analysis of the Russian Academy of Sciences, Russia
Nizar Messai — Université de Tours, France
Tristan Miller — Austrian Research Institute for Artificial Intelligence, Austria
Olga Mitrofanova — St. Petersburg State University, Russia
Alexey A. Mitsyuk — HSE University, Russia
Evgeny Myasnikov — Samara National Research University, Russia
Amedeo Napoli — Loria, CNRS, Inria, and University of Lorraine, France
The Long Nguyen — Irkutsk State Technical University, Russia
Irina Nikishina — Skolkovo Institute of Science and Technology, Russia
Kirill Nikolaev — HSE University, Nizhny Novgorod, Russia
Damien Nouvel — Institut national des langues et civilisations orientales (Inalco University), France
Dimitri Nowicki — Glushkov Institute of Cybernetics of the National Academy of Sciences, Ukraine
Evgeniy M. Ozhegov — HSE University, Perm, Russia
Alexander Panchenko — Skolkovo Institute of Science and Technology, Russia
Polina Panicheva — HSE University, St. Petersburg, Russia

| | |
|---|---|
| Panos Pardalos | University of Florida, USA |
| Marcello Pelillo | University of Venice, Italy |
| Olga Perepelkina | Lomonosov Moscow State University, Russia |
| Georgios Petasis | National Center for Scientific Research "Demokritos", Greece |
| Anna Petrovicheva | Xperience AI, Russia |
| Vladimir Pleshko | RCO LLC, Russia |
| Mikhail Posypkin | Dorodnicyn Computing Centre of the Russian Academy of Sciences, Russia |
| V. B. Surya Prasath | Cincinnati Children's Hospital Medical Center, USA |
| Ekaterina Pronoza | Saint Petersburg State University, Russia |
| Artem Pyatkin | Novosibirsk State University and Sobolev Institute of Mathematics, Siberian Branch of the Russian Academy of Sciences, Russia |
| Irina Radchenko | ITMO University, Russia |
| Delhibabu Radhakrishnan | Kazan Federal University, Russia and VIT University, India |
| Vinit Ravishankar | University of Oslo, Norway |
| Yuliya Rubtsova | Ershov Institute of Informatics Systems, Siberian Branch of the Russian Academy of Sciences, Russia |
| Alexey Ruchay | Chelyabinsk State University, Russia |
| Eugen Ruppert | Universität Hamburg, Germany |
| Christian Sacarea | Babeş-Bolyai University, Romania |
| Aleksei Samarin | St. Petersburg University, Russia |
| Andrey Savchenko | HSE University, Nizhny Novgorod, Russia |
| Friedhelm Schwenker | Ulm University, Germany |
| Oleg Seredin | Tula State University, Russia |
| Tatiana Shavrina | HSE University, Moscow, Russia |
| Andrey Shcherbakov | The University of Melbourne, Australia |
| Sergey Shershakov | HSE University, Moscow, Russia |
| Denis Sidorov | Melentiev Energy Systems Institute, Siberian Branch of the Russian Academy of Sciences, Russia |
| Henry Soldano | Laboratoire d'Informatique de Paris Nord, France |
| Alexey Sorokin | Moscow State University, Russia |
| Andrey Sozykin | Krasovskii Institute of Mathematics and Mechanics, Russia |
| Dmitry Stepanov | Program Systems Institute of Russian Academy of Sciences, Russia |
| Vadim Strijov | Moscow Institute of Physics and Technology, Russia |
| Pavel Sulimov | HSE University, Russia and BetVictor, Gibraltar |
| Rustam Tagiew | German Center for Rail Traffic Research at the Federal Railway Authority, Germany |
| Irina Temnikova | Qatar Computing Research Institute, Qatar |
| Mikhail Tikhomirov | Lomonosov Moscow State University, Russia |
| Martin Trnecka | Palacký University Olomouc, Czech Republic |
| Christos Tryfonopoulos | University of the Peloponnese, Greece |

| Evgenii Tsymbalov | Skolkovo Institute of Science and Technology, Russia |
| Elena Tutubalina | Kazan Federal University, Russia |
| Wil van der Aalst | RWTH Aachen University, Germany |
| Ekaterina Vylomova | The University of Melbourne, Australia |
| Dmitry Yashunin | Harman International, USA |
| Dmitry Zaytsev | HSE University, Moscow, Russia |

## Additional Reviewers

| Vladimir Bashkin | Andrey Rivkin |
| Vadim Fomin | Sergey Sviridov |
| Artem Panin | Marketa Trneckova |

## Organising Committee

| Evgeny Burnaev | Skolkovo Institute of Science and Technology, Russia – AIST 2020 Local Organising Chair |
| Dmitry Ignatov | HSE University, Moscow, Russia – AIST series Head of Organisation |
| Alexander Panchenko | Skolkovo Institute of Science and Technology, Russia – AIST series Head of Organisation |
| Irina Nikishina | Skolkovo Institute of Science and Technology, Russia – AIST 2020 Secretary |
| Ekaterina Artemova | HSE University, Moscow, Russia |
| Ilya Makarov | HSE University, Moscow, Russia |

## Volunteers

| Daryna Dementieva | Skolkovo Institute of Science and Technology, Russia |
| Robiul Islam | Innopolis University, Russia |
| Evgenii Tsymbalov | Skolkovo Institute of Science and Technology, Russia |

# Contents

## Computer Vision

## Social Network Analysis

## Data Analysis and Machine Learning

# Natural Language Processing

# Did You Just Assume My Vector? Detecting Gender Stereotypes in Word Embeddings

Amir Bakarov[✉] [iD]

National Research University Higher School of Economics, Moscow, Russia
amirbakarov@gmail.com

**Abstract.** Recent studies found out that supervised machine learning models can capture prejudices and stereotypes from training data. Our study focuses on the detection of gender stereotypes in relation to word embeddings. We review prior work on the topic and propose a comparative study of existing methods of gender stereotype detection. We evaluate various word embeddings models with these methods and conclude that the amount of bias does not depend on the corpora size and training algorithm, and does not correlate with embeddings performance on the standard evaluation benchmarks.

**Keywords:** Word embeddings · Gender stereotypes

## 1 Introduction

Word embeddings (real-valued word representations produced by neural distributional semantic models) are ubiquitous tools in contemporary NLP. They are mostly treated as black-boxes, and it is unclear how to evaluate them. Recent studies propose various approaches to their evaluation [1,2]. We suppose, though, that certain properties of word embeddings were not considered in recent evaluation studies. One of them is the amount of bias in word vector spaces.

The concept of bias in machine learning commonly refers to prior information [3], but in this work, we will use the notion of **bias** from ethics studies, which is regularities in training data relating to prejudices or stereotypes about a person's physiological features (race, gender, orientation, etc.). Training data historically contains implicit stereotypes and discrimination, so the supervised machine learning models unintentionally capture this bias [4]. It can cause unfair decisions in the model for certain sets of a person's characteristics – for instance, while deciding whether to approve a personal loan [5].

It is usually hard to track such stereotypes in training data for supervised models. For instance, corpus statistics can contain bias that even does not exist in the cognition of the language speakers. The collocation *black sheep* can be more frequent than the collocation *white sheep* in the corpora, but in the real world, black sheep are rarer than white ones [6].

© Springer Nature Switzerland AG 2021
W. M. P. van der Aalst et al. (Eds.): AIST 2020, CCIS 1357, pp. 3–10, 2021.
https://doi.org/10.1007/978-3-030-71214-3_1

Word embeddings tend to frequently capture these unobvious statistics in the data. Such correlations inevitably appear in the resulting models, so the word vectors produce unfair distances between words characterizing men and women. For example, the situation when the word *researcher* is closer to the word *man* than *woman* can be considered as unfair, since there is no reason why women should be less related to research than men. Such situations make the work of systems based on word embeddings "biased": for instance, in the search engine system the query "researchers" can give results more related to men than to women, and then it could be even harder for women to become recognized as researchers. This problem can be solved from the corpus side as well as from the model side, but the main issue is how to evaluate whether the model contains such bias (and how much of it).

The current paper aims to overview recent advances in the detection of gender stereotypes and to propose a comparative analysis of existing gender bias detection methods, finding out whether the amount of bias depends on the diverting factors (corpus size and training algorithm), or embeddings performance on the traditional evaluation benchmarks (word similarity and analogy reasoning). The contribution of our work is to structurize recent studies related to gender bias in word embeddings, to overview existing metrics for gender bias detection, and to compare them with each other. We are the first to propose a comprehensive comparison of gender bias detection metrics for word embeddings.

This work is organized as follows. In Sect. 2 we survey the recent works related to the problem of gender bias in word embeddings and describe existing methods for detecting gender bias. In Sect. 3 we describe our evaluation experiments and make a discussion of the obtained results, while Sect. 4 concludes the paper and reveals our plans on future work on the subject.

## 2   Gender Bias in Word Embeddings

### 2.1   Related Work

In the field of NLP the problem of bias (and, particularly, gender bias) was covered for various tasks, such as text classification [7], machine translation [8], and recommendation systems [9]. We refer an interested reader to a comprehensive review of bias in NLP [10], and narrow the scope of the current paper to *gender* bias in *word embeddings*.

For word embeddings, the problem of bias was firstly introduced in 2015 by a blog post that highlighted the existence of gender unfairness in word vector spaces [11]. Most of the following work was trying to propose algorithms for removing bias in word embeddings: the main effort was to remove bias without hurting their semantic content (i.e., without hurting their performance on the standard benchmarks or downstream tasks).

The first work addressing the problem of measuring and removing gender bias in word embeddings was a work by [12], which proposed two metrics for removing gender bias, and a debiasing method based on the linear transformation of word vector spaces. These metrics were used for evaluation in almost all the following studies. However, [13] noted that human-like semantic biases are reflected through ordinary language, and proposed another metric based on the association tests. [14] suggested another way to generate association tests, supposing that they should be assessed by humans instead of being used automatically.

[15] introduced a triple loss function that penalizes the model for incorrect distances of word pairs related to gender. [16] investigated cultural tracks of gender biases in word embeddings, and concluded that semantic systems of word vector space represent cultural categories rather than biases or distortions. [17] explored semantic shifts from the perspective of bias, and also concluded that the origin of such shifts can reflect the changes in cultural patterns. [18] developed a technique to trace the origin of bias in embeddings back to the original text. [19] suggested using linear projection of all words based on vectors captured by common names. [20] proposed a general framework for unbiasing word embeddings, considering other types of bias rather than only gender. [21] created a neutral-network-based encoder-decoder framework to remove gender biases from gender-stereotyped words and to preserve gender-related information in feminine and masculine words. [22,23] pointed out on the problem of gender bias in contextualized word embeddings, such as ELMo. [24] presented a debiasing method for Hindi, while [25] proposed a technique for the German language. The problem of bias in word embeddings was also investigated in [20,26–31].

## 2.2   Evaluation Metrics

To the best of our knowledge, there are only 3 fully automatic evaluation methods available. We do not consider non-automatic methods that require additional human assessment: for example, the "Analogies exhibiting stereotypes" score, which suggests generating analogies judged by humans [12], or clustering plots, considered as classification of biased neutral words that also need further interpretation [26].

**Occupational Stereotypes (OS)** [12]. This method explores whether the word vectors contain stereotypes connected to occupations, by quantifying a neutral word's vector position on the axis between vectors corresponding to the most representative female and male words (so-called "gender axis"; it was based on the words "she" and "he" in the original study). Given a list of neutral word vectors $N$ (the authors used occupation words, e.g. "programmer"), the projection onto gender axis is computed[1]:

$$proj_x = x \cdot \frac{(v_{female} - v_{male})}{||v_{female} - v_{male}||_2},$$

---

[1] Here and after we will use L2-normalized Euclidian distance in all cases where we measure the distance between the vectors: assuming that the vectors are normalized, the choice between Euclidian distance and cosine similarity does not affect their results [17].

where $x$ is the vector of the neutral (occupation) word, $v_{female}$ and $v_{male}$ are vectors of gender words (e.g. "she" and "he"). The final score is computed as a mean of absolute projection values for a set of occupational words $X$:

$$OS = mean_{x \in X} proj_x = \sum_{x \in X} \frac{proj_x}{|X|}$$

The lower this score, the less bias the model contains. Notably, since bias in other scores has a direct dependency, we use an inverted version of $OS$:

$$IOS = 1 - OS$$

**Gender Difference (GD)** [17]. The method quantifies bias with a so-called "gender vector", calculated as the average of the set of the most representative words of each gender (e.g. "she", "mother", "woman" for female):

$$GD = \sum_{v_m \in M} ||v_m - v_1||_2 - ||v_m - v_2||_2,$$

where $M$ is the set of the neutral word vectors, $v_1$ is the averaged vector of the first gender vectors, and $v_2$ is the average vector of the second gender vectors. The higher the $GD$ score, the more biased the model is.

**Word Embedding Association Test (WEAT)** [13]. The method uses vector similarity to measure the association of two given sets of neutral target words (e.g. occupation words), with two sets of gender attributes words ("she", "mother", "woman", etc.). The score is computed as a probability that a random permutation of the attribute words would produce the observed (or greater) difference in sample means:

$$WEAT = \sum_{x \in X} s(x, A, B) - \sum_{y \in Y} s(y, A, B),$$

where $X$ and $Y$ are two sets of target words of equal size, and the $s(w, A, B)$ statistic measures the association of $w$ with the two sets of attribute words $A$ and $B$:

$$s(w, A, B) = mean_{a \in A} ||w - a||_2 - mean_{b \in B} ||w - b||_2,$$

The higher the $WEAT$ score, the more biased the model.

## 3  Comparison

To empirically compare the aforementioned methods, we evaluate them with a set of pre-trained word embeddings. We used a Word2Vec (continuous skip-gram) model [32] trained on British National Corpus (BNC, 160k tokens), Gigaword (300k tokens), Google News (3M tokens), and Wikipedia (300k tokens), as well as GloVe [33] trained on Gigaword and Wikipedia [34]. For OS we used the dataset of 320 "neutral" words of occupations, along with a dataset of 6 male-female word pairs from [12]. Notably, the "gender" pairs set used in our study is a modified version of the original dataset, since we dropped all pronouns which were out-of-vocabulary in some of our models. Also, unlike the original study, we used projection on the "female"-"male" axis instead of the "she"-"he" axis due to the same reason. These 2 datasets were also used to compute the GD score. For the WEAT score we used the original dataset proposed by the authors of this metric without any changes [13].

Following the experimental setting of [12], we used 4 standard word embeddings evaluation benchmarks: RG [35], RW [36], WS-353 [37] and MSR [38]. The first three measure whether related words have similar embeddings, while the latter measures how well the embedding performs in analogy tasks [39].

The results of our evaluation experiments are presented in Table 1 (best results are highlighted in bold). Surprisingly, the variation in the scores for GD is quite low, unlike the scores for 6 other benchmarks. Considering that a "good" model should have good scores on the standard benchmarks, we note that the amount of bias does not depend on the performance of the model. Henceforth, we can conclude that even a model that is good, from the perspective of standard evaluation benchmarks, can have prejudices and stereotypes.

The results also suggest that while models tend to generalize better on a larger amount of data, it does not mean that a model trained on larger corpora will contain more bias: for instance, according to the WEAT score, the lowest bias have models trained on BNC (the smallest corpora) and GoogleNews (the largest corpora). The amount of bias also does not depend on the training algorithm, and even two different methods for measuring bias can have different scores (the two sets of scores for OS and WEAT have Spearman rank correlation coefficient of $-0.2$).

The code for reproducing the experiments can be found at https://gitlab.com/bakarov/fair-embeddings.

**Table 1.** Evaluation of a set of word embeddings using gender bias metrics (3 first columns) and standard evaluation metrics (4 last columns).

|  | OS | WEAT | GD | RG | RW | WS | MSR |
|---|---|---|---|---|---|---|---|
| GloVe, Gigaword | 0.89 | **0.67** | 0.05 | 0.74 | 0.41 | 0.64 | 0.08 |
| GloVe, Wikipedia | 0.91 | 0.54 | 0.05 | **0.81** | 0.4 | 0.72 | **0.1** |
| Word2Vec, BNC | 0.89 | 0.37 | 0.05 | 0.75 | 0.37 | 0.75 | 0.01 |
| Word2Vec, Gigaword | **0.92** | 0.6 | 0.05 | 0.65 | **0.53** | 0.75 | 0.05 |
| Word2Vec, GoogleNews | **0.92** | 0.35 | **0.06** | 0.75 | 0.51 | 0.74 | 0.04 |
| Word2Vec, Wikipedia | 0.9 | 0.48 | 0.05 | **0.81** | 0.51 | **0.79** | 0.07 |

## 4    Conclusions and Future Work

Our work analyzed the recent methods for measuring gender-oriented biases in word embeddings, and compared the evaluation scores with word embeddings performance on the standard evaluation benchmarks. The results report that the amount of bias does not depend on the performance against these benchmarks, as well as of the training corpora and training algorithms. Results for the different gender bias evaluation methods also do not correlate with each other, so we can conclude that currently, we are unable to unequivocally say which word embeddings model is more biased, and moreover, we cannot reveal the amount of bias from the corpora size or the training algorithm. Any shifts in the embedding model are possible for several reasons. First, shifts are incorporated in the data itself, including collocations of frequency words or cultural phenomena, and second, the training method could be sensitive to certain correlations. Therefore, our study has supported the idea that we cannot capture these correlations (particularly, related to gender) by analyzing external model factors (e.g. training corpora size), and need specific (gender bias detection) methods for this task.

In the future, we suggest extending our research of these methods to the development of new ones, as well as to evaluation experiments in other settings (e.g. other corpora, other training algorithms, or other evaluation benchmarks). Also, in the current study, we used a set of pre-trained models, for which we cannot track the dependence of the results on the random initialization of weights during the training. We plan to train our models and measure the stability of the metric during several training iterations of the same model on the same corpus. Finally, we want to test the dependence of the results of the choices of the gender-specific words, and plan to extend our experiments to other languages (particularly, Russian) by creating datasets for them.

**Acknowledgments.** The reported study was funded by the Russian Foundation for Basic Research project 20-37-90153 "Development of framework for distributional semantic models evaluation".

## References

1. Rogers, A., Hosur Ananthakrishna, S., Rumshisky, A.: What's in your embedding, and how it predicts task performance. In: Proceedings of the 27th International Conference on Computational Linguistics, Santa Fe, New Mexico, USA, Association for Computational Linguistics, pp. 2690–2703, August 2018
2. Senel, L.K., Utlu, I., Yucesoy, V., Koc, A., Cukur, T.: Semantic structure and interpretability of word embeddings. arXiv preprint arXiv:1711.00331 (2017)
3. Bishop, C.M.: Pattern Recognition and Machine Learning. Springer, New York (2006)
4. Torralba, A., Efros, A.A.: Unbiased look at dataset bias. In: 2011 IEEE Conference on Computer Vision and Pattern Recognition (CVPR), pp. 1521–1528. IEEE (2011)

5. Hardt, M., Price, E., Srebro, N., et al.: Equality of opportunity in supervised learning. In: Advances in Neural Information Processing Systems, pp. 3315–3323 (2016)
6. Gordon, J., Van Durme, B.: Reporting bias and knowledge extraction (2013)
7. Wagner, C., Garcia, D., Jadidi, M., Strohmaier, M.: It's a man's Wikipedia? Assessing gender inequality in an online encyclopedia. In: ICWSM, pp. 454–463 (2015)
8. Font, J.E., Costa-jussà, M.R.: Equalizing gender biases in neural machine translation with word embeddings techniques. arXiv preprint arXiv:1901.03116 (2019)
9. Mishra, A., Mishra, H., Rathee, S.: Examining the presence of gender bias in customer reviews using word embedding. arXiv preprint arXiv:1902.00496 (2019)
10. Blodgett, S.L., Barocas, S., Daumé III, H., Wallach, H.: Language (technology) is power: a critical survey of "bias" in NLP. arXiv preprint arXiv:2005.14050 (2020)
11. Schmidt, B.: Rejecting the gender binary: a vector-space operation (2015)
12. Bolukbasi, T., Chang, K.W., Zou, J.Y., Saligrama, V., Kalai, A.T.: Man is to computer programmer as woman is to homemaker? debiasing word embeddings. In: Advances in Neural Information Processing Systems, pp. 4349–4357 (2016)
13. Caliskan, A., Bryson, J.J., Narayanan, A.: Semantics derived automatically from language corpora contain human-like biases. Science **356**(6334), 183–186 (2017)
14. Swinger, N., De-Arteaga, M., NeilThomasHeffernan, I., Leiserson, M.D.M., Kalai, A.T.: What are the biases in my word embedding? CoRR abs/1812.08769 (2018)
15. Zhao, J., Zhou, Y., Li, Z., Wang, W., Chang, K.W.: Learning gender-neutral word embeddings. In: Proceedings of the 2018 Conference on Empirical Methods in Natural Language Processing, pp. 4847–4853 (2018)
16. Kozlowski, A.C., Taddy, M., Evans, J.A.: The geometry of culture: analyzing meaning through word embeddings. arXiv preprint arXiv:1803.09288 (2018)
17. Garg, N., Schiebinger, L., Jurafsky, D., Zou, J.: Word embeddings quantify 100 years of gender and ethnic stereotypes. Proc. Nat. Acad. Sci. **115**(16), E3635–E3644 (2018)
18. Brunet, M.E., Alkalay-Houlihan, C., Anderson, A., Zemel, R.: Understanding the origins of bias in word embeddings. In: International Conference on Machine Learning, pp. 803–811 (2019)
19. Dev, S., Phillips, J.: Attenuating bias in word vectors. In: The 22nd International Conference on Artificial Intelligence and Statistics, pp. 879–887 (2019)
20. Lauscher, A., Glavaš, G., Ponzetto, S.P., Vulić, I.: A general framework for implicit and explicit debiasing of distributional word vector spaces. arXiv preprint arXiv:1909.06092 (2019)
21. Kaneko, M., Bollegala, D.: Gender-preserving debiasing for pre-trained word embeddings. In: Proceedings of the 57th Annual Meeting of the Association for Computational Linguistics, pp. 1641–1650 (2019)
22. Hoyle, A.M., Wolf-sonkin, L., Wallach, H., Augenstein, I., Cotterell, R.: Unsupervised discovery of gendered language through latent-variable modeling. In: 57th Annual Meeting of the Association for Computational Linguistics Meeting of the Association for Computational Linguistics, Association for Computational Linguistics, pp. 1706–1716 (2019)
23. Basta, C., Costa-jussà, M.R., Casas, N.: Extensive study on the underlying gender bias in contextualized word embeddings. Neural Comput. Appl. 1–14 (2020). https://doi.org/10.1007/s00521-020-05211-z
24. Pujari, A.K., Mittal, A., Padhi, A., Jain, A., Jadon, M., Kumar, V.: Debiasing gender biased Hindi words with word-embedding. In: Proceedings of the 2019 2nd International Conference on Algorithms, Computing and Artificial Intelligence, pp. 450–456 (2019)

25. Papakyriakopoulos, O., Hegelich, S., Serrano, J.C.M., Marco, F.: Bias in word embeddings. In: Proceedings of the 2020 Conference on Fairness, Accountability, and Transparency, pp. 446–457 (2020)
26. Gonen, H., Goldberg, Y.: Lipstick on a pig: debiasing methods cover up systematic gender biases in word embeddings but do not remove them. In: Proceedings of the 2019 Conference of the North American Chapter of the Association for Computational Linguistics: Human Language Technologies, Volume 1 (Long and Short Papers), pp. 609–614 (2019)
27. Shin, S., Song, K., Jang, J., Kim, H., Joo, W., Moon, I.C.: Neutralizing gender bias in word embedding with latent disentanglement and counterfactual generation. arXiv preprint arXiv:2004.03133 (2020)
28. Gyamfi, E.O., Rao, Y., Gou, M., Shao, Y.: deb2viz: debiasing gender in word embedding data using subspace visualization. In: Eleventh International Conference on Graphics and Image Processing (ICGIP 2019), vol. 11373, p. 113732F. International Society for Optics and Photonics (2020)
29. Wang, T., Lin, X.V., Rajani, N.F., McCann, B., Ordonez, V., Xiong, C.: Double-hard debias: tailoring word embeddings for gender bias mitigation. arXiv preprint arXiv:2005.00965 (2020)
30. Kumar, V., Bhotia, T.S., Kumar, V., Chakraborty, T.: Nurse is closer to woman than surgeon? Mitigating gender-biased proximities in word embeddings. Trans. Assoc. Comput. Linguist. **8**, 486–503 (2020)
31. Rios, A., Joshi, R., Shin, H.: Quantifying 60 years of gender bias in biomedical research with word embeddings. In: Proceedings of the 19th SIGBioMed Workshop on Biomedical Language Processing, pp. 1–13 (2020)
32. Mikolov, T., Sutskever, I., Chen, K., Corrado, G.S., Dean, J.: Distributed representations of words and phrases and their compositionality. In: Advances in Neural Information Processing Systems, pp. 3111–3119 (2013)
33. Pennington, J., Socher, R., Manning, C.D.: Glove: global vectors for word representation. In: Proceedings of the 2014 Conference on Empirical Methods in Natural Language Processing (EMNLP), pp. 1532–1543 (2014)
34. Kutuzov, A., Fares, M., Oepen, S., Velldal, E.: Word vectors, reuse, and replicability: towards a community repository of large-text resources. In: Proceedings of the 58th Conference on Simulation and Modelling, pp. 271–276. Linköping University Electronic Press (2017)
35. Rubenstein, H., Goodenough, J.B.: Contextual correlates of synonymy. Commun. ACM **8**(10), 627–633 (1965)
36. Luong, M.T., Socher, R., Manning, C.D.: Better word representations with recursive neural networks for morphology. In: Proceedings of the Seventeenth Conference on Computational Natural Language Learning, pp. 104–113 (2013)
37. Agirre, E., Alfonseca, E., Hall, K., Kravalová, J., Pasca, M., Soroa, A.: A study on similarity and relatedness using distributional and wordnet-based approaches. In: Proceedings of Human Language Technologies: The 2009 Annual Conference of the North American Chapter of the Association for Computational Linguistics, pp. 19–27 (2009)
38. Mikolov, T., Yih, W.t., Zweig, G.: Linguistic regularities in continuous space word representations. In: Proceedings of the 2013 Conference of the North American Chapter of the Association for Computational Linguistics: Human Language Technologies, pp. 746–751 (2013)
39. Bakarov, A.: A survey of word embeddings evaluation methods. arXiv preprint arXiv:1801.09536 (2018)

# Detecting Automatically Managed Accounts in Online Social Networks: Graph Embeddings Approach

Ilia Karpov$^{(\boxtimes)}$ ⓘ and Ekaterina Glazkova$^{(\boxtimes)}$ ⓘ

National Research University Higher School of Economics,
Moscow, Russian Federation
karpovilia@gmail.com, catherine.glazkova@gmail.com

**Abstract.** The widespread of Online Social Networks and the opportunity to commercialize popular accounts have attracted a large number of automated programs, known as artificial accounts. This paper (Project repository available at http://github.com/karpovilia/botdetection) focuses on the classification of human and fake accounts on the social network, by employing several graph neural networks, to efficiently encode attributes and network graph features of the account. Our work uses both network structure and attributes to distinguish human and artificial accounts and compares attributed and traditional graph embeddings. Separating complex, human-like artificial accounts into a standalone task demonstrates significant limitations of profile-based algorithms for bot detection and shows efficiency of network structure based methods for detecting sophisticated bot accounts. Experiments show that our approach can achieve competitive performance compared with existing state-of-the-art bot detection systems with only network-driven features.

**Keywords:** Graph embedding · Bot detection · Graph embeddings · Social network analysis · Random walk · Node2Vec · Attri2Vec

## 1 Introduction

The need to quickly search for artificially created accounts is in many practical research tasks. Artificial accounts distort the popularity of groups [6], spread fake news [16,17], and are used for fraud [1] activities. In this work we define an artificial account as an account created and used to generate profit for the owner by violating the rules of a social network. Thus, intuitively, accounts can be divided into two types:

- technical accounts (or bots) created to collect data, increase the number of group members, the popularity of posts etc. Most of these accounts are created and controlled by software. As a rule, these are poorly filled and relatively recently created profiles, which allows to identify them with a sufficiently high accuracy using several profile features.

© Springer Nature Switzerland AG 2021
W. M. P. van der Aalst et al. (Eds.): AIST 2020, CCIS 1357, pp. 11–21, 2021.
https://doi.org/10.1007/978-3-030-71214-3_2

- sophisticated accounts, created and operated semiautomatically by a human, are used primarily for information propagation, marketing and fraud activity. This category also includes hacked and used for fraud accounts.

Most of the existing methods show excellent results in identifying accounts of the first type. However, they do not consider accounts of the second type due to the high complexity of dataset acquisition. At the same time, accounts of the second type are visually and in terms of profile fullness indistinguishable from legitimate user accounts, which makes the task of their classification much more challenging.

The problem of finding artificial accounts of the second type can be reduced to the task of classifying network nodes. Still, its solution requires much more contextual information and learning capacity of the models used than in case of general node classification problem. This makes it appropriate to study the applicability of existing models for describing network nodes and their classification to solve this problem.

This paper has two main contributions as follows:

- First, we introduce the two bot detection dataset, consisting of nodes of three types: real user accounts, technical accounts and manual accounts. Unlike many existing arrays, it contains not only information about the network profile, but also information about the user's friends and groups. Since the array contains data of real people, immediately after collecting the data, all identifiers of users, groups, places of residence were hashed. An important advantage of the proposed array is that it does not contain precalculated walk models, but the initial hashed data themselves for training various models. Thus, it can be used in the future to train and compare new models.
- Second, we investigate two models for constructing network embeddings to solve the problem of classifying network nodes into artificial and natural. As far as we know, we are the first who use a class of the network node attributes simultaneously with the network node embedding to solve this problem.

The use of random walk algorithms optimized for large networks is also of interest. This comparison is especially important given that the problem of identifying real life bots is solved for a graph of $\sim 10^9$ nodes, which imposes significant restrictions on the computational cost of the algorithms used.

The rest of the work is organized as follows. In Sect. 2, we describe existing approaches to bot detection problem and recent advances in network embedding generation. Section 3 is devoted to the proposed model. Section 4 presents our experiments. In Sect. 5, we summarize experiment results and discuss future work.

## 2   Related Work

This section describes the chosen vector representations with attributes approach, the formal criteria to separate simple bots from advanced ones.

## 2.1 Labeled Datasets

Due to the lack of an established definition of the "bot" term, as well as the illegality of the very usage of bots, the formation of a training dataset causes problems. Based on the definition of the above and previously performed works [13,18], the main methods are (1) manual bot labeling – this approach is very resource-intensive and does not work well for sophisticated bots due to the low annotators agreement coefficient, (2) monitoring of suspended users lists – the main disadvantage of this approach are that (a) it is necessary to have time to proactively collect the entire user's network profile before blocking, (b) not all users blocked by a social network due to rules violation are bots (3) Creating accounts that attract bots other than real people, this approach looks very original, but most likely attracts bots using certain communication strategies, which does not fully give an idea of the diversity.

In this work, we use the second approach mainly to form a set of technical bots along with one more previously undescribed strategy. To construct a set of sophisticated accounts, we got access to ∼700 accounts sold by 12 different sellers on special account exchanges. The purchase of accounts is prohibited by the social network rules, all scenarios for using purchased accounts are negative and stay within the proposed definition, which makes monitoring of account exchanges a fairly effective strategy for generating the dataset. At the same time, advanced accounts, as a rule, have a higher price, which allows constructing a cost scale that correlates with the difficulty of identifying such an account. With it, the part of suspended users from the first set can be attributed to sophisticated accounts using the $k$-nearest neighbors method.

## 2.2 Existing Approaches

Researchers designed many bot detection approaches in order to prevent the spread of social media bots. Those approaches can be classified into 3 main categories: graph-based, feature-based and anomaly-based.

The anomaly-based approaches assume that a user with odd behaviour is most likely to be a malicious user. For example, [14] took into assumption that human behaviors exhibit strong heterogeneity and a bot's behavior is less complex than a human's behavior which leads to a smaller entropy of bot actions. They analyse time-interval entropy value and clustered the accounts on Sina Weibo into humans, bots and Cyborgs. Another work [9] created empty accounts and managed to construct a bot dataset with the users that interacted with these empty accounts.

There are several feature-based methods where authors identified a set of behaviour features which describes user actions, interactions, timestamps and may include text counting without deeply analyzing textual content. Then they train several supervised machine learning classifiers like SVM-NN in [8] or Random Forest Classifier in [5]. The sequence of user's actions can also be used as input to CNN-LSTM algorithm [3].

Besides behaviour features researchers use content-based features such as tweets. For example in [11] authors use sentiment analysis and build Contrast Pattern-Based classifier. Besides tweets user's nicknames also can be analysed. Thus, [2] classified user names (strings) into random and non-random and it can be used as additional feature for bot detection classifiers.

Graph-based bot detection methods rely on analyzing the structure of social network. The majority of them make the assumptions that real users form one strongly connected community and refuse to interact with artificial users [4,19,20]. These assumptions mean that there is a sparse cut between the legitimate and bot parts. [18] describe an approach to bot detection using a stacking ensemble with classifier trained on top of a combination of node embeddings from friendship graph, models trained on user's text and bag of subscribers. They use graph embedding obtained from LINE while in our work we compare various different graph embeddings.

## 2.3   Graph Embeddings

Given a weighted graph $G = (V, E, A)$, where $V$ is the set of nodes, $E$ is the edge set and $A$ is the adjacency matrix of a graph $G$ such as $A_{ij} = weight(v_i, v_j)$ if $(v_i, v_j) \in E$, otherwise $A_{ij} = 0$, the goal is to learn a function $V \rightarrow R^d$ that that maps each vertex to a $d$-dimensional ($d \ll |V|$) vector that captures its network connectivity structure. Resulting embedding can be used as input vector for the classification problem.

After the successful application of skip-gram [12] model to the problems of modeling text sequences, this method began to be actively used to model the graph nodes based on linear sequences obtained as a result of the graph random walks. The main advantage of using neural network methods in graph problems is that they do not need the whole training set during the training phase. They optimize by reading the data by batches step by step and correcting continuously which theoretically leads to training on extremely large datasets. In practice most of the frameworks require precomputed transition probabilities, that should be stored in memory.

Attri2Vec [22] generates random walks to capture structural context. After random walks are generated, attri2vec uses the representation $f(x_i)$ of node $v_i$ with attribute $x_i$ in the new attribute subspace to predict its context nodes collected from random walk sequences. In this way, network structure is seamlessly encoded into the new attribute subspace by allowing nodes sharing similar neighbors to be located closely to each other.

Node2Vec [7] is the classical second-order random walk based on the transition parameters $p$ and $q$, where $p$ controls the probability to return back to the starting node of the walk and $q$ – to walk away from the starting node of the walk. Given $p = 1$ and $q = 1$ the algorithm models unparameterized random walk.

# 3   Proposed Approach

VKontakte Social Network provides a significant amount of information about the user that can be converted into features. The features can be divided into five groups:

- Descriptive profile characteristics. These features include gender, age, marital status, date of birth, number of friends, groups, subscribers, country and city of residence, school, university, user work, site, number of posts on the user's wall, number of photos and albums, audio and video records, privacy settings, confirmation by phone number, date of account creation; geotags of photos, links to profiles in other networks;
- Text features received from user's messages, comments and statuses. They can include text embeddings, key topics of messages and comments, variety of vocabulary, user language;
- Image features, images on the page and in the user's albums;
- Graph features – user group membership, user friends structure;
- Time features such as user last seen time, publication time, activity at neighbor pages such as likes, comments at friends and subscribed groups;

In this work, we do not use text, image and time features due to the desire to focus on graph features. We still download text and profile information for further research and dataset enrichment.

## 3.1   Dataset Generation

In our work we consider two sources of artificial accounts information: blocked by the VKontakte social network administration ("banned") and placed on a special exchange for the purpose of selling ("sold"). All accounts have one of the following statuses: active, deleted and banned. We consider only accounts with status banned in "banned" list, which excludes users who have independently deleted their accounts from the network from being included in the set ones. This approach has a certain error ratio, since, as noted above, not only accounts participating in increasing the rating of groups and disseminating information are subject to blocking, but also real users blocked by the administration, for example, due to incorrect behavior towards other participants or publications of prohibited materials. We suggest focusing on "sold" accounts to train more precise models.

For obtaining "banned" accounts we used the following procedure:

1. Users profiles with friends and texts were collected during February 2020. Only profiles, that are accessible without the social network registration were collected.
2. In June 2020 profiles from step 1 were checked with friends and friends of friends. Two lists of users with statuses "active" and "deactivated - banned" are collected.

3. Graph based on found users, their connections and their friends and friends of friends was created.
4. The largest connected component of the obtained graph is considered.

When constructing the final graph, we examined 716 sophisticated and 360 technical accounts found from account shops and 2515 blocked users before they were banned. Blocked users were classified as "technical" and "sophisticated" based on their profile fulfillment and friends count. We obtained mutual friends for all nodes (bots and their friends) for the largest connected component, that gave us 322,917 nodes totally. The resulting connected graph consists of 270 technical accounts and 159 sophisticated accounts and 44,667 users with all information about profile, groups and friends; This node framing strategy distorts periphery nodes centrality measures, since their edges with nodes outside the frame that are not taken into account. It can cause an errors in the results, since it is bot egocentric, but a full-rated modeling of the second-order friends graph is complicated by the fact that it contains about 12 million nodes, which does not allow performing random walks in an acceptable time.

Sophisticated and technical accounts were represented in the largest connected component, which made it possible to train one graph embedding model to classify both types of accounts.

For "banned" users, an additional assessment of the profile filling in was carried out based on the profile filling in by such features as the number of friends, subscribers, groups, photo, account verification.

## 3.2    Selected Features

To train the basic model, we used the following account attributes: age, account verification by phone, nickname, website, Facebook profile, Instagram profile, Twitter profile, photo flags, number of subscriptions to public pages, number of videos, number of audio, number of days since the last login to the network, filled status, number of friends, city, gender.

In the case when individual fields were not filled in or were not available for collection, we substituted the median value. In the future, it seems appropriate to fill in these fields as the median value for k nearest neighbors of the user or to apply more advanced approaches to filling described in the work [23].

To train graph embedding, the user's friend graph was used. Features of encoder walking and training will be described in the section below.

## 3.3    Graph Embeddings

We considered two types of embeddings: Node2Vec [7] and Attri2Vec [22]. To obtain Node2Vec embeddings for our graph we used the SNAP framework [10]. For Attri2Vec embeddings we used Attri2Vec authors implementation[1]. More implementation details are presented in section Experiments.

---

[1] https://github.com/daokunzhang/attri2vec.

### 3.4   Classification

The resulting graph embedding features were combined with the user feature vector and used as input parameters of the classification algorithm. The following classification algorithms were considered:

1. Support Vector Classifier (SVC)
2. Random Forest (RF)
3. Logistic Regression (LogReg)

Implementation of algorithms from scikit-learn [15] was used.

## 4   Experiment

### 4.1   Classification on Account Features

Logistic regression classification algorithm was trained based on the accounts features. Features described in Sect. 3.2 were used. For "city" categorical feature we have chosen 185 cities with more than 10 citizens from our dataset and pre-processed the feature with one-hot encoding. "Gender" feature in VK has several possible values (male, female, unknown), we present it as two one-hot encoded features. Overall, with one-hot encoding of categorical features we obtained 204 features for each account.

Account information is usually sparse since accounts have many blank fields. We considered two approaches for empty fields filling. The first approach is to use some constant value. The second is to use average value among the filled field values. With the first approach on all 204 features we obtained ROC AUC 0.785, with the second approach – 0.762.

### 4.2   Classification on Node2Vec Graph Embeddings

We trained 25 Node2Vec models with different $p$ and $q$ parameters. The grid for each of the parameters is $\{0.25, 0.5, 1, 2, 4\}$ (the same as in the original paper [7]). The other parameters are 10 random walks of length 80 per source, context size – 10 and 10 epochs of SGD optimization.

We consider AUC-ROC (Area Under ROC Curve) as the main metric for our binary classification task. We train Logistic Regression Classification algorithm based on the obtained Node2Vec embeddings. The results for technical and sophisticated accounts are presented in Table 1 and 2, respectively.

We also considered SVM and Random Forest classification algorithms trained on the same embeddings as in Table 1, those approaches best results are 0.929 and 0.922 respectively. The best Logistic Regression results outperform SVM and Random Forest results.

Table 2 (sophisticated accounts) shows a significant decrease in quality compared to Table 1 (technical accounts). Graph random walk parameters are the same. The best model for sophisticated accounts classification shows 0.823 ROC-AUC score.

**Table 1.** LogReg Classification ROC AUC results for technical accounts based on Node2Vec embeddings with different $p$ and $q$ parameters, the number of dimensions is 50.

|  | $p = 0.25$ | $p = 0.5$ | $p = 1$ | $p = 2$ | $p = 4$ |
|---|---|---|---|---|---|
| $q = 0.25$ | 0.856 | 0.814 | 0.804 | 0.823 | 0.780 |
| $q = 0.5$ | 0.787 | 0.768 | 0.813 | 0.799 | 0.822 |
| $q = 1$ | 0.863 | 0.812 | 0.847 | 0.829 | 0.808 |
| $q = 2$ | 0.821 | **0.931** | 0.776 | 0.793 | 0.848 |
| $q = 4$ | 0.793 | 0.801 | 0.848 | 0.809 | 0.823 |

**Table 2.** LogReg Classification ROC AUC results for sophisticated accounts based on Node2Vec embeddings with different $p$ and $q$ parameters, the number of dimensions is 100.

|  | $p = 0.25$ | $p = 0.5$ | $p = 1$ | $p = 2$ | $p = 4$ |
|---|---|---|---|---|---|
| $q = 0.25$ | 0.727 | **0.823** | 0.751 | 0.753 | 0.793 |
| $q = 0.5$ | 0.750 | 0.795 | 0.796 | 0.806 | 0.754 |
| $q = 1$ | 0.771 | 0.804 | 0.765 | 0.788 | 0.772 |
| $q = 2$ | 0.747 | 0.742 | 0.808 | 0.764 | 0.779 |
| $q = 4$ | 0.776 | 0.724 | 0.745 | 0.709 | 0.793 |

### 4.3   Joint Graph Structure and Account Features Learning

Attri2Vec approach allows to use account features with graph structure features. We used authors implementation from the original paper [22]. We considered the same length of random walks, the number of random walks and the context size parameters as for Node2Vec embeddings, i.e. 80, 10 and 10, respectively. The other training parameters are 1M samples and 0.025 initial learning rate value.

We considered several types of the mapping options for constructing embeddings from attributes in Attri2Vec: ReLU mapping, Fourier mapping and Sigmoid mapping. The best results were shown by Sigmoid mapping. The best obtained results for technical and sophisticated accounts are presented in Table 3.

**Table 3.** LogReg Classification ROC AUC results based on Attri2Vec embeddings.

|  | Technical accounts | Sophisticated accounts |
|---|---|---|
| AUC ROC | **0.988** | 0.684 |

The obtained Attri2Vec results outperform Node2Vec results for technical accounts. Results on the sophisticated accounts are worse. This can be caused by more complex structure and profile fields completion of sophisticated accounts and large number of gaps in account features of sophisticated accounts.

### 4.4 Classification on Concatenated Node2Vec Embeddings and Account Features

We considered classification based on concatenated Node2Vec best embeddings and account features. We used Logistic Regression Classification algorithm. The best results on technical and sophisticated accounts are presented in Table 4.

**Table 4.** LogReg Classification ROC AUC results based on concatenated Node2Vec embeddings and account features.

|  | Technical accounts | Sophisticated accounts |
|---|---|---|
| AUC ROC | 0.911 | **0.867** |
| Zegzhda et al. [21] | – | 0.73 |
| Skorniakov et al. [18] | – | 0.820 |

Concatenation of Node2Vec embeddings and account features outperforms account features results for both datasets and gives the best obtained results on sophisticated dataset. Graph based methods show the best results on technical accounts dataset, but fail to classify sophisticated accounts without profile features.

## 5 Conclusions

In this work, we proposed simple and advanced sets for the bots detection problem. We showed the effectiveness of using graph embedding and combining it with user attribute features to solve this problem. The best ROC AUC score of 0.988 for technical accounts and 0.867 for sophisticated accounts shows the complexity difference in proposed tracks. Unfortunately, we cannot directly compare our results with existing approaches because most works do not provide the necessary datasets and code repositories. We took the best results of the bot detection from the paper by Zegzhda et al. [21] and Skorniakov et al. [18] as most close to our work. Proposed approach outperforms both approaches by 4.5% of AUC[2]. Achieved result is sufficient for application purposes, greater results can be reached with the following strategies:

- use of text embeddings since a significant part of artificial accounts performs the function of promoting certain goods or disseminating information, which can be used for classification;
- significant number of accounts hide their friends, but leave open groups that can be used to model a user as a bipartite graph node;
- network modeling as a temporal network is of interest, taking into account such characteristics as the joint appearance of accounts on the network.

---

[2] We did not re-implement the method and only give the result published by the author.

Resulting anonymized dataset, pretrained embeddings and evaluating scripts can be found at project repository[3].

**Acknowledgements.** This work was supported by Russian Academic Excellence Project "5-100" and through computational resources of HPC facilities provided by NRU HSE. We thank Daria Musatkina, Maksim Smirnov and Matvey Osmolovsky for their assistance with data processing and meaningful discussion.

# References

1. Apte, M., Palshikar, G.K., Baskaran, S.: Frauds in online social networks: a review. In: Özyer, T., Bakshi, S., Alhajj, R. (eds.) Social Networks and Surveillance for Society. LNSN, pp. 1–18. Springer, Cham (2019). https://doi.org/10.1007/978-3-319-78256-0_1

2. Beskow, D.M., Carley, K.M.: Its all in a name: detecting and labeling bots by their name. Comput. Math. Organ. Theory **25**(1), 24–35 (2018). https://doi.org/10.1007/s10588-018-09290-1

3. Cai, C., Li, L., Zeng, D.: Detecting social bots by jointly modeling deep behavior and content information. In: Proceedings of the 2017 ACM on Conference on Information and Knowledge Management, pp. 1995-1998. ACM (2017)

4. Danezis, G., Mittal, P.: Sybilinfer: detecting Sybil nodes using social networks. In: 16th Network and Distributed System Security Symposium (NDSS), pp. 1–15 (2009)

5. David, I., Siordia, O.S., Moctezuma, D.: Features combination for the detection of malicious twitter accounts. In: 2016 IEEE International Autumn Meeting on Power, pp. IEEE1-6 (2016)

6. Gaurav, M., Srivastava, A., Kumar, A., Miller, S.: Leveraging candidate popularity on twitter to predict election outcome. In: Proceedings of the 7th Workshop on Social Network Mining and Analysis. SNAKDD 2013. Association for Computing Machinery, New York (2013). https://doi.org/10.1145/2501025.2501038

7. Grover, A., Leskovec, J.: node2vec: scalable feature learning for networks. In: Proceedings of the 22nd ACM SIGKDD International Conference on Knowledge Discovery and Data Mining, KDD 2016, pp. 855–864. Association for Computing Machinery, New York (2016). https://doi.org/10.1145/2939672.2939754

8. Khaled, S., El-Tazi, N., Mokhtar, H.M.: Detecting fake accounts on social media. In: IEEE International Conference on Big Data, pp. 3672-3681. IEEE (2018)

9. Lee, K., Eoff, B.D., Caverlee, J.: Seven months with the devils: a long-term study of content polluters on Twitter. In: Fifth International AAAI Conference on Weblogs and Social Media, pp. 185–192 (2011)

10. Leskovec, J., Sosič, R.: SNAP: a general-purpose network analysis and graph-mining library. ACM Trans. Intell. Syst. Technol. (TIST) **8**(1), 1 (2016)

11. Loyola-González, O., Monroy, R., Rodríguez, J., López-Cuevas, A., Mata-Sánchez, J.I.: Contrast pattern-based classification for bot detection on Twitter. IEEE Access **7**, 45800–45817 (2019)

12. Mikolov, T., Chen, K., Corrado, G., Dean, J.: Efficient estimation of word representations in vector space (2013)

---

[3] Bot detection dataset as well as the project code are available at http://github.com/karpovilia/botdetection.

13. Morstatter, F., Wu, L., Nazer, T.H., Carley, K.M., Liu, H.: A new approach to bot detection: striking the balance between precision and recall. In: Proceedings of the 2016 IEEE/ACM International Conference on Advances in Social Networks Analysis and Mining, ASONAM 2016, pp. 533–540. IEEE Press (2016)
14. Pan, J. and Liu, Y., Liu, X., Hu, H.: Discriminating bot accounts based solely on temporal features of microblog behavior. Physica A **450**, 193–204 (2016)
15. Pedregosa, F., et al.: Scikit-learn: machine learning in Python. J. Mach. Learn. Res. **12**, 2825–2830 (2011)
16. Ratkiewicz, J., et al.: Truthy: mapping the spread of astroturf in microblog streams. In: Proceedings of the 20th International Conference Companion on World Wide Web, WWW 2011, pp. 249–252. Association for Computing Machinery, New York (2011). https://doi.org/10.1145/1963192.1963301
17. Shao, C., Ciampaglia, G.L., Varol, O., Flammini, A., Menczer, F.: The spread of fake news by social bots, 96, 104. arXiv preprint arXiv:1707.07592 (2017)
18. Skorniakov, K., Turdakov, D., Zhabotinsky, A.: Make social networks clean again: graph embedding and stacking classifiers for bot detection. In: 2nd International Workshop on Rumours and Deception in Social Media (RDSM) (2018)
19. Yu, H., Gibbons, P.B., Kaminsky, M., Xiao, F.: SybilLimit: a nearoptimal social network defense against sybil attacks. In: IEEE Symposium on Security and Privacy (SP 2008), pp. 3–17 (2008)
20. Yu, H., Kaminsky, M., Gibbons, P.B., Flaxman, A.D.: Sybilguard: defending against sybil attacks via social networks. IEEE/ACM Trans. Netw. **16**, 576–589 (2008)
21. Zegzhda, P.D., Malyshev, E.V., Pavlenko, E.Y.: The use of an artificial neural network to detect automatically managed accounts in social networks. Autom. Control Comput. Sci. **51**(8), 874–880 (2017). https://doi.org/10.3103/S0146411617080296
22. Zhang, D., Yin, J., Zhu, X., Zhang, C.: Attributed network embedding via subspace discovery. CoRR abs/1901.04095 (2019). http://arxiv.org/abs/1901.04095
23. Žnidaršič, A., Doreian, P., Ferligoj, A.: Treating missing network data before partitioning. In: Advances in Network Clustering and Blockmodeling, pp. 189–224 (2019)

# Semantic Recommendation System for Bilingual Corpus of Academic Papers

Anna Safaryan[1], Petr Filchenkov[1(✉)], Weijia Yan[1], Andrey Kutuzov[2], and Irina Nikishina[3]

[1] National Research University Higher School of Economics, Moscow, Russia
anna.safaryan-813@yandex.ru, psfilchenkov@edu.hse.ru,
renatayanweijia@gmail.com
[2] University of Oslo, Oslo, Norway
andreku@ifi.uio.no
[3] Skolkovo Institute of Science and Technology (Skoltech), Moscow, Russia
Irina.Nikishina@skoltech.ru

**Abstract.** We tested four methods of making document representations cross-lingual for the task of semantic search for the similar papers based on the corpus of papers from three Russian conferences on NLP: Dialogue, AIST and AINL. The pipeline consisted of three stages: preprocessing, word-by-word vectorisation using models obtained with various methods to map vectors from two independent vector spaces to a common one, and search for the most similar papers based on the cosine similarity of text vectors. The four methods used can be grouped into two approaches: 1) aligning two pretrained monolingual word embedding models with a bilingual dictionary on our own (for example, with the VecMap algorithm) and 2) using pre-aligned cross-lingual word embedding models (MUSE). To find out, which approach brings more benefit to the task, we conducted a manual evaluation of the results and calculated the average precision of recommendations for all the methods mentioned above. MUSE turned out to have the highest search relevance, but the other methods produced more recommendations in a language other than the one of the target paper.

**Keywords:** Semantic similarity · Semantic search · Scientific literature search · Document representations · Cross-lingual representations

## 1 Introduction

Only a couple of decades ago, search engines were mostly based on literal occurrences of the query words in the documents. Nowadays, we are witnessing the development of another approach: semantic search, in general terms, is a search with meaning, that can be extracted from a query, some data or an ontology. It enables a search engine user to find relevant pieces of information irrespective of the mentioned formal criteria, i.e. the search results may not contain the query words. The exact definition of the term "semantic search" is quite ambiguous, and we refer the reader to the survey [7] for more details.

W. M. P. van der Aalst et al. (Eds.): AIST 2020, CCIS 1357, pp. 22–36, 2021.
https://doi.org/10.1007/978-3-030-71214-3_3

Performing semantic search in a multilingual corpus of documents is a challenging task. Even within one language there is a huge variety of grammatical, cultural and pragmatic aspects which make the meaning ambiguous. Transferring all these aspects to another language seems extremely confusing. A good start to test multilingual semantic search is therefore to perform it in a corpus of particular domain. Narrowing the corpus to two languages could also facilitate the task. That is why this research deals with the bilingual corpus of academic papers. But its results can be extended to other corpora and languages.

There is a vast number of academic corpora such as a corpus of Wikipedia articles on science and technology [16], a corpus of academic journal papers [9] or any academic sub-corpus of national corpora (BNC[1], RNC[2], etc.). We test several approaches on a similar bilingual corpus of academic papers associated with the RusNLP project [18][3]. RusNLP is a search engine for academic papers presented in Russian NLP conferences: Dialogue, AIST and AINL. This project currently operates with papers in English, although there are still some papers written in Russian.

The research question of the paper stems from the practical issue of using cross-lingual word embedding models for semantic recommendation system on the corpus of academic papers. We have tested the following models: result of a simple word-by-word translation; result of projecting vectors from a Russian model vector space to an English one [14]; Multilingual Unsupervised and Supervised Embeddings (MUSE) [10]; a supervised model based on VecMap framework [2] (more details are given in the Sect. 3.2). Since only one of them was off-the-shelf, we would like to discover, whether it is enough to take pre-aligned word embeddings to create a decent recommendation system or it is better to take independent pretrained monolingual models and try to map them into common vector space in different ways. To the best of our knowledge, there is no other research comparing various approaches to cross-lingual document recommendation on a bilingual corpus of academic papers. Additionally, the most promising approach can be implemented for the RusNLP web service.

The paper is organised as follows: a brief overview of the existing papers on cross-lingual retrieval methods is given in Sect. 2; a more detailed description of the dataset and methods implementation is provided in Sect. 3; the evaluation of implemented cross-lingual search methods is outlined in Sect. 4.

## 2   Related Work

At the moment, there are many described attempts to implement a cross-lingual retrieval on academic papers [6,22]. However, none of them employs cross-lingual word embedding models—a tool that reflects the similarity of words in different languages. Word embedding models infer distributed representations of words in

---

[1] https://www.english-corpora.org/bnc/.

[2] http://www.ruscorpora.ru.

[3] https://nlp.rusvectores.org.

a low dimensional continuous space. Such models may be based on neural networks predicting the context words or low-rank approximations of word-context matrices [19]. At the core of word embeddings lies the distributional hypothesis that can be formulated as follows: the meanings of words used in similar contexts tends to be the same. We can notice that monolingual distributional models are widely employed in various NLP domains [26]. Cross-lingual models are closely connected with the monolingual ones and have a solid number of architectures to be tested. A confusion may occur as they are often based on a combination of different approaches, and the number of publications, meanwhile, grows. [21] is an attempt to provide a thorough survey of cross-lingual word representations. The authors systematise the main types of both mono- and cross-lingual word embedding models, describe their internal structure and working principles.

Supervised learning of cross-lingual embeddings often involves aligning monolingual models: a transformation of vector spaces on the basis of a multilingual dictionary that allow us to make corresponding word embeddings from different language close to each other. One of the first methods was linear transformation, proposed in [14]. Among state-of-the-art alignment methods there are such algorithms as Multilingual Unsupervised and Supervised Embeddings (MUSE) [10], Word Embedding Mapping [2], and Multilingual BERT [20]. A supervised learning may have some restrictions but still is in a focus of attention [17].

Weakly supervised cross-lingual embeddings are another possible method that can be used to deal with the lack of parallel data. A smaller dictionary is the main difference that distinguishes this approach from supervised learning. The paper [1] proposes a bilingual word embedding model with a 25-word dictionary. In [28], we can observe that even as few as 10 seed words are enough to make an alignment between embeddings of two different languages to perform multilingual POS tagging.

Many recent studies propose unsupervised learning of the cross-lingual word embeddings [12,25]. The key advantage of such models is their capability to be trained completely or almost without parallel data. Low-resource languages sometimes leave no choice but to use unsupervised models. In [3], the authors argue that strict unsupervised training without any parallel data is rather impractical. Nevertheless, they acknowledge theoretical scientific value of further research in this direction.

In the main experiment to which this paper is devoted, only word-level supervised methods to align monolingual embeddings were used in addition to the baseline without any cross-lingual embeddings. Their more detailed description is provided below:

- **Linear transformation** of vectors from the one vector space to another. The least squares method is used to calculate a matrix that, when applied to a word vector from the source language, transforms it into a vector which is as close as possible to the word vector of its translation in the target language, i.e., into the target vector space [14].
- **Multilingual Unsupervised and Supervised Embeddings (MUSE)** uses singular value decomposition (SVD) and iterative Procrustes to use the

composition of two resulting non-identity matrices as the projection matrix of the source language word vector into the target language vector space [10].

- **Bilingual Word Embedding Mappings (VecMap)** also uses SVD, but its goal differs from the MUSE and Linear transformation ones. The non-identity matrices resulting from SVD are applied separately to the word vectors of the source and target languages to project them into a new common vector space [2].

In the supervised versions of the MUSE and the VecMap algorithms, matrices for SVD are compiled using a bilingual dictionary.

Several papers dive deeper into the matter of document vectorisation [11,27]. But it is not the focus of this paper. In our experiment, each document was represented as an averaged vector of embeddings of the words included in the text. Some other papers focus on cross-lingual retrieval based on word embedding models [13,24], though they are not dedicated to domain-specific task of academic papers recommendation. Our experiment aims to evaluate different cross-lingual models on academic papers in particular.

## 3   Experimental Setup

As mentioned in Sect. 1, the goal of the experiment was to find out which approach is more beneficial for the cross-lingual recommendation system task: using pre-aligned cross-lingual models or aligning pretrained monolingual embeddings into a shared vector space yourself. To do this, it was decided to build a bilingual semantic recommendation system for the papers written in English and Russian using cross-lingual embeddings obtained with both approaches. To conduct the experiment, the workflow was organised in three stages:

1. **Preprocessing.** All texts were lemmatised with attaching the part-of-speech (POS) tags to words and deleting words with the functional POS tags. For both languages we used pretrained UDPipe [23] models[4]: the Russian model was trained on the SynTagRus corpus, and the English model was trained on the ParTUT corpus, both with the Universal Dependencies ver. 2.4 up.
2. **Generation of a text representation.** The words were converted to numeric vectors using four cross-lingual word embedding models obtained by two approaches (one model was off-the-shelf and the other three were pairs of monolingual embeddings aligned by the authors). In order to proceed to the next stage, each text vector was composed as the normalised average vector of all words in the text. The methods used to obtain the models will be described in more detail in Sect. 3.2.
3. **Search for the closest papers.** After mapping the vectors to the same space, nearest neighbours for the target text vector were selected by multiplying the vector by a matrix of all text vectors. The dot product of a vector and a matrix row can be interpreted as cosine similarity between them. Since each row of the matrix corresponds to a particular text in the corpus, the closest papers to the target vector were found.

---

[4] https://ufal.mff.cuni.cz/udpipe/models.

## 3.1   Data

The RusNLP dataset includes 1,983 papers from three major Russian NLP conferences from their beginning till 2019: Dialogue, AIST, AINL. While the language of the latter two conferences is English, Dialogue also accepts papers in Russian. Their number is decreasing over the years: for the Dialogue, the number of papers in Russian in 2007, 2013, and 2019 is 93, 46, and 25, respectively, which is 97%, 56%, and 40% of the total number of papers in that year; there are no papers in Russian in the AIST proceedings since 2014; AINL has never accepted papers in Russian. Despite this, we can only get a complete picture of the Russian NLP community publications by taking into account papers in all languages. For more information about the corpus, see Table 1 and [4,18].

**Table 1.** RusNLP corpus statistics

| Conference | Since | Texts | Russian | English |
|---|---|---|---|---|
| Dialogue | 2000 | 1,785 | 1,424 | 361 |
| AIST | 2012 | 91 | 21 | 70 |
| AINL | 2015 | 96 | 0 | 96 |
| **Total texts** | | **1,983** | **1,445** | **527** |

## 3.2   Methods

To perform cross-lingual search for the closest papers, it is necessary to map (to align) monolingual word representations for different languages into a common vector space.

Thus, four methods of making document representations cross-lingual were tested (three of them are described in more detail in the Sect. 2). Among these methods we can naturally distinguish two approaches:

1) Align Russian and English vector spaces, using pretrained monolingual Skip-gram [15] word embedding models[5,6] and a bilingual dictionary with approximately 25,600 word pairs (but only 13,400 pairs are contained in the embedding models, mentioned above) from the Facebook repository[7]. Since embedding models contain lemmatised words with the POS tags attached, the bilingual dictionary was also converted to the same format using UDPipe. So, three alignment methods were applied to the same data. It is important to note that the first method (Translation) does not use any alignment in the strict sense, it is a simple word-by-word translation using only one monolingual embedding model, while the others align two independent vector spaces to create cross-lingual word embeddings.

---

[5] http://vectors.nlpl.eu/repository/20/3.zip.

[6] http://vectors.nlpl.eu/repository/20/182.zip.

[7] https://github.com/facebookresearch/MUSE#ground-truth-bilingual-dictionaries.

- **Translation** of texts from Russian to English with the dictionary and vectorisation by the English model. This method was chosen as a baseline because it is simple, however is not cross-lingual in its strict sense. Preprocessed Russian texts are translated word-by-word using the bilingual dictionary preprocessed by the same UDPipe model. They are then vectorised by the English word embedding model, thus being mapped to a shared vector space with the original English texts. This method ignores words that are not present in the bilingual dictionary, while other methods can handle them if they are present in embedding models.
- **Linear transformation** of vectors from the Russian model vector space to the English one trained on the bilingual dictionary (hereafter **Linear projection** or **Projection**).
- **VecMap** alignment algorithm, applied to the same Russian and English Skip-gram word embedding models using the same bilingual dictionary.

2) Using pretrained, already pre-aligned in common vector space and therefore cross-lingual word embeddings.

- **MUSE** aligned word embedding models[8], provided off-the-shelf by Facebook.

In addition to the difference in approaches for obtaining, the models differ in the type of embeddings they are based: the monolingual Skip-grams used for alignment contain lemmatised words with attached POS tags, while the MUSE models are based on fasttext [5] embeddings with unpreprocessed words. All of them are pretrained on the Wikipedia. This study uses only pretrained monolingual vector spaces: this decision was made because our dataset is not big enough to train specialised word embeddings from scratch, although this may have impact on the recommendations quality.

Figure 1 shows the algorithms of all methods of making document representations cross-lingual based on word embeddings.

## 3.3   Evaluation Setup

To evaluate the quality of search for the most similar papers, annotators with expertise in the field and knowledge of both languages were found through crowdsourcing, provided with guidelines and asked to evaluate the outputs from each method by specifying how many recommended papers are relevant to the target one. Each output sample for a particular target paper was evaluated by three annotators, allowing us to calculate for each method not only the average ratio of relevant recommendations, but also the inter-rater agreement: the Krippendorff's alpha coefficient [8]. 15 annotators took part in the evaluation, but most of it was carried out by the authors of this paper, so the annotation can be trusted. We randomly selected 20 papers in Russian and 20 papers in English from the RusNLP corpus as the target papers, and find the closest five papers for each of

---

[8] https://github.com/facebookresearch/MUSE#multilingual-word-embeddings.

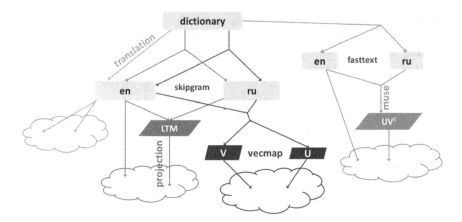

**Fig. 1.** Methods of making document representations cross-lingual based on word embeddings. *en* and *ru* are pretrained, but not aligned in the common vector space word embedding models; *LTM* is the abbreviation for the linear transformation matrix; $U$ and $V$ are resulting matrices of singular value decomposition of $Y \cdot X^T$, where $Y$ is the matrix of the English vectors and $X$ is the matrix of the Russian vectors.

them using cross-lingual models, generated with four methods to map vectors of Russian and English words to the same vector space.

To evaluate all resulting models without taking into account the specificity of the RusNLP dataset, they were additionally tested on another dataset. For this purpose, we selected 54 pairs of articles from the Russian and English Wikipedia with parallel titles (that is, they are marked in Wikipedia as written about the same subject in different languages) from several fields: biology, chemistry and medicine; history and culture; flora and fauna; other. For each article it was automatically evaluated whether the article with a parallel title was included into the top-1, top-5, and top-10 recommendations.

## 4    Results and Discussion

### 4.1    Quantitative Results

We evaluated the quality of each method using average precision (the average number of relevant recommended papers given by each method). As for recall, it could not be evaluated since the RusNLP dataset is too large for manual verification.

By observing the results presented in Table 2(a), we can tell that compared with Translation (54.5%), Projection (54.5%) and VecMap (54.2%), MUSE has the highest search relevance (58.5%). It is worth pointing out that all methods except MUSE used Skip-gram embeddings with lemmatised and POS tagged words in the model, and MUSE used Fasttext embeddings trained on a non-lemmatized corpus. Though the gap between their results is not that large, only the values for the first three methods are fully comparable. However, all the results are important from a practical point of view.

**Table 2.** RusNLP experimental results for target papers in both languages

(a) Average precision

| Method | Precision |
|--------|-----------|
| Translation | 54.5 |
| Projection | 54.5 |
| VecMap | 54.2 |
| MUSE | **58.5** |

(b) Inter-rater agreement

| Method | Krippendorff's $\alpha$ |
|--------|-------------------------|
| Translation | **0.347** |
| Projection | 0.262 |
| VecMap | 0.163 |
| MUSE | 0.170 |

Most of the recommendations turned out to be in the language of the target paper; this will be discussed in detail in Subsect. 4.3.

As already mentioned in the Sect. 3.2, the models differ not only in the methods used for obtaining them, but also in the type of word embeddings. It could be noticed, that MUSE, which used fasttex with unpreprocessed words, outperformed other models based on Skip-grams and containing lemmatised words with POS tags.

Within the same alignment method for each target paper we also evaluated the inter-rater agreement of the annotators using the Krippendorff's alpha coefficient [8]. The closer the $\alpha$ is to 1, the higher is the agreement. The values, presented in Table 2(b) are generally low. The highest Krippendorff's $\alpha$ is only 0.347 (Translation), while the $\alpha$ of MUSE, which has the highest average precision, is only 0.17. Such low consistency means that the annotators had very different opinions about the recommendations produced by MUSE. Thus, this method cannot be considered as the best one with full confidence, especially when the absolute difference in average precision with other methods is small. The following difficulties in the annotation process may have caused such a bias in the annotators agreement:

1. **Ambiguity of the guidelines.** The annotating guidelines could have been not clear enough, which may have caused misunderstandings.
2. **Not paper-specific evaluation.** For every five recommendations under the same target, the annotator was asked how many of them were relevant, but not whether each paper is relevant or not.
3. **Size of the annotation forms.** The forms were too long and too time-consuming to fill in, therefore the annotators could fill them out less carefully.

It could be noted that the presence of cross-lingual results did not affect the quality of the evaluation, since the knowledge of both English and Russian was a requirement for the annotators.

## 4.2 Examples of Relevant Recommendations

In this section, the relevance of examples will be analysed from our point of view, but this analysis is unavoidably subjective, which is proven by the low inter-rater agreement.

Here is a search target example where, according to the annotators, MUSE has outperformed other alignment methods. The title of the target paper is *'Semantic Role Labelling with Neural Networks for Texts in Russian'*. Five recommendations from the approach using MUSE are listed below:

1. *Semantic Role Labelling for Russian Language Based on Russian FrameBank*
2. *Classification Models for RST Discourse Parsing of Texts in Russian*
3. *Exploiting Russian Word Embeddings for Automated Grammeme Prediction*
4. *Methods for Semantic Role Labelling of Russian Texts*
5. *Wear the Right Head: Comparing Strategies for Encoding Sentences for Aspect Extraction*

The first and the fourth recommendations are truly devoted to the semantic role labeling task and obviously can be considered relevant to the target paper. The second and the fifth papers also seem to be appropriate: they are both devoted to the topics closely related to the semantic role labelling task. Moreover, the second paper also evaluates different neural models as the target one. The third paper is dedicated to the applicability of word embeddings to the prediction of classifying grammemes and seems quite irrelevant to the target. Therefore, the estimated precision for this example is 80% which is even higher than the average precision given by the annotators (66.6%).

Another example is the one where the most recommendations were not in the language of the target paper. It has been proposed by the Translation algorithm. The target paper *'Разработка формализма для описания сегментных морфологических процессов в германских языках и его компьютерная реализация'*. The recommendations from the system are the following:

1. *Morphological Analyser and Generator for Russian and Ukrainian Languages*
2. *Part-of-Speech Tagging: the Power of the Linear SVB-Based Filtration Method for Russian Language*
3. *К проблеме лемматизации несловарных словоформ*
4. *Особенности лексико-морфологического анализа при извлечении информационных объектов и связей из текстов естественного языка*
5. *Grammatical Dictionary Generation Using Machine Learning Methods*

The first and the fourth papers introduce morphological analysers or generators, which is particularly relevant to the target. The other ones can also be appropriate as they are devoted to the tasks closely related to morphology (POS tagging, lemmatisation and automatic grammatical dictionary compilation).

According to the annotators, the average precision for this example is 40%, although we can still detect relevant features in each recommended paper. Evaluation ambiguity may occur as there is a bias in what is considered relevant. The annotation guidelines had a general explanation of relevance but there is still a possibility that it could be interpreted differently.

### 4.3   Cross-lingual Recommendations

First of all, it should be noted that our main aim was to get the most relevant recommendations regardless of their language. Nevertheless, the difference between recommendations with regards to cross-linguality seem curious and worth some analysis.

As already mentioned, in the vast majority of cases the results were given in the language of the target. This can be partially explained by the specifics of the corpus: papers in Russian (especially from the Dialogue conference) tend to be more often devoted to theoretical aspects of linguistics, while papers about Natural Language Processing and computational linguistics are usually written in English. Thus, for some topics there could be even no papers in the other language to recommend. However, we can say that all results are cross-lingual, but for some methods, recommendations in the other language appear in positions higher than for others. Only Translation and Linear projection managed to yield cross-lingual recommendations among top-5 most similar papers. However, among the recommendations of MUSE and VecMap, there were also cross-lingual ones, but at lower positions. The frequency distribution of cross-lingual recommendations for positions up to 50 is shown in Fig. 2(a). The positions are grouped in bins of five, so the maximum frequency is 200 occurrences: 40 target papers × 5.

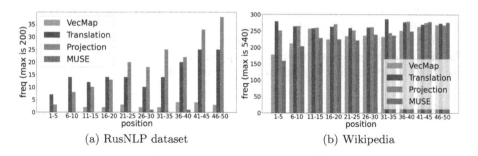

(a) RusNLP dataset                (b) Wikipedia

**Fig. 2.** Distribution of cross-lingual recommendations by positions

In other cases, we assume that a large number of words in papers are deleted during vectorisation because they are not present in word embedding models and/or in the bilingual dictionary (more details about the coverage are presented in the Subsect. 4.5). Since the Translation method relies on both resources, more words were lost, and the search for nearest neighbours relied mainly on common words (including words from the examples given in the paper) and frequent terms of syntax and morphology like 'analysis', 'text', 'sentence', 'part', 'form' etc., while specific words are thrown out. The remaining words are translated using the dictionary, vectorised by the same model, and as a result they are converted to the same vector. Since the average representations of the text are generated from mostly identical sets of words, the vectors are very similar. With the other methods, the texts probably contained many words which were present in only one embedding model (either English or Russian) and had no translations in the other one, so even in the aligned cross-lingual space, the nearest neighbours for these words were words of the same language.

Thus, the reason for the more cross-lingual recommendations in the top positions of the Translation results is that this method has a very limited vocabulary, which makes all papers very close to each other, regardless of the topic. However, we do not yet know the explanation for the difference between the results of Projection and VecMap, which are very similar algorithmically and are trained on the same resources.

It is useful to remember that since our goal is to find the most relevant papers that are not necessarily in a different language, we should not assume that the more cross-lingual results the model offers, the better it is.

### 4.4   Testing on the Wikipedia Dataset

We conducted an additional experiment to evaluate all resulting models without taking into account the specificity of the RusNLP dataset.

For testing on the Wikipedia, the results are presented in the Table 3. It should be noted that while for the RusNLP dataset the recommendation systems were evaluated manually using precision, the metric for this dataset is recall, since it was automatically evaluated whether the 'correct' article (one with a parallel title) had been included into the top 1, top 5, and top 10 recommendations. The gold standard in this experiment is a list of articles that are marked in Wikipedia as written about the same subject in different languages. The metric for inter-rater agreement (Krippendorff's $\alpha$) is not relevant for this case, since the evaluation was performed automatically.

**Table 3.** Wikipedia experimental results for target papers in both languages

| Method | Recall@1 | Recall@5 | Recall@10 |
|---|---|---|---|
| Translation | 51.85 | 87.96 | 95.37 |
| Projection | **56.48** | **91.67** | 97.22 |
| VecMap | 38.89 | 85.19 | 99.07 |
| MUSE | 34.26 | 90.74 | **100.00** |

In this experiment, MUSE has the lowest recall@1, but the highest recall@10: parallel paper in the other language always occurs in the top-10 recommendations.

On the Wikipedia dataset, all methods turn out to recommend cross-lingual results in high positions, although, as in the experiment on academic papers, Translation and Projection outperform VecMap and MUSE in this respect (Fig. 2(b)). As for MUSE, recommendations in another language are distributed mostly across low positions.

These results are not quite comparable to the actual results on academic papers not only because different metrics (precision and recall) are used, but also since the corpora contain texts of different genres. However, they allow us to get an idea of the quality of all methods without taking into account the specificity of the RusNLP dataset.

So, we can say that the results of the experiment on academic papers were confirmed. MUSE outperformed the other methods as measured by recall@10, which is arguably more important for the recommendation task than recall@1, because it is desirable not to miss any of the relevant papers.

## 4.5   Analysing Coverage

We also calculated the token and vocabulary coverage for the papers. Table 4 shows percentage of tokens from the text length and the percentage of unique words from the text vocabulary taken into account when vectorising by each model. The values of some models are equal, since the main difference between them is in the vector space aligning methods, not in the model dictionaries. Thus, these results characterise not the methods, but the dictionaries of embedding models used.

**Table 4.** Coverage (%)

| Method | English texts | | | Russian texts | | |
|---|---|---|---|---|---|---|
| | Tokens | Vocab | Dict size | Tokens | Vocab | Dict size |
| Translation | 71.53 | 63.15 | 296,630 | 53.91 | 47.99 | 19,118 |
| Projection | 71.53 | 63.15 | 296,630 | **89.30** | **85.57** | 248,978 |
| VecMap | 71.53 | 63.15 | 296,630 | **89.30** | **85.57** | 248,978 |
| MUSE | **89.30** | **83.21** | 200,000 | 86.58 | 82.84 | 200,000 |

The highest number of words was excluded when vectorising Russian texts with the Translation method, because it used words which had been included in the intersection of the bilingual dictionary and the model dictionary. However, this method had most recommendations in another language in the top-10, and the average precision was the same as for VecMap. At the same time, MUSE, which received the best average precision and gave the least cross-lingual recommendations, had the highest coverage. It can be assumed that the percentage of coverage affects the precision and the amount of cross-lingual recommendations.

## 5    Conclusions and Future Work

The research question of this paper stemmed from the practical issue of choosing an approach to make document representations cross-lingual for the semantic recommendation system. Hence, the aim was to find the best option for our task between off-the-shelf solution and aligning presumably high-quality pretrained monolingual embedding models on our own. It turned out that the first approach is good enough.

MUSE has the best average precision (58.5%). Other three methods were slightly worse, however, Translation and Projection seem to provide more cross-lingual recommendations than MUSE and VecMap. The fact that most of recommended papers are in the same language as the target one can be explained by algorithmic constraints of all four methods or by the corpus structure.

The results cannot be considered fully consistent as inter-rater agreement is low. It indicates that the annotators were inclined to disagree on the average precision of the four methods. So, further evaluation should be done with several changes: less ambiguous annotating guidelines, paper-specific evaluation of pair relevance, shorter annotation forms. It might be useful to use a ranking task with NDCG as a metric, rather than a binary score, but inter-rated agreement might then be even lower.

Additional testing on a bilingual set of Wikipedia articles demonstrated that models performance may vary depending on the number of recommendations. This experiment confirmed the ambiguity of the results. Nevertheless, in general it showed the same picture as the experiment on the RusNLP dataset.

In practice, as the result of the presented experiments, MUSE outperforms other methods in average precision and demonstrates the largest vocabulary coverage for both languages, therefore can be selected for incorporation into the RusNLP web service until further results are obtained.

Vocabulary coverage is a factor that could affect the results, and we plan to pay attention to it in the future. The results may also depend on the fact that the word embedding models used in the experiment were pretrained on Wikipedia. Presumably, training specialised models on in-domain texts would improve the quality of recommendations. Currently this is not possible to train solely on the RusNLP dataset, since it is too small, but it can be considered as a topic for further experiments. Moreover, while the present experiment was limited to using only word-level embeddings, it would be interesting to apply text-level vectorization methods to the task.

## References

1. Artetxe, M., Labaka, G., Agirre, E.: Learning bilingual word embeddings with (almost) no bilingual data. In: Proceedings of the 55th Annual Meeting of the Association for Computational Linguistics (Volume 1: Long Papers), Vancouver, Canada, pp. 451–462. Association for Computational Linguistics, July 2017

2. Artetxe, M., Labaka, G., Agirre, E.: Generalizing and improving bilingual word embedding mappings with a multi-step framework of linear transformations. In: Proceedings of the Thirty-Second AAAI Conference on Artificial Intelligence, pp. 5012–5019 (2018)
3. Artetxe, M., Ruder, S., Yogatama, D., Labaka, G., Agirre, E.: A call for more rigor in unsupervised cross-lingual learning. In: Proceedings of the 58th Annual Meeting of the Association for Computational Linguistics, pp. 7375–7388. Association for Computational Linguistics, Online, July 2020
4. Bakarov, A., Kutuzov, A., Nikishina, I.: Russian computational linguistics: topical structure in 2007–2017 conference papers. In: Proceedings of Dialogue-2018, Online Papers. ABBYY (2018)
5. Bojanowski, P., Grave, E., Joulin, A., Mikolov, T.: Enriching word vectors with subword information. Trans. Assoc. Comput. Linguist. 5, 135–146 (2017)
6. Celli, F., Keizer, J.: Enabling multilingual search through controlled vocabularies: The AGRIS approach. In: MTSR (2016)
7. Klusch, M., Kapahnke, P., Schulte, S., Lécué, F., Bernstein, A.: Semantic web service search: a brief survey. KI - Künstliche Intelligenz 30, 139–147 (2015)
8. Krippendorff, K.: Content Analysis: An Introduction to Its Methodology. Sage Publications, Thousand Oaks (2018)
9. Kwary, D.A.: A corpus and a concordancer of academic journal articles. Data Brief 16, 94–100 (2018)
10. Lample, G., Conneau, A., Ranzato, M., Denoyer, L., Jégou, H.: Word translation without parallel data. In: International Conference on Learning Representations (2018)
11. Lau, J.H., Baldwin, T.: An empirical evaluation of doc2vec with practical insights into document embedding generation. ArXiv abs/1607.05368 (2016)
12. Litschko, R., Glavas, G., Ponzetto, S.P., Vulic, I.: Unsupervised cross-lingual information retrieval using monolingual data only. In: The 41st International ACM SIGIR Conference on Research and Development in Information Retrieval (2018)
13. Litschko, R., Glavas, G., Vulic, I., Dietz, L.: Evaluating resource-lean cross-lingual embedding models in unsupervised retrieval. In: Proceedings of the 42nd International ACM SIGIR Conference on Research and Development in Information Retrieval (2019)
14. Mikolov, T., Le, Q.V., Sutskever, I.: Exploiting similarities among languages for machine translation (2013)
15. Mikolov, T., Sutskever, I., Chen, K., Corrado, G.S., Dean, J.: Distributed representations of words and phrases and their compositionality. In: Advances in Neural Information Processing Systems, pp. 3111–3119 (2013)
16. Minguillón, J., Lerga, M., Aibar, E., Lladós-Masllorens, J., Meseguer-Artola, A.: Semi-automatic generation of a corpus of Wikipedia articles on science and technology. Profesional De La Informacion 26, 995–1004 (2017)
17. Moshtaghi, M.: Supervised and nonlinear alignment of two embedding spaces for dictionary induction in low resourced languages. In: Proceedings of the 2019 Conference on Empirical Methods in Natural Language Processing and the 9th International Joint Conference on Natural Language Processing (EMNLP-IJCNLP), Hong Kong, China, pp. 823–832. Association for Computational Linguistics, November 2019
18. Nikishina, I., Bakarov, A., Kutuzov, A.: RusNLP: semantic search engine for Russian NLP conference papers. In: van der Aalst, W.M.P., et al. (eds.) AIST 2018. LNCS, vol. 11179, pp. 111–120. Springer, Cham (2018). https://doi.org/10.1007/978-3-030-11027-7_11

19. Pilehvar, M.T., Camacho-Collados, J.: Embeddings in Natural Language Processing. Morgan and Claypool Publishers (2020)
20. Pires, T., Schlinger, E., Garrette, D.: How multilingual is multilingual bert? ArXiv abs/1906.01502 (2019)
21. Ruder, S., Vulić, I., Søgaard, A.: A survey of cross-lingual word embedding models. J. Artif. Intell. Res. **65**, 569–631 (2019)
22. Stanković, R., Krstev, C., Obradović, I., Trtovac, A., Utvić, M.: A tool for enhanced search of multilingual digital libraries of e-journals. In: Proceedings of the Eighth International Conference on Language Resources and Evaluation (LREC 2012), Istanbul, Turkey, pp. 1710–1717. European Language Resources Association (ELRA), May 2012
23. Straka, M., Straková, J.: CoNLL 2017 shared task: multilingual parsing from raw text to universal dependencies. In: Proceedings of the CoNLL 2017 Shared Task: Multilingual Parsing from Raw Text to Universal Dependencies, Vancouver, Canada. Association for Computational Linguistics, August 2017
24. Wang, Z., et al.: Estimation of cross-lingual news similarities using text-mining methods. J. Risk Financ. Manage. **11**, 8 (2018)
25. Xu, R., Yang, Y., Otani, N., Wu, Y.: Unsupervised cross-lingual transfer of word embedding spaces. In: Proceedings of the 2018 Conference on Empirical Methods in Natural Language Processing, Brussels, Belgium, pp. 2465–2474. Association for Computational Linguistics, October– November 2018
26. Young, T., Hazarika, D., Poria, S., Cambria, E.: Recent trends in deep learning based natural language processing. IEEE Comput. Intell. Mag. **13**, 55–75 (2018)
27. Zhang, W., Li, Y., Wang, S.: Learning document representation via topic-enhanced LSTM model. Knowl. Based Syst. **174**, 194–204 (2019)
28. Zhang, Y., Gaddy, D., Barzilay, R., Jaakkola, T.: Ten pairs to tag - multilingual POS tagging via coarse mapping between embeddings. In: Proceedings of the 2016 Conference of the North American Chapter of the Association for Computational Linguistics: Human Language Technologies, San Diego, California, pp. 1307–1317. Association for Computational Linguistics, June 2016

# Prediction of News Popularity via Keywords Extraction and Trends Tracking

Alexander Pugachev[1]([✉]), Anton Voronov[2]([✉]), and Ilya Makarov[1]([✉])

[1] HSE University, Moscow, Russia
avpugachev@edu.hse.ru, iamakarov@hse.ru
[2] Sberbank of Russia, Moscow, Russia
voronov.a.dm@sberbank.ru

**Abstract.** In the last years, news agencies have become more influential in various social groups. At the same time, the media industry starts to monetize online distributed articles with contextual advertising. However, the efficiency of online marketing highly depends on the popularity of news articles. In our work, we present an alternative and effective way for article popularity forecasting with two–step approach: article keywords extraction and keywords-based article popularity prediction. We show the benefits of this technique and compare with widely used methods, such as Text Embeddings and BERT–based methods. Moreover, the work provides an architecture of the model for dynamic keyword tracking trained on the newest dataset of Russian news articles with more than 280k articles and 22k keywords for the popularity of forecasting purposes.

**Keywords:** Online news popularity forecasting · Keyword extraction · Popularity prediction · BERT · Text embedding

## 1 Introduction

Since the last century, the way how people find out news has changed significantly. Half a century ago, people discovered news from newspapers, magazines, radio, television, and other media resources. Nowadays, the Internet has become the central resource for consuming information. Today people prefer to receive news about events, politics, sports, economy, and other topics from different digital news sources, online aggregators, and mobile applications. People always want to stay informed and get the news as soon as possible. News agencies publish hundreds of articles daily and continuously keep the reader up to date with events. Today, each person can follow global trends and tendencies in real–time, keep up to date with meaningful events, comment on, and discuss the news that he is interested.

For the last several years, the news agencies' websites traffic increased significantly[1]. Any website on the Internet can be considered a platform for

---

[1] https://www.liveinternet.ru/stat/lenta.ru/summary.html?lang=en.

© Springer Nature Switzerland AG 2021
W. M. P. van der Aalst et al. (Eds.): AIST 2020, CCIS 1357, pp. 37–51, 2021.
https://doi.org/10.1007/978-3-030-71214-3_4

advertising, and the news websites are no exception. The higher website traffic provides higher benefits for owners. It is necessary to reduce costs and increase web advertisement efficiency and display ads according to a particular news article's popularity.

Forecasting news articles' popularity task has many challenges, mainly because it is often difficult to find a suitable definition and measure popularity. The most appropriate metric for measuring the article's popularity related to the advertising display is the number of article web page views. From this point of view, it is useful to know in advance will the particular news article become popular and decide whether to show the advertisement on its web page or not.

This paper investigates approaches for solving the news article's popularity forecasting problem, considering the number of articles views as the base metric. For this purpose, we have collected a dataset consisting of news articles from one of the most famous Russian news agencies, "RIA News"[2] over a long time. We present an article popularity prediction pipeline, which consists of two steps: relevant keywords extraction and keywords–based popularity prediction. We compare this approach to the most popular ones based on text embeddings and BERT [4] model.

The paper is structured as follows. Section 2 presents an overview of related works; Sect. 3 introduces the dataset; Sect. 4 describes the proposed method. The experiment results present in Sect. 5. Finally, Sect. 6 summarizes the conclusion.

## 2   Related Work

Although news popularity prediction is a relatively novel problem, it is in many researchers' scope of interest. Various studies consider different ways to measure news article popularity and related data to make predictions.

Balali et al. [2] took the number of comments as the base metric of the article's popularity. For building a reliable predictive model, authors extracted both textual (article title, body, comment tree) and non–textual (number of views, likes, dislikes) information from article web pages. Lee et al. [9] proposed an unsupervised keyword extraction technique that can be used for tracking news topics over time. Authors introduced six weighted TF–IDF variants and applied keyword extraction for several news domains (e.g., politics, business) separately. Keneshloo et al. [7] took into account that the life span of a news article is relatively short. It is more useful and valuable to predict an article's early popularity rather than its long-term popularity. Their study authors defined an article's popularity as the number of page views within the first 24 h after publication. The proposed model tracks the article for 30 min upon publication, extracts a range of temporal, social, and contextual features, and makes accurate forecasting. Xia et al. [18] analyzed base roles such as people, organizations, and political parties to understand the main trends in the news. Gayberi et al. [5] considered

---

[2] https://ria.ru/.

the task of popularity prediction of posts in social networks taking into account user, post and statistical features, and image object detection related features.

Piotrkowicz et al. [14] thoroughly studied the wide variety of textual features (morphological, semantic, syntactic, etc.) of articles' titles and built a reliable predictive model based on a support vector machine. Lamprinidis et al. [8] considered multi–task learning approach for news headlines popularity prediction. The main objective for researchers was to build a machine learning model based on the title of a news article that predicts whether users will click on this title or not. Lu et al. [10] performed an analysis of people's behavior when reading online news articles. They found that users are more likely to click on low–quality articles because of their high title persuasion. Voronov et al. [16] developed a language–independent model with an Online Machine Learning approach. Researchers trained and evaluated models on Russian and Chinese articles considering different text preprocessing methods according to these languages' morphology and syntax.

Some researches preferred to use sentiment analysis to predict the popularity of the content. In the work of Wang et al. [17] the popularity indicates by the sentiment analysis of the popular internet trends. However, it can predict only the class of the sentiment for most observed users, but it can not predict the final number of views. The model can indicate a person's attitude to a particular topic, but not the retention of the article in the news feed.

In contrast to the previous works, we consider articles' relevant keywords as the basis for making predictions. Using keywords, we can obtain a concise representation of a particular news article and track trends' dynamics.

## 3   Dataset

### 3.1   Data Collection

There exist many news articles datasets publicly available on the Internet, the majority of them are English–language. One of the most appropriate for our study Russian–language dataset is "News dataset from Lenta.Ru"[3]. However, it does not contain any information about articles' number of views. Since no other dataset with the latest news and sufficient for models training number of samples, it was decided to collect a suitable dataset on our own with the "RIA News" news agency as the base source of news articles. "RIA News" is one of the world's largest news agencies, according to LiveInternet, it has more than 62.9 million visitors per month[4]. We took articles for the period from May 2006 to March 2020.

### 3.2   Dataset Information

Each of the articles has a title, body, a set of relevant keywords, information about its publication time, and views count. An element from the collected

---

[3] https://www.kaggle.com/yutkin/corpus-of-russian-news-articles-from-lenta.

[4] https://www.liveinternet.ru/stat/RS_Total/Riaru_Total/summary.html.

corpora represents a JSON object. An example from the dataset presented in Table 1, all the fields have been translated from Russian to English.

**Table 1.** Example of one item from the collected dataset.

| Key | Value |
|---|---|
| Title | It became known who Meghan Markle wants to play when she returns to the movies |
| Link | https://ria.ru/20200302/1566826292.html |
| Time | 21:53 02.03.2020 |
| Topic | World |
| Views | 2108 |
| Article | The media discovered that the Duchess of Sussex Meghan Markle asked her agent nick Collins to find her a role in the Marvel superhero blockbuster, the Mirror reports. The source reports that Prince Harry's wife is confident that participating in such a major project will help her revive her acting career... |
| Tags | Culture, Movies and TV series, Culture news, Meghan Markle |

There are 23,254 unique keywords in the collected dataset. The keywords distribution (sorted by descending order of occurrence) among the articles as well as the distribution of the number with respect of unique keywords for each article shown in Figs. 1 and 2 respectively. Both histograms are presented on a logarithmic scale. Approximately 98% of articles have less than 8 related tags.

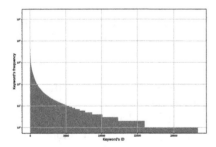

**Fig. 1.** Keywords distribution among all articles.

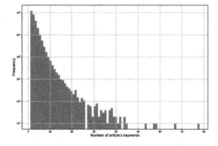

**Fig. 2.** Distribution of number of article's keywords.

The distribution of the number of views and the distribution of articles' lengths in terms of the number of words are shown in logarithmic scale on Figs. 3 and 4 respectively.

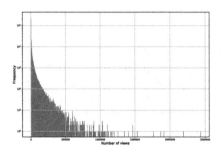

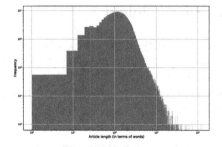

**Fig. 3.** Number of views distribution.      **Fig. 4.** Article lengths distribution.

### 3.3   Data Preprocessing

The data preprocessing procedure included deleting duplicate articles, items with no title, text, or set of tags, and samples that do not have a time of publication or number of views. After the preprocessing step, the number of articles decreased from approximately 335k to 296k.

The keywords preprocessing is also worth mentioning. Firstly, several most popular tags were removed from each of the articles. Many keywords are close semantically in the collected dataset and are similar in meaning but different in spelling. Several examples of these cases provide in Table 2, the names of the keywords are translated from Russian to English.

**Table 2.** Examples of similar keywords.

| Keyword #1 | Keyword #2 |
| --- | --- |
| Analytics—Religion and worldview | News – Religion and worldview |
| G20 Summit in Argentina | G20 Summit |
| Khanty–Mansi Autonomous Okrug | Khanty–Mansi AO |
| Space | Space—RIA Science |

To reduce the number of unique keywords by replacing a pair of similar keywords with a single one, we took the most frequently encountered keyword pairs and using Yandex. Toloka[5] obtained a labeled dataset of keyword pairs. The assessors marked whether a suggested pair of keywords are similar in meaning or not. If the majority of assessors marked two keywords as similar, then these keywords were merged into one. There were involved approximately 350 assessors in keyword similarity labeling. After merging similar keywords, the number of unique tags among all articles decreased from 23,254 to 22,919.

---

[5] https://toloka.yandex.com/.

# 4  Methodology

The popularity of a specific news article directly depends on the topic of the article. A set of relevant keywords often describes one specific topic. Therefore, based on keywords, the article's topic can be determined and predicted. The proposed model involves first extracting relevant keywords from a news article and then predicting the number of views based on the predicted keywords.

## 4.1  Keyword Extraction

For the keyword extraction pipeline, we combined the One–Shot Learning approach for building a custom vectorizer and k–Nearest Neighbors multi–label classification algorithm for obtaining relevant keywords.

**Vectorizer Model.** Despite the existence of several well–established text embedding methods such as Word2Vec [11], FastText [6], and ELMo [13], we implemented our vectorizer model taking into consideration the specificity of our task and data. Before treating a text sample to the model, it goes through the tokenization procedure the same way as when working with the BERT model. The RuBERT[6] vocabulary used for tokenizing.

We built a Siamese Neural Network, which produces multi–dimensional embedding vectors close to each other for two similar texts. For dissimilar samples, it produces two vectors with a large distance between them. The model particularly takes input a triplet of tokenized texts: an Anchor article, one that is similar to it (called Positive), and another one that is dissimilar to it (called Negative). Before training, all the articles were sorted in ascending order by publication time (oldest articles were in the beginning). While training for a particular Anchor article, there were randomly taken one Positive and one Negative article among 2000 closest by date of published articles. The rule for determining the similarity between a pair of articles was the following:

- Two articles are considered *similar* if they have no less than a half common keywords
- Two articles are considered *dissimilar* if they have strictly less than a half common keywords

The key feature of such architecture consists in specific loss called Triplet Loss [15]:

$$\mathcal{L}(s_a, s_p, s_n) = \max\left\{||s_a - s_p|| - ||s_a - s_n|| + \alpha, 0\right\} \tag{1}$$

where $s_a$, $s_p$, $s_n$ are vector embeddings for Anchor, Positive and Negative samples, respectively, $||\cdot||$ is Euclidean distance and $\alpha$ is hyperparameter (was set to 0 for all experiments). The architecture of vectorizer model is depicted at Fig. 5.

---

[6] http://docs.deeppavlov.ai/en/master/features/models/bert.html.

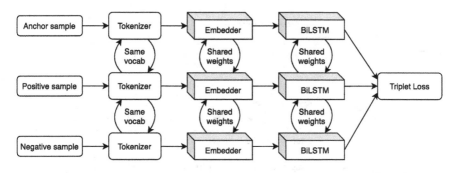

**Fig. 5.** Vectorizer model architecture.

**Multi–label kNN.** In order to predict a set of relevant keywords for a specific article, we construct a multi–dimensional vector space from embeddings of previously published articles. There are found $k$ its nearest neighbors for the article according to cosine similarity metrics, the Top–$n$ most frequent keywords among these neighbors considered as predicted keywords for this article. The value 25 for the $k$ parameter and value 3 for the $n$ parameter were found to be optimal during experiments.

## 4.2   Popularity Forecasting

To predict the popularity of a news article regarding the number of views based on the relevant keywords, we applied the Tracking Window strategy with gradient boosting.

**Tracking Window.** For a certain moment of time $t$ the popularity of a keyword can be calculated as follows:

$$P_{(t,K)} = \frac{\sum V_{d_t \cap K}}{N_{d_t \cap K}} \tag{2}$$

where $P_{(t,K)}$ is the popularity of keyword $K$ at the moment $t$, $\sum V_{d_t \cap K}$ is the sum of all views of articles for which keyword $K$ is relevant for the time moment $t$, and $N_{d_t \cap K}$ is a number of articles for the moment $t$ for which keyword $K$ is relevant.

During the experiments, one day was taken as a time moment $t$. To predict the impact of a specific keyword's popularity on the number of views, we consider the keyword's tendencies for $N$ past moments using the Tracking Window mechanism.

The main steps of processing articles' keywords into a Tracking Window matrix shows in Fig. 6. First, for each article's keyword, calculates its average popularity for the past $N$ days. The Tracking Window ($N$ parameter) was set to 30 (days), including the day of publication of a specific article, which allows determining general and local trends more accurately. After that, the Top–3

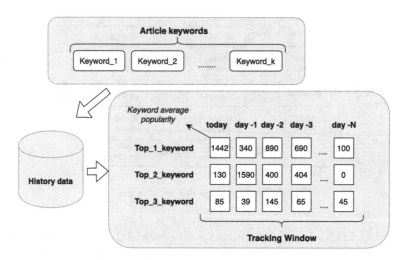

**Fig. 6.** Processing article's keywords into the Tracking Window matrix.

article's keywords were taken according to the average popularity and merged into a $3 \times N$ matrix. The obtained matrix represents the trends of each keyword individually and, at the same time, shows the overall trend for a news article. If a specific article had less than 3 keywords, the matrix's corresponding values for missing keywords were set to 0 for each of the $N$ days.

**Number of Views Forecasting.** Gradient boosting was chosen as a model for predicting the number of views due to its ability to analyze data sequences and tendencies, which are reflected in the Tracking Window. Moreover, the approach itself implies the continuous model actualization according to changing trends or news resource. Algorithms based on decision trees, in particular gradient boosting, are well suited for such requirements. We used one of the most widely used implementations of gradient boosting through our study, CatBoost [12].

Before treating the Tracking Window matrix to CatBoost, it was converted to a one–dimensional vector. According to the Top–3 keywords and separated by a "−1" value, each of the rows was situated. The obtained vector was treated to the CatBoost regression model, which outputs the number of views. The full forecasting pipeline for obtaining the number of views based on the Tracking Window matrix is presented in Fig. 7.

### 4.3 Full Prediction Pipeline

The proposed popularity prediction pipeline has presented in Fig. 8. Firstly the article's vector embedding is obtained using a trained vectorizer model. Then, the kNN multi–label classifier used for extracting 3 most relevant keywords. Finally, considering predicted keywords and news articles for the past 30 days, the

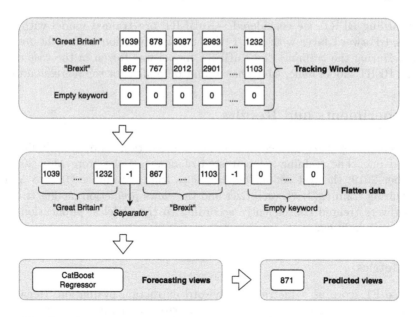

**Fig. 7.** Number of views forecasting based on Tracking Window. The Tracking Window matrix is turned into a one–dimensional vector and then treated to CatBoost Regression model which outputs the number of views.

Tracking Window matrix is built and treated to the trained CatBoost Regressor model for predicting views.

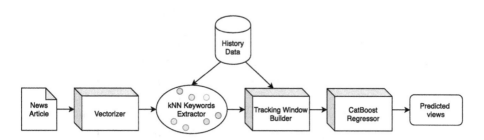

**Fig. 8.** Full popularity forecasting pipeline.

### 4.4 Comparable Approaches

To show the keyword extraction step's usefulness, we compare our approach to the most popular ones nowadays. We considered the BERT model and a neural network with trainable custom embeddings and two fully connected layers with Dropout. These models took as input tokenized articles and directly predicted the number of views without considering relevant keywords.

Regarding BERT, we considered a RuBERT pre–trained model with 128–unit Feed Forward layer with ReLu [1] activation function before the model's output. During the training, the RuBERT model was frozen in the case of the original BERT model issue, which has a limit of quality for the regression task[7].

## 5   Experiments and Results

We evaluated keyword extraction and views prediction models, including the proposed one. The training set consisted of news articles from May 2006 to November 2019; the test set included articles from November 2019 to March 2020. During training, both vectorizer and the views prediction models training samples were treated consequently according to the articles' publication date without any shuffling.

### 5.1   Metrics

We chose F1 score as a metric for keyword extraction evaluation. Regarding views prediction models, the Mean Absolute Error (MAE) metric was used during each of them' training. For the views prediction evaluation, we took Root Mean Squared Logarithmic Error (RMSLE) [3]. This metric is considered robust to outliers and penalizes the underestimation more severely than the overestimation. Also, we evaluated models in terms of percentile rank concerning Absolute Logarithmic Error.

**Table 3.** Keyword prediction performance.

| ID | kNN space set | Text | Similar keywords | F1 |
|----|---------------|------|------------------|-----|
| 1 | May 2006—Nov 2019 | Full article | Merged | 0.445 |
| 2 | | | Not merged | 0.435 |
| 3 | | First paragraph | Merged | 0.439 |
| 4 | | | Not merged | 0.431 |
| 5 | Jan 2019—Nov 2019 | Full article | Merged | **0.465** |
| 6 | | | Not merged | 0.452 |
| 7 | | First paragraph | Merged | 0.455 |
| 8 | | | Not merged | 0.446 |

---

[7] https://github.com/google-research/bert/issues/462.

## 5.2   Keyword Extraction

The most critical information of the news article concentrates on its first paragraph. Hence, we conducted several experiments where instead of the full articles' text, we took only their first paragraphs. Also, we evaluated keyword prediction depending on the set of articles in the kNN space. Two sets were considered: from May 2006 to November 2019 and from January 2019 to November 2019. Finally, we trained and evaluated the keywords prediction model before and after merging similar keywords.

Results for keyword extraction are presented in Table 3. It can be clearly noticed that after merging semantically close keywords, the overall performance increased. We can also see that the number of articles located in the kNN space plays a significant role. Using only the most recent articles, instead of all ever published, the keyword prediction performance gets better according to F1 score. Ultimately, in Table 3, we can notice that the quality of predictions using the first paragraphs is not significantly lower compared to the use of full articles' texts. The hypothesis confirms that the articles' first paragraphs contain the most important information for prediction.

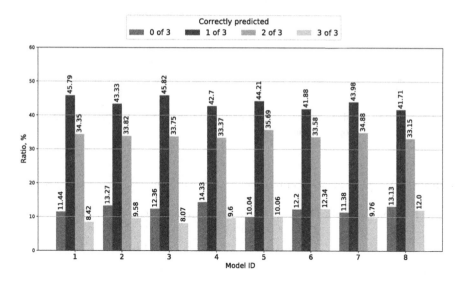

**Fig. 9.** Distribution of the number of correctly predicted keywords. IDs correspond to models in Table 3.

For each keyword prediction model we present the fraction of correctly predicted keywords, shown in Fig. 9. According to the diagrams, with a probability of 90%, our best model predicts at least 1 of 3 keywords correctly. With probability, more than 45% at least 2 of 3 keywords are predicted correctly.

## 5.3 Popularity Forecasting

We conducted two experiments on the popularity forecasting evaluation for the proposed approach: based on predicted and ground truth keywords. In the first case, for each article from the training set, we predict keywords based on previous articles for the last 2 months and train the CatBoost Regressor model for predicted keywords. In the second case, we took ground truth keywords for each article from the training set. The evaluation results in terms of a percentile rank and RMSLE metric for each of the mentioned models are presented in Fig. 10 and Table 4, respectively. We provide some examples of views predictions in Table 5.

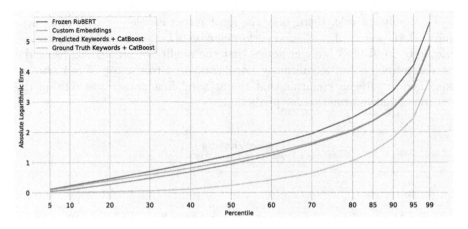

**Fig. 10.** Evaluation of number of views prediction in terms of percentile rank with respect to Absolute Logarithmic Error.

**Table 4.** Models evaluation in terms of Root Mean Squared Logarithmic Error.

| RuBERT | Custom embeddings | Predicted keywords + CatBoost | Ground truth keywords + CatBoost |
|--------|-------------------|-------------------------------|----------------------------------|
| 2.019  | 1.708             | 1.672                         | 1.042                            |

From Table 5 we can see that our model behaves differently depending on the article's topic and trends preceding its publication. For instance, examples 2, 3, and 9 show that in cases where news is preceded by related events, predictions are made more accurately. However, in cases when an event happens suddenly, we can get both high–quality (examples 1, 4) and low–quality predictions (examples 5, 7).

**Table 5.** Number of views prediction examples.

| # | Article's title | Ground truth number of views | Predicted number of views | | CatBoost | |
|---|---|---|---|---|---|---|
| | | | RuBERT | Custom embeddings | Predicted keywords | Ground truth keywords |
| 1 | In Sydney, a group of baboons escaped from the Royal hospital.[a] | 623 | 3215 | 1643 | 1584 | 583 |
| 2 | The Samsung S10's price reached minimum on the eve of the new products' release.[b] | 6865 | 1602 | 2899 | 1792 | 7029 |
| 3 | The Ministry of Health gave recommendations on the prevention of coronavirus.[c] | 11205 | 1282 | 1885 | 3701 | 11265 |
| 4 | Aeroflot received a plane with beds and a bar.[d] | 303 | 2627 | 1759 | 739 | 214 |
| 5 | Magnitude 5.5 earthquake jolts Japan.[e] | 1675 | 2456 | 504 | 812 | 879 |
| 6 | Boeing did not receive orders for planes in January for the first time since 1962.[f] | 2126 | 1262 | 3329 | 4566 | 2053 |
| 7 | The neural network colored the video of 1896 about Moscow.[g] | 10502 | 5778 | 7115 | 2422 | 4786 |
| 8 | Ukrainian citizen refused to evacuate from China without her dog.[h] | 6439 | 4705 | 5539 | 3564 | 20158 |
| 9 | A Serbian pensioner went on foot to Moscow for the parade on May 9.[i] | 9142 | 3725 | 2360 | 4051 | 11289 |
| 10 | Jack Ma donated money for the development of remedy against coronavirus.[j] | 783 | 775 | 918 | 1582 | 2134 |

[a] https://ria.ru/20200225/1565191567.html
[b] https://ria.ru/20200212/1564602625.html
[c] https://ria.ru/20200130/1564059279.html
[d] https://ria.ru/20200304/1568142823.html
[e] https://ria.ru/20200212/1564594675.html
[f] https://ria.ru/20200211/1564568950.html
[g] https://ria.ru/20200228/1565339495.html
[h] https://ria.ru/20200220/1565032131.html
[i] https://ria.ru/20200216/1564846656.html
[j] https://ria.ru/20200214/1564787083.html

# 6   Conclusion

This paper presents an approach for predicting news articles' popularity based on articles' keywords. This work shows that the keyword-based prediction is a more flexible tool than the text-based prediction. Using keywords as a base feature, we can estimate the real history of trends' popularity and make better predictions. Experiments show that the prediction of articles' views based on keywords gives the maximum quality on the test dataset. Moreover, the proposed model has potential for improvement: due to limited resources, we trained CatBoost on 10 million iterations, the model did not finish training while the other models met at a certain minimum. The most promising improvement, in our opinion, is to reduce the number of unique keywords, which can lead to significant growth in quality. It is worth noting that the proposed model shows acceptable quality up to 80% of cases.

This work also shows that the deep linguistic model does not show good quality in this task because news articles already have a compressed semantic representation, which indicates certain words' frequencies, which are, to some extent, a representation of keywords. However, a large pre–trained linguistic model is tied to the search for linguistic meaning for a specific time moment, which could be useless for predicting articles' popularity.

# References

1. Agarap, A.F.: Deep learning using rectified linear units (ReLU). CoRR abs/1803.08375 (2018)
2. Balali, A., Asadpour, M., Faili, H.: A supervised method to predict the popularity of news articles. Computación y Sistemas **21**, 703–716 (2018)
3. Desai, K.V., Ranjan, R.: Insights from the Wikipedia contest (IEEE contest for data mining 2011). arXiv preprint arXiv:1405.7393 (2014)
4. Devlin, J., Chang, M.W., Lee, K., Toutanova, K.: Bert: Pre-training of deep bidirectional transformers for language understanding. arXiv preprint arXiv:1810.04805 (2018)
5. Gayberi, M., Oguducu, S.G.: Popularity prediction of posts in social networks based on user, post and image features. In: Proceedings of the 11th International Conference on Management of Digital EcoSystems, MEDES 2019, pp. 9–15. Association for Computing Machinery, New York (2019)
6. Joulin, A., Grave, E., Bojanowski, P., Mikolov, T.: Bag of tricks for efficient text classification. CoRR abs/1607.01759 (2016)
7. Keneshloo, Y., Wang, S., Han, E.H., Ramakrishnan, N.: Predicting the popularity of news articles. In: SDM (2016)
8. Lamprinidis, S., Hardt, D., Hovy, D.: Predicting news headline popularity with syntactic and semantic knowledge using multi-task learning. In: Proceedings of the 2018 Conference on Empirical Methods in Natural Language Processing, Brussels, Belgium, pp. 659–664. Association for Computational Linguistics, October–November 2018
9. Lee, S., Kim, H.J.: News keyword extraction for topic tracking. In: 2008 Fourth International Conference on Networked Computing and Advanced Information Management, pp. 554–559, October 2008
10. Lu, H., Zhang, M., Ma, W., Shao, Y., Liu, Y., Ma, S.: Quality effects on user preferences and behaviorsin mobile news streaming. In: The World Wide Web Conference, WWW 2019, pp. 1187–1197. Association for Computing Machinery, New York (2019)
11. Mikolov, T., Chen, K., Corrado, G., Dean, J.: Efficient estimation of word representations in vector space. CoRR abs/1301.3781 (2013)
12. Ostroumova, L., Gusev, G., Vorobev, A., Dorogush, A.V., Gulin, A.: CatBoost: unbiased boosting with categorical features. arXiv preprint arXiv:1706.09516 (2018)

13. Peters, M., et al.: Deep contextualized word representations. In: Proceedings of the 2018 Conference of the North American Chapter of the Association for Computational Linguistics: Human Language Technologies (2018)
14. Piotrkowicz, A., Dimitrova, V., Otterbacher, J., Markert, K.: Headlines matter: using headlines to predict the popularity of news articles on Twitter and Facebook. In: ICWSM, pp. 656–659, May 2017
15. Schroff, F., Kalenichenko, D., Philbin, J.: FaceNet: a unified embedding for face recognition and clustering. In: Proceedings of the IEEE Conference on Computer Vision and Pattern Recognition, pp. 815–823, June 2015
16. Voronov, A., Shen, Y., Mondal, P.K.: Forecasting popularity of news article by title analyzing with BN-LSTM network. In: Proceedings of the 2019 International Conference on Data Mining and Machine Learning, ICDMML 2019, pp. 19–27. Association for Computing Machinery, New York (2019)
17. Wang, C., Xiao, Z., Liu, Y., Xu, Y., Zhou, A., Zhang, K.: SentiView: sentiment analysis and visualization for internet popular topics. IEEE Trans. Hum. Mach. Syst. **43**(6), 620–630 (2013)
18. Xia, C., Zhang, H., Moghtader, J., Wu, A., Chang, K.W.: Visualizing trends of key roles in news articles. arXiv preprint arXiv:1909.05449 (2019)

# Methods for Verification of Sentiment Frames

Irina Matueva and Natalia Loukachevitch[(✉)]

Lomonosov Moscow State University, Moscow, Russia
iramatueva@gmail.com, louk_nat@mail.ru

**Abstract.** The paper describes the approaches to verification of sentiment frames in the RuSentiFrames lexicon describing sentiment connotations related to specific Russian predicates. Two approaches for verification were used: 1) analysis of specific sentences from Russian National
Corpus, 2) via crowdsourcing platform Yandex.Toloka. The idea was to
find similarities and differences between the annotations made by the
experts in RuSentiFrames and by non-experts from Yandex.Toloka, thus
verifying the RuSentiFrames descriptions. The first approach showed
that implicit information influences greatly on the author's attitude and
that the context plays crucial role. The second approach showed mostly
the agreement between the expert's and non-expert's annotations in
case of relations between the participants in sentiment frames, but the
author's attitudes were estimated differently in some cases.

**Keywords:** Sentiment analysis · Frame · Crowdsourcing

## 1 Introduction

In various sentiment analysis tasks, the approaches based on specialized sentiment lexicons are still quite significant. For many languages, general sentiment
lexicons have been created [12,20,21]. At the same time, every sentiment analysis domain requires a specialized sentiment lexicon, which can be created with
tuning a general lexicon or extracting sentiment words from domain-specific text
collections [6,8].

Most sentiment vocabularies are presented as lists of words and expressions
with scores of their sentiment [21]. Some vocabularies provide also additional
characteristics of the word sentiment called as 'strength'. Also sentiment scores
can be assigned to specific senses of ambiguous words [1]. However, for more
accurate extraction of sentiment attitudes cited or expressed in texts, it is not
enough to have a simple sentiment list with sentiment scores assigned to words
and expressions [5,16].

For example, Maks and Vossen [14] analyse the following sentence:
"Damilola's killers were boasting about his murder". It is evident that this sentence conveys that the killers are positive towards the murder, and the author is
negative towards to the killers. In fact, both type of information can be inferred

© Springer Nature Switzerland AG 2021
W. M. P. van der Aalst et al. (Eds.): AIST 2020, CCIS 1357, pp. 52–64, 2021.
https://doi.org/10.1007/978-3-030-71214-3_5

from the use of verb *to boast*. The authors of [3] infer that Russia is negative towards Saakashvili from the sentence "Russia criticized Belarus for permitting Georgian President Mikheil Saakhashvili to appear on Belorussian television", which is mainly based on interaction of two words: *to criticize* and *to permit*.

These observations make it necessary to create structured sentiment lexicons, which provide more detailed description of sentiments, attitudes and effects, associated with specific words [4,10,13]. However, such detailed representations can be subjective; additional efforts are needed for their verification.

For more detailed description of sentiment and effects associated with specific words and expressions of the Russian language, sentiment lexicon RuSentiFrames has been created [13]. It was already used for automatic creation of a training collection for deep learning approaches to improve the extraction of attitudes between entities mentioned in texts [19] and for unsupervised extraction of sentiment relations between political entities [13]. Some procedures on evaluation of the created resource have been already implemented, but verification procedures of structured sentiment resources should be further investigated.

In this paper we consider approaches to verify descriptions made in the sentiment lexicon RuSentiFrames created for the Russian language. The verification methods are applied to specially selected words to check experts' decisions. The methods include the check of the expert descriptions on sentences from Russian National Corpus, and also crowdsourcing verification via Yandex.Toloka based on specially designed sentences. Using crowdsourcing does not mean that we prefer opinions of native speakers more than experts, but we try to reveal inconsistencies or accidental mistakes of experts.

## 2   Related Work

Our work concerns two directions in natural language processing: creating structured sentiment lexicons with more complicated structure than word lists with scores, and crowsourcing related to sentiment analysis tasks, especially when the quality of expert and native speaker responses is compared.

### 2.1   Structured Sentiment Resources

Maks and Vossen [14] present a lexicon model for subjectivity description of Dutch verbs. They describe the subjectivity relations that exist between the participants of the verbs and from a speaker (a writer), expressing multiple attitudes for each verb sense. The authors estimated inter-annotator agreement in established relations and found that scores of attitudes between participants have quite high agreement.

In [18] the authors stressed that it is important to extract implied sentiments and proposed the approach to description of so-called connotation frames for transitive verbs. The description includes three participants' roles: agent, theme, and writer. The frame includes the attitude polarity of participants to each other (positive, negative, or neutral), the effect of the situation to agent or theme, mental states and values of the participants. For experiments, 1000 most frequent English verbs were extracted from a corpus. These verbs were provided with five example sentences constructed from most frequently seen Subject-Verb-Object triples in Google syntactic ngrams and were annotated by crowdsourcers. The obtained values were averaged.

In [10] the authors introduce Sentiframes, a German verb resource for verb-centered sentiment inference. Their model specifies verb polarity frames that capture the polarity effects on the fillers of the verb's arguments in a sentence with that verb frame. By 2017, 1500 verb-polarity frames for 1100 verbs were described [11]. Some nominalizations (for example, *destruction*) have been also considered as frame entries. Authors also introduced an algorithm to apply the verb resource to treebank sentences.

In [4] the authors consider the problem of analyzing implicit sentiments. Some such sentiment attitudes are conveyed mentioning of events, which positively or negatively affect the object: malefactive (BadFor) and benefactive (GoodFor) events (effects). To study the inference of implicit sentiments, the authors extract GoodFor and BadFor words from FrameNet and WordNet. Examples of GoodFor verbs are *to encourage, to promote*. BadFor verbs examples are *to assault, to assail, to lower*. They also created rules for inference, for example, if the writer expresses a positive sentiment toward a BadFor event, then it is possible to infer that the writer is positive toward the agent of the event and negative toward the theme.

## 2.2 Crowdsourcing in Sentiment Analysis

Crowdsourcing in sentiment analysis is executed through online crowdsourcing platforms, which can be paid or free. In [2] the authors analyzed both approaches. To implement crowdsourcing on a fee basis, they used the CrowdFlower platform to classify students' comments about the learning process. For crowdsourcing on a free basis, the authors used the Pybossa platform. The study showed that crowdsourcing analysis of sentiment on both paid and free platforms is much more accurate than automatic sentiment analysis algorithms, but not as accurate as manual expert annotation.

In [7] contributors had to label term groups in three different main classes, emotion, intensifier and none, without instructions through the CrowdFlower platform. The authors compared several interfaces and also annotations performed by experts and non-expert annotators. The cost of hiring two experts was equal to the cost of hiring 19 participants at CrowdFlower. The authors concluded the crowd is capable of producing and evaluating a quality pure emotion lexicon without gold standards. However, it was also found that spam is very common and quality assessment should be implemented.

In [17] the authors examined the problem of multi-class annotation of pictures in the crowdsourcing platform MTurk with the help of experts and non-expert annotators. In the end, having measured several statistics, they concluded that with a good guide for annotators they can get a significantly better dataset.

Thus, because non-expert annotators are ordinary people and can be found in large numbers, crowdsourcing is an effective way to collect accurate data.

In the work [15] the author identifies several types of sentences that are difficult to annotate with sentiment. These are, for example, sentences describing the success or failure of one of the parties over another, sentences with sarcasm, requests, or rhetorical questions. To improve the consistency of annotation, the author proposed two new types of questionnaires (general and target-oriented). Both questionnaires were applied for annotation of tweets via a crowdsourcing platform.

## 3   RuSentiFrames Lexicon

Russian Sentiment Lexicon RuSentiFrames describes sentiments and connotations conveyed with a predicate word in form of sentiment frames [13,19]. Sentiment frame is a set of positive or negative associations (connotations) related to a predicate word or expression. A predicate usually describes a situation with some participants. The types of connotations that are conveyed in sentiment frames are as follows:

- attitude of the author of the text towards mentioned participants,
- positive or negative sentiment between participants,
- positive or negative effects on participants,
- positive or negative mental states of participants related to the described situation.

To designate roles of a predicate, the approach of PropBank [9] is accepted when semantic arguments are numbered, starting from zero. For a particular predicate , Arg0 is generally the argument exhibiting features of a Prototypical Agent of the situation, while Arg1 is a Prototypical Object.

Initial descriptions in RuSentiFrames are created by experts. To justify their descriptions, expert have to analyze specific sentences mentioning a word under analysis [13]. All the assertions in RuSentiFrames are provided with the score of confidence, which currently has two values: 1, if an expert believes that this assertion is true almost always, or 0.7, if the assertion is considered as default. It is difficult to obtain more fine-grained scores from experts. Assertions about neutral sentiments, effects and states of participants are not described.

For example, verb хвастаться (to boast) is associated with the following frame (Example 1). The frame indicates, that the first participant of the situation

---

**Example 1:** Frame "Хвастаться" (Boast)

"roles":   {"a0": "who boasts",
            "a1": "about what"}
"polarity": {["a0", "a1", "pos", 1.0],
            ["author", "a0", "neg", 1.0]},
"state":   {["a0","+", 1.0],

---

(a0) is positive towards the theme of boasting (participant a1). The participant a0 is in positive mood, but the author is negative towards a0. Thus, the frame explains the interpretation of the example from [14] mentioned in the introduction section. Also we can see a mix of positive and negative sentiments associated with the same word.

Another example is the frame for words such as *позволить* (to permit) (Example 2). It does not seem that this word has significant sentiment strength, but it has several positive associations, including positive relations between participants, positive effects, positive private state. But the author's attitude is not conveyed.

---

**Example 1:** Frame "Позволить" (Permit)
| | |
|---|---|
| "roles":    | {"a0": "who permits", |
|             | "a1": "what permitted"} |
|             | "a2": "whom is permitted"} |
| "polarity": | {["a0", "a1", "pos", 0.7], |
|             | ["a2", "a0", "pos", 0.7], |
|             | ["a0", "a2", "pos", 0.7], |
|             | ["a2", "a1", "pos", 1.0]}, |
| "effect":   | {["a1", "+", 1.0], |
|             | ["a2", "+", 1.0]}, |
| "state":    | {["a2","+", 1.0]}, |

---

The created frames are associated with a set of related words and expressions, which have the same attitudes and connotations. The set of lexical units of a frame can include: single words (mainly verbs and nouns); idioms (*вешать лапшу на уши* – to hang noodles on the ears—to lie); light verb constructions (*нанести вред* – inflict harm)[13].

Currently, RuSentiFrames contains 311 frames with more than 7K associated frame entries. Table 1 presents the distribution of frame entries according to sentiments between main participants of the situation and from the author to the participants in the current version of RuSentiFrames.

**Table 1.** Distribution of frame entries according to sentiments.

| Attitude | Sentiment | Number |
|---|---|---|
| a0 to a1 | Pos | 2,252 |
| a0 to a1 | Neg | 2,802 |
| Author to a0 | Pos | 1,178 |
| Author to a0 | Neg | 1,571 |
| Author to a1 | Pos | 1,429 |
| Author to a1 | Neg | 815 |

# 4 Verification Procedures

The created sentiment frames need evaluation and possibly correction. The procedures of verification can be implemented using manual analysis of sentiments conveyed in specific sentences, via crowdsourcing, or with the use of distributional methods. Currently, only sentence analysis and crowdsourcing were implemented for RuSentiFrames verification.

## 4.1 Previous Experiments for Checking Sentiment Frames

Previously, the following experiments on evaluation of RuSentiFrames were implemented.

In the first experiment, two experts described frames for selected words in parallel using their intuition and text examples. In the second experiment, one expert created frames and gave only roles (without connotations) to an annotator. The annotator gathered 10 random non-duplicate sentences for each word from different topics of the current news flow. The task of the annotator was to assign positive or negative scores to each role of the word mentioned in a sentence under analysis. The obtained scores were averaged. The average scores and connotations were compared with the original frame of the word.

The agreement in both experiments is estimated as 0.76–0.78 of the harmonic mean of relative intersections between both annotations, which can be considered as a quite high value [13]. It was found that the most agreement is met in polarity of relations between participants, and also in estimation of effects. But the author's position towards the participants is most dependent on context and subjectivity of an expert.

## 4.2 New Verification Procedures

In new experiments, the sentiment connotations of words were evaluated using two approaches: 1) analysis of specific sentences; 2) crowdsourcing via the Yandex.Toloka service[1].

In previous evaluation experiments, predicate words for verification were selected randomly. In new procedures of RuSentiFrames evaluation, words for verification were specially chosen. For sentence analysis, words which have positive or negative author's attitudes described in RuSentiFrames were selected. For crowdsourcing, word pairs close in meanings but different in some aspects (usage, single word or phrase, connotations) were chosen for comparison.

## 4.3 Sentence Analysis

It was previously found [13] that the author's attitude mostly depends on the context. To verify the sentiment frames based on the analysis of specific sentences, we compiled a list of sentiment predicates having positive or negative author's attitude to the participants of this situation according to RuSentiFrames.

---

[1] https://toloka.yandex.ru/.

The set of words with described negative author's attitudes includes the following words: *дохнуть (to die)*, *грозить (to threaten)*, *убить (to kill)*, *ухудшить (to worsen)*, *похитить (to abduct)*, *нарушить (to violate)*, *осудить (to condemn)*, *ухмыляться (to grin)*, *ябедничать (to sneak)*. Examples of selected words with positive author's attitudes are as follows: *предотвратить (to prevent)*, *почить (to rest)*, *карать (to punish)*, *ладить (to get along)*, *воодушевить (to inspire, вразумить (to reason)*, *постигать (to comprehend)*, *развивать (to develop)*.

For each sentiment predicate we selected example sentences from Russian National Corpus using the following principles:

- number of sentences: 10 example sentences for each sentiment predicate,
- uncomplicated construction: the sentiment predicate should be in the main clause of the sentence, in active voice, without negation, without introductory constructions,
- sentiment of other words: sentences should not contain evident negative or positive words; according to [5] if a participant in the situation is represented by a negative lexical unit, then regardless of the sentiment orientations of other components of the sentiment frame, the attitude towards this participant is negative,
- cases of irony, humor and sarcasm were not included due to interpretation problems,
- if the syntactic valences of the sentiment predicate are filled with anaphoric expressions, we replace them with antecedents.

Thus, sentences like *'The situation was only worsened by the press and television programs'* were included in the evaluation. Sentences like *'Nothing could be improved or worsened by haste'* were not included.

Author's attitude scores were calculated manually on a numerical scale from *'very negative' (−2)* to *'very positive' (+2)*. The obtained numbers were averaged, and this averaged result was compared with the original sentiment of frame components in RuSentiFrames.

As a result, there is not large numerical difference between relations in sentiment predicates in RuSentiFrames and the ones, obtained from the sentence analysis in Russian National Corpus in case of:

- verbs with the meaning of deprivation of life: *убить ('to kill')*, *дохнуть ('to die')*, *почить ('to rest')*
- verbs with meaning of damage: *ухудшить ('to worsen')*, *нарушить ('to break')*
- verbs with the meaning of impact on a negative object: *предотвратить (to prevent)*, *обезвредить (to neutralize)*
- verbs with the lexical meaning *'express disapproval'*: *ябедничать (to sneak)*, *осудить (to condemn)*

**Table 2.** Annotation of some predicates in RuSentiFrames and in the sentence analysis

| Verb | Attitude | RuSentiFrames score | Sentence score |
|------|----------|---------------------|----------------|
| ябедничать (to sneak) | (author, a0) | −1.0 | −0.5 |
| ухудшить (to worsen) | (author, a0) | −0.7 | −0.4 |
| нарушить (to violate) | (author, a0) | −1.0 | −0.5 |
| ухмыляться (to grin) | (author, a0) | −0.7 | −0.4 |
| предотвратить (to prevent) | (author, a0) (author, a1) | +0.7 −0.7 | 0.5 −0.35 |
| обезвредить (to neutralize) | (author, a0) (author, a1) | +1 −1.0 | 0.35 −0.5 |

Some sentiment predicates that were provided with the positively assessed author's attitudes showed significant difference from those obtained from the sentence analysis, for example:

- *карать* (*'to punish'*): positive author's attitude towards the actor (a0) assigned with experts was not confirmed,
- *ладить* (*'to get along'*): the author of the statement usually remains neutral towards the participant with whom one gets along (a2), but in frames the positive attitude to that participant was described. Only the main actor (a0) was positively characterized by the author as the most active participant.

The found inconsistencies should be corrected in a new version of RuSentiFrames. Table 2 shows some examples of initial RuSentiFrames annotations and results of sentences analysis. All scores are given on the scale [−1, 1].

### 4.4 Verification via Crowdsourcing

To verify sentiment frames with a crowdsourcing experiment, we selected word pairs that correspond to one of the following criteria:

- words are similar in meaning but have different sentiment connotations, specifically some differences in author's attitudes towards participants, and, therefore, these words are assigned to different frames. For example, sentiment predicate *укокошить* (*'finish off as to kill'*) differs from *'убить'* (*'to kill'*) in that in addition to negative author's attitude towards the main participant, there is a negative author's attitude towards the second participant who was killed;

- words are similar in meaning but have difference in polarity of relations between the participants. For example, in sentiment predicate *'смеяться'* (*'to laugh'*) the one who is laughing will show negative attitude towards the one who does it. But in *'насмехаться'* (*'to mock'*) this negative attitude will be more intense;
- synonymous words, assigned to the same frame, but different in language register to understand the difference in scores, for example *одобрить – похвалить (to approve – to praise)*. This group also includes cases when a synonymous entry is a phrase: *надеяться/возлагать надежду (to hope), доверять/оказывать доверие (to trust)*;
- antonym words. We would like to understand, if scores of connotations obtain opposite values from respondents or in sentences: *запретить/разрешать (to prohibit/to allow), улучшить/ухудшить (to improve/to worsen), нарушать/соблюдать (to violate/comply)*.

To conduct a crowdsourcing experiment in the Yandex.Toloka platform we selected 89 sentiment predicates according to the above-mentioned principles. Then we created sentences with these predicates, and questions about attitudes between the entities in the sentiment frame. Respondents were asked to answer these questions. The sentences were constructed artificially, using neutral names as Ivanov or Petrov, so as not to cause specific associations among annotators.

The sentences looked like this: *Иванов воздал Петрову по заслугам.(* *'Ivanov paid Petrov what he deserved'*). Then the following questions were asked:

- Как Иванов относится к Петрову?('How does Ivanov feel about Petrov?')
- Как Петров относится к Иванову? ('How does Petrov feel about Ivanov?')
- Как автор относится к Иванову? ('What is the author's attitude towards Ivanov?')
- Как автор относится к Петрову? ('What is the author's attitude towards Petrov?')

Respondents were asked to choose one answer from the options, that were ranked from *very negative(−2)* to *very positive(+2)*, as well as in the analysis of specific sentences.

To identify the optimal settings, we included in each experiment several test sentences containing an obvious positive/negative relationship between entities, such as: *Ivanov loves Petrov*. It is obvious that Ivanov has the positive attitude towards Petrov. Those respondents who correctly answered the test sentences were accepted. Thus, it is possible to block wrong annotators by collecting high-quality data without intervening the experiment process itself.

In total, 100 respondents participated in each experiment. The respondents' answers were averaged and compared with the initial expert assessment in RuSentiFrames.

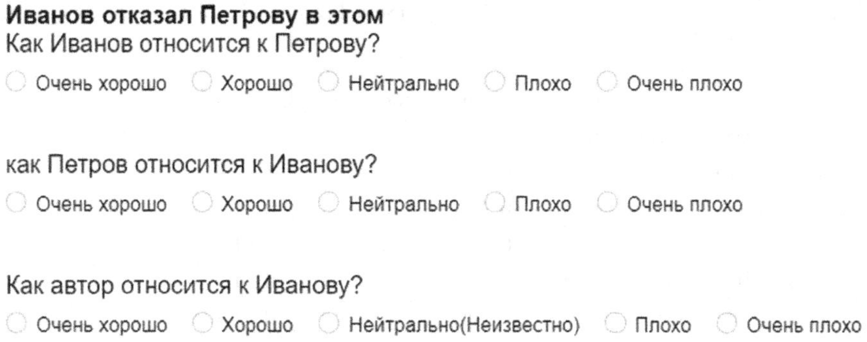

**Иванов отказал Петрову в этом**
Как Иванов относится к Петрову?

○ Очень хорошо   ○ Хорошо   ○ Нейтрально   ○ Плохо   ○ Очень плохо

как Петров относится к Иванову?

○ Очень хорошо   ○ Хорошо   ○ Нейтрально   ○ Плохо   ○ Очень плохо

Как автор относится к Иванову?

○ Очень хорошо   ○ Хорошо   ○ Нейтрально(Неизвестно)   ○ Плохо   ○ Очень плохо

**Fig. 1.** Example of a sentence and questions asked about this sentence in Yandex.Toloka.

As a result, the annotations of the experts in RuSentiFrames and non-expert annotators showed the agreement in the relationships of the participants of the sentiment predicate towards each other. For example, in the pair *наказать-карать (to punish)* respondents annotated the subject's attitude in *наказать* towards the object as negative as well as in *карать* and it coincides with the expert annotations in RuSentiFrames.

However, significant differences were found for some predicates. Respondents also estimated the author's attitude to the first participant (a0) of verb *карать* negatively, which agrees with previously mentioned results of sentence analysis. In predicates *упустить (to miss), опоздать (to be late)* the main participant does not have positive attitudes towards what is missed or where the participant is late, according to respondents. The RuSentiFrames experts assigned positive scores in default (0.7) to these attitudes.

Also the author's attitude towards the main participant in high-style words like *умереть-почить (to die), понять-постичь (to understand), жаловаться-роптать (to complain)* did not change to a more positive unlike RuSentiFrames data. Most often respondents chose the answer *neutral* when classifying the author's attitude towards the participants in case of low frequency predicates as for the words: *изобличить (to expose), постичь (to comprehend)*.

If the predicate was a colloquial word, then the polarity of author and participants was estimated as more intense. For example, in pairs: *умереть-дохнуть (to die), убить-укокошить (to kill), ругать-охаять (to scold), отказать-отбрить (to refuse)*.

Table 3 shows the annotation of some predicates made by the experts in RuSentiFrames and the results of crowdsourcing experiment. All scores are given on the scale [−1, 1]. Table 4 shows the average deviation of crowdsourcing scores from expert scores.

**Table 3.** Annotation of some predicates in RuSentiFrames and in the crowdsourcing experiment

| Verb | Attitude | RuSentiFrames score | Crowdsourcing score |
|---|---|---|---|
| ябедничать | (author, a0) | −1 | −0.61 |
| (to sneak) | (author, a1) | 0 | −0.1 |
| ухудшить | (author, a0) | −0.7 | −0.34 |
| (to worsen) | (author, a1) | 0 | −0.24 |
| нарушить | (author, a0) | −1 | −0.24 |
| (to violate) | (author, a1) | 0 | −0.34 |
| ухмыляться | (author, a0) | −0.7 | −0.32 |
| (to grin) | | | |
| убить | (author, a0) | −0.7 | −0.93 |
| (to kill) | (author, a1) | 0 | −0.12 |
| укокошить | (author, a0) | −0.7 | −0.73 |
| to finish off | (author, a1) | −0.7 | −0.28 |

**Table 4.** Average deviations of crowdsourcing scores from expert scores

| Parameter | (a0, a1) | (a1, a0) | (author, a0) | (author, a1) |
|---|---|---|---|---|
| Deviation | 0.41 | 0.49 | 0.31 | 0.2 |

## 5   Conclusion

In this paper we presented the approaches to verification of sentiment frames in the RuSentiFrames lexicon describing sentiment connotations related to specific Russian predicates. Two approaches for verification were used: 1) analysis of specific sentences from Russian National Corpus, 2) via crowdsourcing platform Yandex.Toloka.

The results of annotation of sentiment frames in specific sentences and in the crowdsourcing experiment were compared with the original annotation data made by experts in RuSentiFrames. As a result, context and implicit information from the context had a great influence on the sentiment in the analysis of specific sentences.

We see the prospects of the study in the possibility of creating crowdsourcing experiment with the help of non-professionals to annotate relations in the sentiment frame, but in the context of different sentences. Thus, it would be possible to study the problem of implicit information in sentiment analysis more deeply, which seems us to be a promising task not only for system of sentiment frames, but also for improving the quality of sentiment analysis on the whole.

**Acknowledgments.** The reported study was funded by RFBR according to the research project № 20-07-01059.

# References

1. Baccianella, S., Esuli, A., Sebastiani, F.: SentiWordNet 3.0: an enhanced lexical resource for sentiment analysis and opinion mining. In: LREC-2010, pp. 2200–2204 (2010)
2. Borromeo, R.M., Toyama, M.: Automatic vs. crowdsourced sentiment analysis. In: Proceedings of the 19th International Database Engineering & Applications Symposium, pp. 90–95 (2015)
3. Choi, E., Rashkin, H., Zettlemoyer, L., Choi, Y.: Document-level sentiment inference with social, faction, and discourse context. In: Proceedings of the 54th Annual Meeting of the Association for Computational Linguistics (Volume 1: Long Papers), pp. 333–343 (2016)
4. Choi, Y., Deng, L., Wiebe, J.: Lexical acquisition for opinion inference: a sense-level lexicon of benefactive and malefactive events. In: Proceedings of the 5th Workshop on Computational Approaches to Subjectivity, Sentiment and Social Media Analysis, pp. 107–112 (2014)
5. Deng, L., Wiebe, J.: Sentiment propagation via implicature constraints. In: Proceedings of the 14th Conference of the European Chapter of the Association for Computational Linguistics, pp. 377–385 (2014)
6. Hamilton, W.L., Clark, K., Leskovec, J., Jurafsky, D.: Inducing domain-specific sentiment lexicons from unlabeled corpora. In: Proceedings of the Conference on Empirical Methods in Natural Language Processing. Conference on Empirical Methods in Natural Language Processing, EMNLP-2016, p. 595 (2016)
7. Haralabopoulos, G., Simperl, E.: Crowdsourcing for beyond polarity sentiment analysis a pure emotion lexicon. arXiv preprint arXiv:1710.04203 (2017)
8. Huang, S., Niu, Z., Shi, C.: Automatic construction of domain-specific sentiment lexicon based on constrained label propagation. Knowl. Based Syst. **56**, 191–200 (2014)
9. Kingsbury, P.R., Palmer, M.: From TreeBank to PropBank. In: LREC, pp. 1989–1993. Citeseer (2002)
10. Klenner, M., Amsler, M.: Sentiframes: a resource for verb-centered German sentiment inference (2016)
11. Klenner, M., Tuggener, D., Clematide, S.: Stance detection in Facebook posts of a German right-wing party. In: Proceedings of the 2nd Workshop on Linking Models of Lexical, Sentential and Discourse-level Semantics, pp. 31–40 (2017)
12. Koltsova, O.Y., Alexeeva, S., Kolcov, S.: An opinion word lexicon and a training dataset for Russian sentiment analysis of social media. Comput. Linguist. Intellect. Technol. Mater. DIALOGUE **2016**, 277–287 (2016)
13. Loukachevitch, N., Rusnachenko, N.: Sentiment frames for attitude extraction in Russian. In: Proceedings of the International Conference on Computational Linguistics and Intellectual Technologies (Dialogue-2020) (2020)
14. Maks, I., Vossen, P.: A lexicon model for deep sentiment analysis and opinion mining applications. Decis. Support Syst. **53**(4), 680–688 (2012)
15. Mohammad, S.: A practical guide to sentiment annotation: challenges and solutions. In: Proceedings of the 7th Workshop on Computational Approaches to Subjectivity, Sentiment and Social Media Analysis, pp. 174–179 (2016)
16. Neviarouskaya, A., Prendinger, H., Ishizuka, M.: SentiFul: a lexicon for sentiment analysis. IEEE Trans. Affect. Comput. **2**(1), 22–36 (2011)

17. Nowak, S., Rüger, S.: How reliable are annotations via crowdsourcing: a study about inter-annotator agreement for multi-label image annotation. In: Proceedings of International Conference on Multimedia Information Retrieval, pp. 557–566 (2010)
18. Rashkin, H., Singh, S., Choi, Y.: Connotation frames: a data-driven investigation. In: Proceedings of the 54th Annual Meeting of the Association for Computational Linguistics, pp. 311–321 (2016)
19. Rusnachenko, N., Loukachevitch, N., Tutubalina, E.: Distant supervision for sentiment attitude extraction. In: Proceedings of International Conference on Recent Advances in Natural Language Processing (RANLP 2019), pp. 1022–1030 (2019)
20. Waltinger, U.: GermanPolarityClues: a lexical resource for German sentiment analysis. In: LREC, pp. 1638–1642 (2010)
21. Wilson, T., Wiebe, J., Hoffmann, P.: Recognizing contextual polarity in phrase-level sentiment analysis. In: Proceedings of Human Language Technology Conference and Conference on Empirical Methods in Natural Language Processing, pp. 347–354 (2005)

# Investigating the Robustness of Reading Difficulty Models for Russian Educational Texts

Ulyana Isaeva[1] and Alexey Sorokin[1,2]($\boxtimes$)

[1] Moscow State University, Moscow, Russia
ya.uvi23@yandex.ru, alexey.sorokin@list.ru
[2] Moscow Institute of Physics and Technology, Dolgoprudny, Russia

**Abstract.** Recent papers on Russian readability suggest several formulas aimed at evaluating text reading difficulty for learners of different ages. However, little is known about individual formulas for school subjects and their performance compared to that of existing universal readability formulas. Our goal is to study the impact of the subject both in terms of model quality and on the importance of individual features. We trained 4 linear regression models: an individual formula for each of 3 school subjects (Biology, Literature, and Social Studies) and a universal formula for all the 3 subjects. The dataset was created of schoolbook texts, randomly sampled into pseudo-texts of size 500 sentences. It was split into train and test sets in the ratio of 75 to 25. As for the features, previous papers on Russian readability do not provide proper feature selection. So we suggested a set of 32 features that are possibly relevant to text difficulty in Russian. For every model, features were selected from this set based on their importance. The results obtained show that all the one-subject formulas outperform the universal model and previously developed readability formulas. Experiments with other sample sizes (200 and 900 sentences per sample) prove these results. This is because feature importances vary significantly among the subjects. Suggested readability models might be beneficial for school education for evaluating text relevance for learners and adjusting those texts to target difficulty levels.

**Keywords:** Readability assessment · Text reading difficulty · Russian · Educational texts · Robustness · Text simplicity

## 1 Introduction

Text complexity [2], reading difficulty [1], readability [2] and comprehensibility [15] are text characteristics which are often undistinguished in the literature related to text complexity. They all refer to measuring to what extent the structure of a text affects the effort that the reader needs to understand the text. This characteristic is one of the crucial criteria for moderating course materials of educational programs since the content of educational material (e.g. textbooks)

© Springer Nature Switzerland AG 2021
W. M. P. van der Aalst et al. (Eds.): AIST 2020, CCIS 1357, pp. 65–77, 2021.
https://doi.org/10.1007/978-3-030-71214-3_6

influences the results of education. The concept is studied at the intersection of psychology and linguistics [11].

For more than 40 years there have been held numerous studies aiming at defining a proper set of parameters which could become a basis for a universal automated tool for measuring the readability of texts in Russian. In recent years a step forward was made by a series of surveys introduced by V. D. Soloviev, M. I. Solnyshkina and their coauthors [4,12,13]. These papers share the authors' experience of using a wide variety of features for evaluating the complexity of Russian school textbooks on Social Studies and provide several Russian readability linear formulas which showed good quality in application to textbooks and outperformed the older formulas.

In this work, we provide a pilot study of the extensibility of the above-mentioned models (trained on texts on Social Studies) to textbooks on other school subjects: Biology and Literature. We suggested some new features which may be useful for evaluating the readability of textbooks. After feature selection, we trained individual linear models for each of the three subjects. Another model was trained on a mixed dataset of texts on all the 3 subjects. These models outperformed the existing readability models for Russian. We provide an analysis of the selected features, which demonstrated that adding new morphological and syntactic features results in significant improvement in model performance. Also, we show that there is little coincidence in feature sets for different subjects, which supports the assumption that readability models trained on texts on one subject are not always extensible to other subjects.

## 2   Related Work

The history of readability formulas begins in the XX century. One of the first formulas was introduced in 1949 by Flesch [3]. It is based on two text parameters: 'average sentence length' and 'average syllables per word'. This formula was adjusted to use for many other languages.

First Russian readability formulas developed by Matskovsky and Mikk [8,9] aimed at evaluating a learner's ability to understand texts. At that time a narrow set of features was used: it consisted mostly of features related to sentence and word length, the familiarity of the words to a reader, and their abstractness [11]. One of the most important steps in the direction of modern view at measuring readability was made in the late 1970s with the development of the theory of information by Claude Shannon. A big shift was made towards using not only the above-mentioned quantitative features but also a wide range of qualitative features such as parts of speech and syntax phrases [11].

With the development of computational tools began the modern period of study of text readability. The one who contributed greatly to the development of modern Russian readability formulas was I. Oborneva, who provided an adaptation of the Flesch Reading Ease formula to the Russian language [10]. To do this she carried out a contrastive study of the average length of English and Russian words using 'Dictionary of the Russian Language' edited by S. Ozhegov and

'English-Russian Dictionary' edited by V. Muller. I. Oborneva analyzed a hundred of fictional English texts with Russian translations and found the average word lengths to be 3.29 and 2.97 syllables for Russian and English respectively. The formula is as follows:

$$FRE = 206.836 - (1.52 \times ASL) - (65.14 \times ASW) \tag{1}$$

In 2018 Oborneva's formula was developed further by V. Solovyev and his colleagues. In the paper [13] the formula was applied to the Russian Readability Corpus constructed of school textbooks on Social Studies and its coefficients were adjusted. The authors suggested an extended number of features and computed the values of correlation coefficients between each feature and the target variable. The results obtained showed that the most correlated features, i.e. most effective for evaluating text readability of textbooks on Social Studies are 'frequency of content words' and 'number of adjectives per sentence'. Considering this, the authors suggested a new formula. It represents a modification of the Flesch-Kincaid-Grade (FKG) [6] formula with a new feature – 'number of adjectives per sentence'. The modified formula allows to predict the grade of the text fragment with the RMSE of $0.51^1$ on their dataset:

$$FKG(ASL, ASW, ADJ) = 1.59 + (0.23 \times ASL) - (1.48 \times ASW) + (3.97 \times ADJ) \tag{2}$$

Another paper by the same group of authors [4] covers an investigation of the impact of 24 lexical, syntactic, and frequency features on readability assessment. This work provides a correlation analysis of features that demonstrated that features traditionally used for evaluating readability ('average sentence length' and 'average words per sentence') have the highest degree of correlation with the target variable (grade level). Also, among the most correlated parameters, there are such features as 'average number of coordinating chains' and 'average number of participial constructions'. This proves that information about syntactic structure can be made use of when assessing text readability. Using the ridge regression approach, the authors selected a subset of 16 features that have the maximum influence on the target variable. However, they did not evaluate the performance of model on the selected set of features.

Despite this, most questions concerning the complexity of Russian texts remain unsolved. One of these questions is the robustness of developed formulas across domains and genres. We address this problem in the current study.

---

[1] We replicate the calculation of RMSE for [12] model and our dataset and obtain much larger error. Since the exact test set and model predictions are not available, we have no explanation of such difference.

## 3   Methodology

### 3.1   Data Collection and Sampling

For our task, we collected a corpus of Russian schoolbook texts. Some of them were obtained by OCR processing and some were taken from the database provided by the authors of [4,12,13] at https://kpfu.ru/slozhnost-tekstov-304364.html. The dataset size and structure are provided in Table 1. Since the size of our corpus was not large enough to make statistically significant conclusions, we had to utilize a text sampling technique. That is, we evaluated our model, not on contiguous sentences or paragraphs, but random samples from the text. Note that this approach was also used in the previous studies on the KPFU dataset.

**Table 1.** Dataset size (in sentences) and structure.

| Grade/Subject | Biology | Literature | Social studies | Total |
|---|---|---|---|---|
| 5 | 4836 | – | 2326 | 7162 |
| 6 | 4995 | 10525 | 2949 | 18469 |
| 7 | 6798 | 5003 | 3942 | 15743 |
| 8 | 3424 | 15995 | 7122 | 26541 |
| 9 | 6650 | 21075 | 6465 | 34190 |
| 10 | 14834 | 8010 | 20822 | 43666 |
| 11 | – | 6780 | 22218 | 28998 |
| Total | 41537 | 67388 | 68544 | 174769 |

The most controversial issue in text sampling was the sample size. In [12] sample size 500 was suggested as one which allows obtaining readability assessment for a sample close to that of the text from which the sample was taken. This value was derived empirically, so we carried out our own investigation of the impact of sample size on readability prediction accuracy. We took 500 as the basic value for the sample size and also performed several experiments with sample sizes 200 and 900 to ensure robustness.

### 3.2   Evaluation of Recent Readability Formulas for Russian

Our initial assumption is that existing readability formulas trained on texts on a particular school subject may appear to underperform when applied to texts on other subjects. So, we applied the formula (2) to texts on Biology, Literature, Social Studies, and to the entire dataset (texts on 3 subjects). The results are provided in Table 2. It is quite expectable that the RMSE values vary across the subjects, also, it is important to notice that the model intended for texts on Social Studies shows better performance on Literature texts.

Considering this, further in this study we will treat differently the subsets (texts on different subjects) of our dataset and train individual independent

**Table 2.** Model performance across different subjects (500 sentences per sample).

| Subject | Biology | Literature | Social studies | Total |
|---------|---------|------------|----------------|-------|
| RMSE | 1.36 | 1.03 | 1.15 | 1.22 |
| $R^2$ | 0.43 | 0.52 | 0.46 | 0.49 |

models for them. Also, we will build a general model for the entire dataset and compare its performance to that of the individual models to find the answer to our research question about readability models extensibility (universality).

### 3.3 Introducing a Feature Set

The number of features used in readability models and observed in the theoretical papers differs widely, starting from the first Flesch-Kincaid Grade with 2 features to large studies like [2], where the total number of features viewed in the paper is 87. Generally, the more features are suggested as possibly relevant to the target variable, the better performance may be achieved by the final model based on selected features provided the features are relevant to the problem under consideration.

Thus, we developed a set of 32 features which might be useful for assessing the readability of schoolbook texts. To the best of our knowledge, a wider set of features for Russian readability models had never been observed, though a comparable number of the features (26) appeared in [4]. The features are divided into 3 groups based on their nature: the first group consists of features traditionally used for text readability estimation, the second is for features based on morphological features of the words, and the third one contains averaged data about syntactic characteristics of the sentences. All morphosyntactic information was obtained using the UDPipe [14] tagger and parser.

– Group 1. Traditional features:

1. Average number of words per sentence.
2. Average number of characters per word.
3. Average number of syllables per word.
4. Average frequency of words in the text.
5. Average Juilland coefficient [5].

– Group 2. Features based on POS-tags: computed as a proportion of words with the feature among all words in a sentence:

6. Finite verbs.
7. Non-finite verbs.
8. Verbs in active voice.
9. Verbs in passive voice.
10. Active participles.

11. Passive participles.
12. Adverbs.
13. Adverbial participles.
14. Nouns.
15. Full adjectives.
16. Short adjectives.
17. Personal pronouns.
18. Other pronouns.
19. Prepositions.
20. Conjunctions.
21. Negative particles.
22. Particles.
23. Content words.

– Group 3. Syntactic features: computed for every sentence and averaged for a text:

24. Number of clauses.
25. Syntax tree depth.
26. Length of adverbial participle phrases.
27. Length of participle phrases.
28. Number of embedded NPs with genitive case.
29. Ratio of subordinate clauses.
30. Number of relative clauses.
31. Ratio of infinite clauses.
32. Maximum distance between a node and its descendant.

### 3.4    Feature Selection

Feature selection is one of the most important steps in model building, since non-informative or duplicate predictors may adversely affect the model quality. Moreover, the more features there are in the model, the more complex it becomes, which requires more computing resources [7].

As indicated above, the suggested set of 32 features is a hypothetical one and is subject to a thorough feature selection procedure. This section will cover the feature selection methods which we applied to our features. These are univariate feature selection, recursive feature elimination, selection by model coefficients, and decision trees. Univariate feature selection gave the best results, so we will provide the results only for this method here.

Since all the feature values are numerical and the target variable is numerical too, it is convenient to view the task of readability assessment as a regression task. We do not aim at obtaining the predicted readability level as an integer, because float numbers are not less illustrative in this case. This assumption is supported by the fact that first readability formulas were built as linear models and so were many modern ones [4, 10, 12].

The selected subsets of features formed a basis for 4 linear regression models: one for every subject and one universal model trained on the texts on all the 3 subjects. We used the Ordinary Least Squares Regression algorithm implementation provided at the ScikitLearn library (LinearRegression). To evaluate the models, we computed 2 parameters: root mean squared error (RMSE) and determination coefficient ($R^2$). RMSE was chosen as the basis for model comparison.

**Univariate Feature Selection.** The univariate feature selection method used in this work aims at evaluating the predictors from the initial feature set based on some criterion and selecting a subset of features that will form the basis of the model. We chose 2 selection criteria: mutual information and F-score function between a predictor and the target variable since they are both commonly used for linear regression tasks. Figure 1 presents a dependency between the number of features selected by the method and the performance of a model based on this number of features. These results were obtained by training a model for every number of features and validating using 10-fold cross-validation.

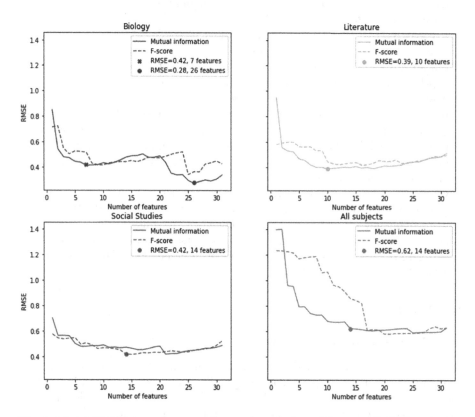

**Fig. 1.** Model RMSE dependency on the number of selected features. Solid lines represent results obtained with mutual information criteria and dashed line represent that of F-score function. The values of RMSE are negative to visualize the effect as 'the higher – the better'.

It can be observed that for every model there is a too small and a too large number of features with which the model's quality decreases. Approximately from 1 to 7 features the error decreases significantly with each added predictor and reaches a plateau (with a slight decrease) until it grows again when the model starts to overfit due to an excess of features.

Since we seek to find a model with the best possible quality and at the same time face a restriction on the number of features, for every model we chose a point at the start of the plateau with the least RMSE. These points are marked with circles in Fig. 1. There was the only exception for the Biology model because after the plateau there is a notable decrease in error. That is, the best point is 26 features, being much larger than 7 features (which is the first peak point marked with a cross on Fig. 1). We set this model to be based on 7 features for the sake of interpretability.

Summarizing, the following features were kept in the final model:

- **Biology (7 features)**: average number of words (1), average number of syllables per word (3), verbs in active voice (8), personal pronouns (17), syntax tree depth (25), number of embedded noun phrases (28), maximum linear distance between parent and child (32).
- **Literature (10 features)**: average number of words (1), average number of characters per word (2), finite verbs (6), verbs in active voice (8), nouns (14), full adjectives (15), personal pronouns (17), content words (23), syntax tree depth (25), number of embedded noun phrases (28).
- **Social studies (14 features)**: average number of words (1), average number of characters per word (2), average number of syllables per word (3), average word frequency (4), non-finite verbs (7), verbs in active voice (8), passive participles (11), adverbs (12), personal pronouns (17), other pronouns (18), negative particles (21), particles (22), syntax tree depth (25), number of embedded noun phrases (28).
- **General (14 features)**: average number of words (1), average number of characters per word (2), average number of syllables per word (3), average word frequency (4), average Juilland coefficient (5), finite verbs (6), verbs in active voice (8), adverbs (12), nouns (14), personal pronouns (17), content words (23), syntax tree depth (25), number of embedded noun phrases (28), ratio of infinite clauses (31).

## 4   Results

### 4.1   Evaluation of the Models

To evaluate the models' quality, we created 10 random splits of our dataset into train and test sets in proportion 75/25. The estimates presented in Table 3 are averaged over these 10 evaluations. The sample size is set to default 500 sentences. For comparison, we also evaluate the model 2 on our data.[2]

---

[2] We do not evaluate the adjusted FRE-formula (1), [10] as it predicts the abstract readability score, not the school grade.

**Table 3.** Model performance across different subjects (500 sentences per sample) compared to formula 2.

| Subject | Biology | Literature | Social studies | All subjects |
|---|---|---|---|---|
| Opt. number of features | 7 | 10 | 14 | 14 |
| RMSE (our model) | 0.52 | 0.45 | 0.49 | 0.77 |
| RMSE (2) | 1.36 | 1.03 | 1.15 | 1.22 |

As it was stated above, we validated these results on different sample sizes. The values of RMSE depending on sample size are presented in Table 4.

**Table 4.** Model performance across different subjects evaluated with different sample sizes (RMSE values).

| Subject | Biology | Literature | Social studies | All subjects |
|---|---|---|---|---|
| Sample size = 200 | 0.48 | 0.54 | 0.39 | 0.59 |
| Sample size = 500 | 0.54 | 0.44 | 0.43 | 0.63 |
| Sample size = 900 | 0.35 | 0.35 | 0.33 | 0.54 |

It can be observed that within all the experiments the universal models tend to be inferior compared to the single-subject models. However, introducing separate models for different subjects or domains may be inefficient under some circumstances, so it is necessary to define some criteria to assess a model independently from other models and decide on whether a model is acceptable. We set this RMSE threshold to be 0.5 because in terms of defining whether a text is appropriate for students of a particular grade such an error seems to be acceptable. Of course, this assumption needs a more proper investigation in future works.

## 4.2    Analysis of Selected Features

To analyze which features were selected we plotted a bar chart that presents the coefficients from the 4 linear models according to the selected features. An important thing to notice here is that the so-called traditional features (1–3) turned out to be among the most 'powerful' in terms of readability prediction.

Four features (#3, #8, #17, and #25) were selected for all the four models. So we decided to measure the contribution of the other selected features to each of the models over these 4 features. The results are present in Table 5.

**Table 5.** Model performance across different subjects evaluated with different sample sizes (RMSE values).

| Subject | Biology | Literature | Social studies | All subjects |
|---|---|---|---|---|
| Number of features | 4 | 4 | 4 | 4 |
| RMSE | 0.65 | 0.59 | 0.61 | 1.13 |
| Number of features | 7 | 10 | 14 | 14 |
| RMSE | 0.52 | 0.45 | 0.49 | 0.77 |

Notice that for all 3 individual subjects the model on 4 features achieves performance comparable to the one of the optimal model. However, for the entire dataset the relative gap between these two classes of models occurs to be more significant. It implies that feature weights vary considerably between the universal models and the subject-specific models.

## 5 Discussion

The formulas suggested in this paper differ significantly from existing readability formulas, first of all, in terms of the number of predictors. A modern formula applied to texts on different subjects showed noticeable variance in the values of RMSE which is chosen as the main parameter to evaluate and compare readability models. The new models developed separately for each of the 3 subjects (Biology, Literature, and Social Studies) outperformed the previous formulas.

This was explained during the analysis of the features selected by the univariate feature selection method. The observation showed that there is not much coincidence in feature sets for different subjects and thus a significant decrease in model performance is expected when the model trained on one subject is applied to texts on other subjects. Though there is a common "core" of 4 features that are kept for all the domains, even their weights vary significantly across subjects.

Also, an increase in models' quality is explained by newly added features which were chosen according to the specific characteristics of texts on each of the subjects. Although traditional features such as 'average sentence length' and 'average syllables per word' turned out to be the most informative for assessing text readability, we managed to obtain a significant improvement in models' performance using features referring to linguistic characteristics of a text: morphology and syntax. Further improvement may be achieved by taking into account lexical characteristics but we leave it for future work. We also expect that lexical characteristics tend to be more domain-dependent than more abstract features (Fig. 2).

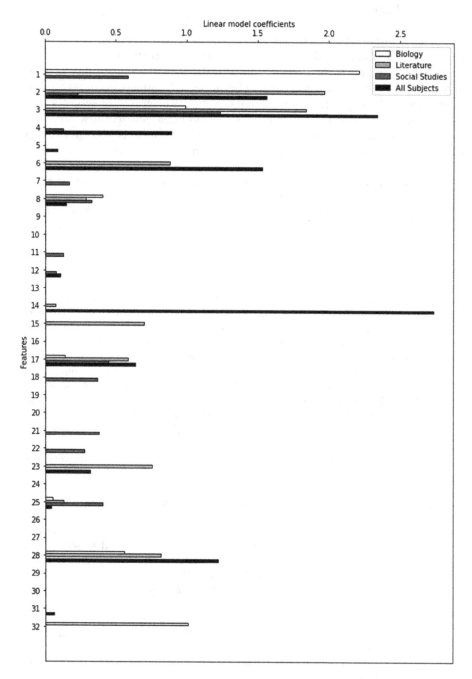

**Fig. 2.** Linear models' coefficients for the selected features. The features which have 0 were not selected to any of the four models.

## 6 Conclusion

The results obtained in this pilot study represent the next step in the development of readability models for Russian educational texts. To the best of our knowledge, there have not been any investigations of the universality of readability models for educational texts. The presented findings encourage future studies on readability formulas for different subjects or groups of subjects.

The main result of the paper is that the most relevant features vary from subject to subject, nevertheless 'traditional features' (such as average sentence length or word length in syllables) are present in all feature sets.

**Acknowledgements.** We thank V. Solovyev, M. Solnyshkina, V. Ivanov et al. for publishing their database of schoolbook texts that we used in our study. This contributed greatly to the findings obtained in our work. We are also very grateful to anonymous AIST reviewers whose thorough comments helped to improve the paper.

## References

1. Collins-Thompson, K., Callan, J.P.: A language modeling approach to predicting reading difficulty. In: Proceedings of the Human Language Technology Conference of the North American Chapter of the Association for Computational Linguistics: HLT-NAACL 2004, pp. 193–200 (2004)
2. Dell'Orletta, F., Wieling, M., Venturi, G., Cimino, A., Montemagni, S.: Assessing the readability of sentences: which corpora and features? In: Proceedings of the Ninth Workshop on Innovative Use of NLP for Building Educational Applications, pp. 163–173. Association for Computational Linguistics (2014). https://doi.org/10.3115/v1/W14-1820. http://aclweb.org/anthology/W14-1820
3. Flesch, R., Gould, A.J.: The Art of Readable Writing, vol. 8. Harper, New York (1949)
4. Ivanov, V., Solnyshkina, M., Solovyev, V.: Efficiency of text readability features in Russian academic texts. In: Komp'juternaja Lingvistika I Intellektual'nye Tehnologii, pp. 267–283 (2018)
5. Juilland, A., Chang-Rodríguez, E.: Frequency dictionary of Spanish words. Technical report (1964)
6. Kincaid, J.P., Fishburne Jr., R.P., Rogers, R.L., Chissom, B.S.: Derivation of new readability formulas (automated readability index, fog count and Flesch reading ease formula) for Navy enlisted personnel. Technical report, Naval Technical Training Command Millington TN Research Branch (1975)
7. Kuhn, M., Johnson, K.: Applied Predictive Modeling. Springer, New York (2013). https://doi.org/10.1007/978-1-4614-6849-3
8. Matskovsky, M.: Problems of readability of printed material. In: Semantic Perception of Speech Messages in Conditions of Mass Communication, Nauka, pp. 126–142 (1976)
9. Mikk, Y.: On factors of comprehensibility of educational texts. Ph.D. thesis, Tartu University (1970)
10. Oborneva, I.: Automatic assessment of the complexity of educational texts on the basis of statistical parameters. Ph.D. thesis, Moscow State Pedagogical University (2006)

11. Solnyshkina, M.I., Kiselnikov, A.S.: Text complexity: study phases in Russian linguistics. Vestnik Tomskogo gosudarstvennogo universiteta. Filologiya **6**(38), 86–99 (2015). https://doi.org/10.17223/19986645/38/7
12. Solovyev, V., Ivanov, V., Solnyshkina, M.: Assessment of reading difficulty levels in Russian academic texts: approaches and metrics. J. Intell. Fuzzy Syst. **34**(5), 3049–3058 (2018)
13. Solovyev, V., Solnyshkina, M., Ivanov, V., Batyrshin, I.: Prediction of reading difficulty in Russian academic texts. J. Intell. Fuzzy Syst. **36**(5), 4553–4563 (2019)
14. Straka, M., Straková, J.: Tokenizing, POS tagging, lemmatizing and parsing UD 2.0 with UDPipe. In: Proceedings of the CoNLL 2017 Shared Task: Multilingual Parsing from Raw Text to Universal Dependencies, pp. 88–99 (2017)
15. Zakaluk, B., Samuels, S.: Issues related to text comprehensibility: the future of readability. Revue québécoise de linguistique **25**(1), 41–59 (1996)

# A Multi-task Learning
# Approach to Text Simplification

Anna Dmitrieva[1,2](✉) and Jörg Tiedemann[1]

[1] University of Helsinki, Helsinki, Finland
{anna.dmitrieva,jorg.tiedemann}@helsinki.fi
[2] National Research University Higher School of Economics, Moscow, Russia

**Abstract.** We propose a multi-task learning approach to reducing text complexity which combines text summarization and simplification methods. For the purposes of this research, two datasets were used: the Simple English Wikipedia dataset for simplification and the CNN/DailyMail dataset for summarization.

We describe several experiments with reducing text complexity. One experiment consists in first training the model on summarization data, then fine-tuning it on simplification data. Another experiment involves training the model on both datasets simultaneously while augmenting source texts with a task-specific tag that shows the model which task (summarization or simplification) needs to be performed on a given text. Models with a similar architecture were also trained on each dataset separately for comparison. Our experiments have shown that the multi-task learning approach with task-specific tags is more effective than the fine-tuning approach, and the models trained for both tasks simultaneously can perform as good at each of them as the models that were trained only for that specific task.

**Keywords:** Text simplification · Text complexity · Abstractive summarization

## 1 Introduction

Text simplification is the process of reducing grammatical and lexical complexity of a text, while retaining its information content. In broad terms, text simplification can involve lexical and grammatical simplification, clarifying modification, when some concepts are explained in more detail, or summarization, when unnecessary or repeating fragments of text are being removed. Automatic text summarization, however, is a complex natural language processing task on its own. An automatic text summarization system generates a shorter version, or a summary, of a text which contains the most important information about a topic. A summary generated by an automatic text summarizer should consist of the most relevant information in a document and at the same time, it should occupy

W. M. P. van der Aalst et al. (Eds.): AIST 2020, CCIS 1357, pp. 78–89, 2021.
https://doi.org/10.1007/978-3-030-71214-3_7

less space than the original document [5]. Text summarization can be extractive, when relevant sentences are simply extracted from the text, or abstractive, when the summary aims to capture the essence of the text, sometimes involving rewriting selected information.

These two tasks are sometimes combined in the development of systems for text complexity reduction. Saggion [18] lists multiple examples of such systems. In one of them, sentence simplification was used as a part of a multi-document extractive summarization algorithm. Sentences were simplified before clustering and selecting relevant sentences from each cluster [18]. In another system [14], lexical simplification is used during summary generation to replace difficult words. However, the usage of summarization techniques to aid simplification systems has not yet been explored. Considering summarization as a process that makes a text simpler, we propose a framework where summarization assists text simplification in a joint learning or a multi-task setup.

In this work, we propose to reduce text complexity using methods and datasets for text simplification and summarization. We aim to develop a system that in the future can be used for languages where datasets for training simplification models are not available. We believe that in most languages, summarization-specific datasets are easier to build than simplification datasets: for example by aligning Wikipedia article summaries (short paragraphs that preface the main body of text in an article) with the text of the article. Simplification-specific datasets, on the other hand, are difficult to build, since a lot of languages such as Russian do not have "simple" versions of regular media sources, like Simple Wiki or news/information written in simple language. Besides, both simplification and abstractive summarization require a good understanding of text semantics, so finding a proper way to preserve the meaning of texts will benefit both tasks.

## 2   Related Work

The field of automatic text simplification dates back to the middle of the 1990s [11]. Early work on automatic text simplification mainly aimed at grammar and style simplification in English. In the beginning, text simplification systems were mostly rule-based. One of the first works which dealt with automatic text simplification belongs to Chandrasekar and Srinivas [4]. Their research focused on syntax simplification techniques. The main goal was to speed up and reduce the error rate for other NLP tasks, such as syntactic parsing and machine translation.

Today the task of text simplification is often viewed as a monolingual translation problem [26]. Once viewed as such, a natural approach is to use sequence-to-sequence (seq2seq) neural machine translation (NMT) models [13]. The seq2seq NMT systems attempt to build a single large neural network such that every component is tuned based upon training sentence pairs [26]. While models of this kind tend to have drawbacks such as copying directly from the original sentence, making the output unnecessarily complex [13], there are solutions that can help to make up for these kinds of problems. Using deep neural networks allows for creating simple and fluent modifications that preserve the meaning of the original sentence [27]. State-of-the-art text simplification systems are able to generate shorter and simpler sentences according to Flesch-Kincaid grade level and human judgments, and provide a comprehensive analysis using human evaluation and a qualitative error analysis. Text simplification systems can also be trained on unlabeled data in an unsupervised manner [23].

There are multiple approaches to text summarization. The majority of methods that are used today are methods of extractive summarization. However, in recent years, with the rise of neural text generation techniques, abstractive techniques are also becoming increasingly popular [16]. For example, in [6] a completely abstract model biased with a powerful content selector is proposed. In this research, a data-efficient content selector is used to select phrases from a source document that should be part of the summary. This approach not only can be used for obtaining accurate summaries, but also can easily be transferred to a new domain [6]. Finally, sometimes extractive and abstractive approaches can be combined in one summarization system. For example, in [16] an extractive summarization model is trained end-to-end using abstractive summaries.

All systems mentioned above in this section were trained on and work with English language data. As mentioned before, simplification datasets are more difficult to produce than summarization datasets, therefore for the majority of languages the text simplification problem is understudied. Some examples of non-English text simplification research include the creation of the Spanish simplification corpus as part of the Simplext project [3] and developing a complex system for extracting texts with high readability from a pre-crawled corpus and performing text adaptation for learners of Russian as a foreign language [9].

## 3   Data

For the purposes of this research, two different datasets were used: one designed for summarization and one for simplification. We used data aligned at the document level instead of sentence-to-sentence alignment approach commonly used in simplification. First, this alignment is common for text summarization datasets as it allows models to learn to omit whole sentences and parts of the text. Second, for languages without large simplification datasets, it would be easier and less time-consuming to create a simplification dataset aligned at the document level. Based on these prerequisites, the Simple Wikipedia dataset [10] was chosen for simplification and the CNN/DailyMail dataset [15] for summarization. Both of these datasets are frequently used for their respective tasks: for example, CNN/DailyMail is used for training and testing summarization models in [6] and [16] and Simple Wiki is used as part of a simplification dataset in [27]. Thus we can conclude that, although these datasets have some drawbacks, they are sufficient for correct training of our models.

**Simplification.** There are not many datasets dedicated specifically to the purpose of text simplification. For approximately five years, Simple Wikipedia has been the main source of data in this field [26]. Today, some new simplification datasets have emerged, including Newsela [26]. Although less noisy, they tend to provide only sentence-to-sentence alignments. That is why for the purposes of this research the Simple Wikipedia dataset was chosen. It comprises texts from Simple English Wikipedia and corresponding articles from English Wikipedia [10]. The data was downloaded from Wikipedia in May 2013 and comprises all Simple Wiki articles that at that time also had a corresponding article in English Wikipedia. This dataset has versions with sentence-to-sentence and document-to-document alignment. The document aligned version consists of full texts separated into sentences, each sentence provided with the article name and paragraph number. For the purposes of this research, we aligned the documents text-to-text, so the "regular" version of each text corresponded to its simpler version with the same name. This dataset has some flaws including noise and incorrect alignments (for instance, sometimes the last sentence of one article becomes the first sentence in another), but otherwise, it is the largest openly available dataset that suits the purpose of this research.

**Summarization.** For this task, we chose the CNN/DailyMail news articles dataset[1]. CNN/DailyMail comprises multi-sentence human-generated abstractive summaries of news articles aligned with full articles [15]. It is a version of the dataset that was presented in a shared task of question answering [7]. Articles were collected starting in April 2007 for CNN and June 2010 for the Daily Mail, both until the end of April 2015. Validation data is from March, test data is from April 2015 [7].

The parameters of the datasets used in this research are listed in Table 1.

**Table 1.** Parameters of datasets.

|  | CNN/DailyMail | Simple Wiki | All |
|---|---|---|---|
| **Documents** | | | |
| Train set | 287227 | 58775 | 346002 |
| Validation set | 13368 | 500 | 13868 |
| Test set | 11490 | 500 | 11990 |
| **Tokens (words)** | | | |
| Train set source | 199570815 | 84975889 | 284546704 |
| Train set target | 14050660 | 7867296 | 21917956 |
| Validation set source | 9092698 | 691540 | 9784238 |
| Validation set target | 73173 | 66381 | 139554 |
| Test set source | 7901203 | 726250 | 8627453 |
| Test set target | 597406 | 80049 | 677455 |

---

[1] https://github.com/harvardnlp/sent-summary.

# 4   Preprocessing

**Training:** both datasets described above come with tokenization that inserts spaces between words and punctuation. For most of the experiments described here, that format was preserved. In addition, all texts were converted to lowercase and non-ASCII symbols were removed.

*Sentence Boundary Tagging:* For some of the models, an additional tokenization method was applied to the data. It has been shown that on the CNN/DailyMail dataset the summarization models can perform better if sentence boundary tagging is applied to the target text beforehand [6] like this: ‹t› w1 w2 w3 . ‹/t›. We decided to try this technique on both datasets and compare the performance of the models.

*Tagging Source Sentences with Task-specific Tags:* In addition to training models separately on different datasets, we also attempted to train on both datasets simultaneously, merging them together and shuffling. We use a method introduced before for multilingual neural machine translation. In this method, an artificial token is added to the input sequence to indicate the required target language [8]. In our research, the artificial token in front of each text in the joined dataset indicates if the text needs to be summarized of simplified. This token is added in the front of the source sentence, so hereafter we call these tokens "front tags". Our front tags are ‹2sum› for summarization and ‹2simp› for simplification.

*SentencePiece:* As an experiment, another way of preprocessing using SentencePiece technology[2] was tested on the Simple Wiki dataset. SentencePiece is a text tokenizer that implements a unigram language model and subword units like byte-pair encoding. For this experiment, all texts in the dataset were detokenized using the Moses toolbox[3]. After that, SentencePiece unigram and byte-pair encoding models were trained and used on the Simple Wiki texts. This tokenization technique was applied only on Simple Wiki and the joined datasets.

We use the OpenNMT-py toolkit [12] for preprocessing our data, as well as for training and testing the models. We follow the approach of See et al. [19] also used by Gehrmann et al. [6] for automatic summarization on CNN/DailyMail and truncate the source texts to 400 tokens and the target texts to 100 tokens. Although such truncation might seem brutal, in [19] it was proven that at least for the CNN/DailyMail dataset it can actually raise the performance of the summarization model. We also used a dynamic dictionary and shared vocabulary to ensure that source and target sentences are aligned and use the same dictionary, which is needed for copy attention (an implementation of pointer-generator networks that considers copying words from the source sequences [6,24]).

---

[2] https://github.com/google/sentencepiece.
[3] https://github.com/moses-smt/mosesdecoder.

**Evaluation:** Before evaluation, output texts produced by the models, as well as source and target test texts, were detokenized with the help of the Moses decoder or the SentencePiece decoder if the corresponding encoding was used. All additional tags were removed. Source and target texts were also cropped so that their length matches the size of the texts used for training: 400 words for source texts, 100 for target.

## 5    Models

For our experiments, we used the Pointer-Generator model architecture proposed in [19] and also used in [6]. The primary reason for the choice of the model and its parameters was twofold. First, the content selector used in this architecture has been proved to be simple yet accurate and improve the performance of summarization models [6]. Second, due to its data-efficiency, this technique can easily be adjusted to a new domain, which in our case is the simplification task [6]. Since this architecture has already been used on the CNN/DailyMail dataset, we are using similar model parameters to those used for this dataset before.

The model was built using the OpenNMT toolkit. It uses a one-layer LSTM with 512 hidden states and an embedding size of 128. The encoder is a bidirectional LSTM with 512 hidden states (256 in both directions). It uses copy-attention [24] which allows to copy words from the source. The attention mechanism that the model uses was introduced in [2]. Their approach allows for information to be spread throughout the sequence of annotations, which can be selectively retrieved by the decoder accordingly. The standard attention is then reused as copy attention. The loss is modified in a way that makes it divide the loss of a sequence by the number of tokens in it, which was proven by [6] to generate longer sequences during inference. This model uses Adagrad optimizer, no dropout, and gradient-clipping with a maximum norm of 2.

At the inference stage, beam search with a beam size of 10 is used, because it has been found out that bottom-up attention requires a larger beam [6]. Multiple penalties are applied: length penalty is used to encourage longer sequences, coverage penalty is used to avoid repetitions, and repeating trigrams are blocked.

The following setups were tried:

**Experiments on One Dataset.** In order to correctly evaluate the performance of the models trained on the two datasets, the models were also trained on each dataset separately. Moreover, separate models were trained on versions of the datasets with and without sentence boundary tagging. In addition to that, we also experimented with applying different SentencePiece tokenization and comparing the performance of models trained on data preprocessed with SentencePiece to that of models trained on regular data.

**Experiments with Fine-tuning.** In an attempt to train a model that would combine the summarization and simplification operations, we tried a fine-tuning approach. We first trained a model on the summarization dataset and then fine-tuned it on the smaller simplification dataset. For this experiment, we also tried versions of the dataset with and without sentence boundary tagging.

**Experiments on the Joined Dataset.** As mentioned in the above paragraph, some models were trained on both datasets simultaneously. All source texts were augmented with task-specific tags. In one experiment, the model was trained on a joined dataset with the volume of each original dataset preserved. In another experiment, the summarization data was undersampled and the simplification data was oversampled (with some texts being repeated) so that the amount of source simplification data being 117,550 texts and the summarization data being 287,227 texts. In the third experiment, the model mentioned in the first experiment was additionally fine-tuned on simplification data that it had already seen.

## 6 Evaluation

For evaluation we used metrics usually applied to text simplification and text summarization evaluation. The metrics were BLEU, SARI and FKGL (Flesch-Kincaid Grade Level) from the EASSE package [1] and also a pure Python implementation of the ROUGE score[4]. Despite BLEU being widely used for different neural machine translation tasks, its effectiveness on text simplification is sometimes doubted [22], that is why we also use SARI to evaluate the simplification of texts. SARI measures how the simplicity of a sentence is improved based on the words added, deleted and kept by a system [1]. For more extensive evaluation of a text complexity reduction system it is indeed best to also have human assessors, but in this initial study we limit ourselves to the listed metrics.

Since there has not been research featuring simplification or summarization systems that were built using both datasets that we are using simultaneously, or the Simple Wiki dataset aligned text-to-text the way that it is done here, it is difficult to compare most of our results to other work. Therefore, for the most part, we will only compare different results of our models.

**Experiments on One dataset.** Among the models trained on just one dataset, be it CNN/DailyMail or Simple Wiki (not including the joined dataset), the best ROUGE scores were obtained by the model trained on Simple Wiki data preprocessed with SentencePiece BPE tokenization. However, in terms of BLEU, SARI and FKGL scores the model trained on Simple Wiki without specific tokenization seems to perform slightly better, outperforming both BPE and sentence boundary tokenization (see Table 2).

In all tables below "R" stands for the ROUGE score. To make the data easier to read, we limited ourselves to reporting only F-scores of ROUGE-1, 2 and L.

---

[4] https://github.com/pltrdy/rouge.

**Table 2.** Testing the models trained on the Simple Wiki data on the Simple Wiki test set.

| Data type | BLEU | SARI | FKGL | R1 F | R2 F | RL F |
|-----------|------|------|------|------|------|------|
| BPE | 22.89 | 51.11 | 7.77 | 0.46 | **0.31** | **0.47** |
| Tagged | 22.97 | 50.65 | 7.67 | **0.50** | 0.26 | 0.41 |
| Plain | **27.48** | **51.54** | **7.40** | 0.43 | 0.27 | 0.43 |

As for the models trained only on the CNN/DailyMail dataset, the version with sentence boundary tagging seems to give slightly better performance than the plain version in terms of simplicity and readability metrics (see Table 3). Nevertheless, the ROUGE scores of these models were not higher than those of some models described below (for example, compare Table 3 to Table 5).

**Table 3.** Testing the models trained on the CNN/DailyMail data on the CNN/DailyMail test set.

| Data type | BLEU | SARI | FKGL | R1 F | R2 F | RL F |
|-----------|------|------|------|------|------|------|
| Tagged | **13.88** | **42.79** | 9.35 | 0.33 | 0.13 | 0.31 |
| Plain | 13.44 | 42.70 | **9.33** | 0.33 | 0.13 | 0.31 |

**Fine-tuning Experiments.** The output of summarization models fine-tuned on simplification data (both with and without sentence boundary tagging), as seen in Table 4, proved to be inferior in quality to the output of models trained on one dataset. Readability, however, improved on the CNN/DailyMail test set. As can be seen, the presence of sentence boundary tagging does not make a big difference here, although it does slightly affect the performance. The ROUGE scores also were not improved in comparison to some other models (for example, compare Table 4 to Table 3 above):

**Table 4.** Testing the models trained on the CNN/DailyMail data and fine-tuned on the Simple Wiki data on the CNN/DailyMail test set.

| Data type | BLEU | SARI | FKGL | R1 F | R2 F | RL F |
|-----------|------|------|------|------|------|------|
| CNNDM fine-tuned on Simple Wiki | 10.53 | 41.01 | **8.91** | 0.32 | 0.12 | 0.30 |
| Tagged version | **10.96** | **41.32** | 9.50 | **0.33** | **0.13** | **0.31** |

**Experiments on the Joined Dataset.** The models trained on the joined dataset with different configuration described above (see Sect. 5, Experiments on the joined dataset) obtained higher ROUGE scores on the CNN/DailyMail test set compared to all other models. It should be noted that models with an architecture similar to ours trained on this dataset can get ROUGE-2 F-scores as big as 17.25 [6] (our highest score on this scale is 14), and the newest baseline ROUGE-2 F-score for CNN/DailyMail equals 19.24 [17].

**Table 5.** Testing the models trained on the joined data with front tags on the CNN/DailyMail test set.

| Data type | BLEU | SARI | FKGL | R1 F | R2 F | RL F |
|---|---|---|---|---|---|---|
| Joined dataset | 12.38 | 41.52 | **10.08** | 0.34 | 0.14 | 0.32 |
| Undersample CNNDM, oversample SimpleWiki | **13.02** | **41.93** | 10.21 | **0.35** | 0.14 | 0.32 |
| Joined dataset fine-tuned on SimpleWiki | 12.50 | 41.59 | 10.20 | 0.34 | 0.14 | 0.32 |

However, the results shown on the Simple Wiki test set were less high than those described above in Table 2.

Table 6 presents a comparison of three models: a model trained on the joined dataset, a model trained on undersampled summarization data and oversampled simplification data, and a model trained on joined dataset and fine-tuned on simplification data (see Sect. 5, Models). In terms of performance on the joined test set, the model trained on joined data without sampling or fine-tuning seems to have the best simplicity and readability scores. The ROUGE scores, however, did not really differ from one model to another.

**Table 6.** Testing the models trained on the joined data on the joined test set.

| Data type | BLEU | SARI | FKGL | R1 F | R2 F | RL F |
|---|---|---|---|---|---|---|
| Joined dataset | **14.10** | **42.51** | 9.79 | **0.35** | **0.14** | **0.33** |
| Undersample CNNDM, oversample SimpleWiki | 11.63 | 39.86 | 9.94 | 0.33 | 0.13 | 0.31 |
| Joined dataset fine-tuned on SimpleWiki | 13.61 | 42.05 | 10.21 | **0.35** | **0.14** | **0.33** |

The models generally perform better when evaluated on the test set from the same dataset they were trained on than on a test set from a different dataset. In our case, the results of models trained trained on the joined dataset are similarly high with the results of models trained on just one dataset when evaluated on a corresponding test set. For example, the scores obtained from the model trained on the joined data on the Simple Wiki test set were similar to, but not higher than those obtained on the same test set from the model trained specifically on Simple Wiki data (compare Table 7 to Table 2).

It should be noted that, unlike in models trained only on the Simple Wiki dataset, using SentencePiece tokenization did not improve the performance of the models trained on the joined dataset.

**Table 7.** Testing the models trained on the joined data on the Simple Wiki test set.

| Data type | BLEU | SARI | FKGL | R1 F | R2 F | RL F |
|---|---|---|---|---|---|---|
| Joined dataset | **26.13** | **50.18** | **8.37** | 0.43 | **0.27** | 0.43 |
| Undersample CNNDM, oversample SimpleWiki | 24.81 | 49.38 | 8.40 | 0.43 | 0.26 | 0.43 |
| Joined dataset fine-tuned on SimpleWiki | 25.40 | 49.73 | 8.40 | 0.43 | 0.26 | 0.43 |

In order to see if the models understand the semantics of task-specific tags, we compared the performance of the models trained on the joined dataset on regular test sets and on test sets with reversed tags (where ‹2sum› becomes ‹2simp› and vice versa). Evaluation scores somewhat decreased but the difference was less than expected. The length of the output texts, however, has slightly increased on average, which is reflected in the increased readability scores (compare Table 8 to Table 6):

**Table 8.** Testing the models trained on the joined data on the joined test set with reversed front tags.

| Data type | BLEU | SARI | FKGL | R1 F | R2 F | RL F |
|---|---|---|---|---|---|---|
| Joined dataset | **12.79** | **41.92** | 10.56 | **0.35** | **0.14** | **0.33** |
| Undersample CNNDM, oversample SimpleWiki | 11.58 | 38.79 | **10.09** | 0.33 | 0.13 | 0.31 |
| Joined dataset fine-tuned on SimpleWiki | 12.13 | 41.43 | 10.46 | **0.35** | **0.14** | **0.33** |

A look at a small amount of randomly selected output texts confirmed that, on average, the simplified articles are longer than summaries. This is interesting given that all target texts are truncated to the same length during preprocessing. Examining the output also confirmed that the same source text with different tags will be processed differently by a model. This can be illustrated by the following example. Below is an article from the Simple Wiki dataset [10], which is still present in both English Wikipedia[5] and Simple Wiki[6]. These are the versions of the texts that were used for the evaluation of the models.

**Original text:** sofia wistam -lrb- born 15 may 1966 in liding, stockholm county, sweden -rrb- is a swedish television host on tv4 and tv3 and radio talk-show host. She has also worked as a stylist for stars such as carola, jerry williams and tommy nilsson. in 2008 she was also a judge on the talent show sweden's got talent, during this year she also hosted her own radio show on rix fm. during 2009 sofia will host the competition show on swedish television.

**Target text:** sofia wistam -lrb- may 15 1966 -rrb- is a swedish television host and radio talk-show host.

**Output with <2simp> tag:** sofia wistam -lrb- born 15 may 1966 in liding, stockholm county, sweden -rrb- is a swedish television host on tv4 and tv3 and

---

[5] https://en.wikipedia.org/wiki/Sofia_Wistam.
[6] https://simple.wikipedia.org/wiki/Sofia_Wistam.

radio talk-show host. She has also worked as a stylist for stars such as carola, jerry williams and tommy nilsson.

**Output with <2sum> tag:** sofia wistam is a swedish television host on tv4 and tv3 and radio talk-show host. She has also worked as a stylist for stars such as carola, jerry williams and tommy nilsson.

The outputs with different tags are not the same even when the source text is quite short. However, it is hard to pinpoint the exact differences that each tag triggers, not only because the evaluation of such phenomena is generally difficult, but also because the tasks of summarization and simplification seem to be less distinguishable in their nature than, for example, the tasks of translating a text into two different languages.

## 7   Conclusion

In this paper we described the attempts to reduce text complexity using summarization and simplification data and models. We used different approaches such as fine-tuning summarization models on simplification data and training a model to solve both tasks simultaneously. In addition to that, different preprocessing techniques were applied. Our experiments have shown that, in terms of preprocessing, using the BPE tokenization seems to give better results in summarization for models trained on a single dataset. Sentence boundary tagging, although it improved summarization performance on the CNN/DailyMail dataset, did not seem to be effective on other data. As for training the models, using joined datasets with task-specific tags proved to be more effective than training a model for one task and fine-tuning it for another. The models trained on the joined dataset can perform different tasks with the same effectiveness as models trained for one task and tested on the corresponding data. The different tokenization approaches tested did not seem to have a significant effect on the performance of these models.

## References

1. Alva-Manchego, F., Martin, L., Scarton, C., Specia, L.: EASSE: easier automatic sentence simplification evaluation. arXiv preprint arXiv:1908.04567 (2019)
2. Bahdanau, D., Cho, K., Bengio, Y.: Neural machine translation by jointly learning to align and translate. arXiv preprint arXiv:1409.0473 (2014)
3. Bott, S., Saggion, H.: An unsupervised alignment algorithm for text simplification corpus construction. In: Proceedings of the Workshop on Monolingual Text-To-Text Generation, pp. 20–26 (2011)
4. Chandrasekar, R., Srinivas, B.: Automatic induction of rules for text simplification. Knowl.-Based Syst. **10**(3), 183–190 (1997)
5. Gambhir, M., Gupta, V.: Recent automatic text summarization techniques: a survey. Artif. Intell. Rev. **47**(1), 1–66 (2016). https://doi.org/10.1007/s10462-016-9475-9
6. Gehrmann, S., Deng, Y., Rush, A.M.: Bottom-up abstractive summarization. arXiv preprint arXiv:1808.10792 (2018)

7. Hermann, K.M., Kocisky, T., Grefenstette, E., Espeholt, L., Kay, W., Suleyman, M., Blunsom, P.: Teaching machines to read and comprehend. In: Advances in Neural Information Processing Systems, pp. 1693–1701 (2015)

8. Johnson, M., et al.: Enabling zero-shot translation: Google's multilingual neural machine translation system. Trans. Assoc. Comput. Linguist. **5**, 339–351 (2017)

9. Karpov, N., Sibirtseva, V.: Towards automatic text adaptation in Russian. Higher School of Economics Research Paper No. WP BRP, 16 (2014)

10. Kauchak, D.: Improving text simplification language modeling using unsimplified text data. In: Proceedings of the 51st Annual Meeting of the Association for Computational Linguistics (Long papers), vol. 1, pp. 1537–1546, August 2013

11. Keskisärkkä, R.: Automatic text simplification via synonym replacement (2012)

12. Klein, G., Kim, Y., Deng, Y., Senellart, J., Rush, A.M.: Opennmt: open-source toolkit for neural machine translation. arXiv preprint arXiv:1701.02810 (2017)

13. Kriz, R., et al.: Complexity-weighted loss and diverse reranking for sentence simplification. arXiv preprint arXiv:1904.02767 (2019)

14. Lal, P., Ruger, S.: Extract-based summarization with simplification. In: Proceedings of the ACL, July 2002

15. Nallapati, R., Zhou, B., Gulcehre, C., Xiang, B.: Abstractive text summarization using sequence-to-sequence RNNs and beyond. arXiv preprint arXiv:1602.06023 (2016)

16. Nallapati, R., Zhai, F., Zhou, B.: SummaRuNNer: a recurrent neural network based sequence model for extractive summarization of documents. In: Thirty-First AAAI Conference on Artificial Intelligence (2017)

17. Raffel, C., et al.: Exploring the limits of transfer learning with a unified text-to-text transformer. arXiv preprint arXiv:1910.10683 (2019)

18. Saggion, H.: Automatic text simplification. Synth. Lect. Hum. Lang. Technol. **10**(1), 1–137 (2017)

19. See, A., Liu, P.J., Manning, C.D.: Get to the point: Summarization with pointer-generator networks. arXiv preprint arXiv:1704.04368 (2017)

20. Siddharthan, A.: A survey of research on text simplification. ITL-Int. J. Appl. Linguist. **165**(2), 259–298 (2014)

21. Štajner, S., Saggion, H.: Data-driven text simplification. In: Proceedings of the 27th International Conference on Computational Linguistics: Tutorial Abstracts, pp. 19–23 (2018)

22. Sulem, E., Abend, O., Rappoport, A.: BLEU is not suitable for the evaluation of text simplification. arXiv preprint arXiv:1810.05995 (2018)

23. Surya, S., Mishra, A., Laha, A., Jain, P., Sankaranarayanan, K.: Unsupervised Neural Text Simplification. arXiv preprint arXiv:1810.07931 (2018)

24. Vinyals, O., Fortunato, M., Jaitly, N.: Pointer networks. In: Advances in Neural Information Processing Systems, pp. 2692–2700 (2015)

25. Wang, T., Chen, P., Rochford, J., Qiang, J.: Text simplification using neural machine translation. In: Thirtieth AAAI Conference on Artificial Intelligence (2016)

26. Xu, W., Callison-Burch, C., Napoles, C.: Problems in current text simplification research: new data can help. Trans. Assoc. Comput. Linguist. **3**, 283–297 (2015)

27. Zhang, X., Lapata, M.: Sentence simplification with deep reinforcement learning. arXiv preprint arXiv:1703.10931 (2017)

# Federated Learning in Named Entity Recognition

Efim Luboshnikov[1](✉) and Ilya Makarov[1,2](✉)

[1] HSE University, Moscow, Russia
`ealuboshnikov@edu.hse.ru`, `iamakarov@hse.ru`
[2] Big Data Research Center, National University of Science and Technology MISIS,
Moscow, Russia

**Abstract.** This article is devoted to the implementation of the federated approach to named entity recognition. The novel federated approach is designed to solve data privacy issues. The classic BiLSTM-CNNs-CRF and its modifications trained on a single machine are taken as baseline. Federated training is conducted for them. Influence of use of pretrained embedding, use of various blocks of architecture on training and quality of final model is considered. Besides, other important questions arising in practice are considered and solved, for example, creation of distributed private dictionaries, selection of base model for federated learning.

**Keywords:** Federated learning · Federated averaging · Named entity recognition · BiLSTM-CNNs-CRF

## 1 Introduction

In response to the recent increase in incidents involving leaks, illegal use and sale of client data, governments are creating and enacting laws to protect the rights of clients, both people and companies. Such laws impose significant restrictions on the transfer, storage and processing of customer data. Most of the solutions based on machine learning in business are built using client data. Companies directly use the data for training and model validation. With the adoption of laws, it will become more difficult, if not impossible, to continue implementing models.

One possible solution is to use federated learning. The main idea is to share the resources of many independent parts and then benefit all participants. Federated learning is called distributed training of the general model, where data is always stored on clients during training. Despite its recent appearance, federated learning is already being applied in various areas. Projects such as Google Keyboard, joint medical and financial research are good examples. In all these projects, private data was successfully and safely used for model training and evaluation.

This work is devoted to solving the practical problem of extracting named entities in a federated approach. This task more often arises as a subtask of more

© Springer Nature Switzerland AG 2021
W. M. P. van der Aalst et al. (Eds.): AIST 2020, CCIS 1357, pp. 90–101, 2021.
https://doi.org/10.1007/978-3-030-71214-3_8

complex, for example, machine translation or question answering. Most studies in the field of federated learning use multi-layer perceptron, the simplest CNN, which is of little practical interest. NLP tasks are more difficult to implement because of the complexity of general, secure preprocessing and preparation for learning, for example, creating common dictionaries. Besides, the models themselves have much more complicated architecture. In this study, we will try to find answers to several questions in this complex task and take a step towards safe and effective learning.

## 2  Related Work

This work has two logical parts: federated learning and named entity recognition. Analysis of related work, current approaches, benchmarks and results also has two semantic parts. Therefore, at the beginning of this part, we will give an overview of the NER area, and then we will move on to the issue of federated learning, which is the main interest of the work.

*Named Entity Recognition.* Named Entity Recognition (NER) is a machine learning task that consists of identifying and segmenting named objects, then classifying or classifying them according to various predefined classes. This task appears both independently and as a subtask in other tasks, for example: text understanding, information retrieval, automatic text summarization, question answering, natural language understanding, knowledge base construction.

New state-of-the-art results are obtained with the help of deep learning approaches. According to the article [1] there are three key aspects of deep learning success in this area: complex non-linear mapping between network input and output, automatic features creation and end-to-end approach.

Most neural network architectures for this task have a 3-step structure. In the first stage, we get a good representation of input. Not only the word itself is important in the sentence, but also the context surrounding it. So, CNN, RNN, LM or Transformer are used for this purpose in the second step. In the final step, we use decoder to get the predictions themselves - tags for words. By convention, we can call these three steps as input, encoder, decoder.

The work BiLSTM-CNNs-CRF [2] was introduced in 2016 and marked a new trend in the solution. So, BiLSTM-CRF is the most common architecture now. The latest SOTA results were achieved with really huge models. The Bi-directional Transformer in a cloze-style manner [3] reaches (93.5%) on the CoNLL03 dataset. Instead of treating the NER problem as a sequence marking problem, the article [4] suggests formulating it as a machine reader's understanding problem (MRC). As result, models achieve significant amount of performance boost over current SOTA models on nested NER datasets for chinese, english languages. The success of NER systems strongly depends on input representation. The use of pre-trained language model embeddings is prevalent in new developments and research.

*Federated Learning.* The growing popularity of federated learning is primarily due to restrictions on the use of data. GDPR in European Union [5], PDPA in Singapore [6], CCPA [7] in the USA and many other laws and regulations limit the use and transfer of client data. Those who work with this data have to look for solutions. Using the full amount of distributed, private data for training would allow for better models, new tasks, and ultimately, better lives.

The main idea of federativity is cooperation of many independent participants. Federated systems include concepts of federated learning, federated database systems [8] and federated cloud [9]. The idea of federativity is inherent in all of them, but there are some peculiarities. In federated databases there is a focus on distributed data management, in federated cloud there is a focus on resource planning, in federated learning systems (FLS) there is a secure computation between all components of the system.

Research in this area has just begun, so in most works do not experiment with the choice of the optimizer. Default stochastic gradient descent has theoretical grounds and is well researched. Strategies of averaging local model improvements into global ones are studied much more. The simplest and most reliable way is to average local weights of the model changed after local training, the so-called FedAvg. In the article [10] the authors of the algorithm investigate the question of improving the quality of the global model with a federated approach on very simple models. The authors put forward two hypotheses to improve the quality of the global model:

1. Increase the number of epochs on local clients.
2. Increase the number of transfers between local clients and the central server where the global model is stored.

Logical prerequisites are justified by experiments. The authors study optimal parameters, as the increase in the number of epochs leads to the increase in the cost of local calculations, and the growth of transfers leads to an increase in the cost of communication activities.

In works [11,12] the authors, relying on FedAvg, achieve better results with changing the target function for optimization. In FedProx, changes in the local loss function are limited, this approach has theoretical grounds. In the Agnostic FL method, a centralized model optimized for any target class distribution is trained. The main problem for practical application of federated training is efficiency. In [13] authors reduce the size of the reference at each iteration. The work [14] is devoted to finding the necessary number of communication rounds. Trade-off between local computations and weight exchange is still unexplored.

Only one article [15] was found on the subject of this study. Team of authors claim to have created a comprehensive model containing two modules. One of the modules is taught in a federated approach, respectively, after graduation trained weights and coefficients are common to all clients. The second module is trained on local data only. Thus, they offer to use the advantage of federated training to obtain aggregated information on private data distributed among clients, and additional training on local data will allow to take into account the context of a particular client. Unfortunately, this article has no source code and was published quite recently, so there is no possibility to judge about the reliability of the results.

## 3   Dataset and Metrics

The CoNLL03 dataset [16] is based on common themes and has a split into train (14041), valid (3250), test (3453). There are four entities: LOC (location), ORG (organization), PER (person) and MISC (miscellaneous). Original tagging scheme is BIO, but classic experiments results with IOBES is better. So, in all our experiments we use IOBES.

In most cases, accurate estimates are used for NER tasks. In extracting entities, there are two subtasks that need to be solved: defining exactly the boundaries and type of entity. For accurate estimates, only an extraction where boundaries and type are correctly marked is correct. Similar to the classification tasks, the concepts True Positive, False Positive, False Negative, True Negative are introduced. The obtained values are used to calculate metrics: precision, recall, F-measure. These metrics are defined for each class separately. In practical tasks it is always more convenient to focus on one number as a measure of quality. For this purpose, different types of F-measure averaging are used. In macro average we count F-score for each tag, and then take the arithmetic mean. We use micro-average, so, we sum up TP, FP, FN, TN all over the body without tags, and then take the average.

## 4   Baseline Model

Bi-directional LSTM-CNNs-CRF [2] is an strong baseline for NER tasks. There is no special preprocessing, feature-engineering in the model. This model can be trained for any sequence labeling task. Common NER solution are based on three-step architecture as shown on Fig. 1. At the beginning of the model, convolutional layers are used to obtain information from words at the symbol level. Then we combine word and character based views into one common one and send it to BiLSTM. With RNN, we aggregate context information from specific words and characters. BiLSTM outputs are fed into a random field to account the correlation of neighbor labels in order to obtain the most likely sequence of labels for the input sentence.

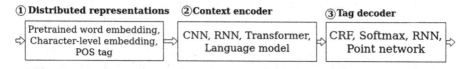

**Fig. 1.** Typical three-step architecture for NER. Used blocks are shown in blue. (Color figure online)

For the baseline word vectors were trained from scratch, not used character level features. Parameters were initialized in the same way and were optimized with stochastic gradient descent (SGD). We set initial learning rate $\eta_0 = 0.015$, momentum $m = 0.9$. Also, we used multiplicative learning rate scheduler with factor $f = 0.95$. The model achieves 85.33% F1 quality for NER on CoNLL03 dataset. In this article [17] more than $50,000$ experiments have been conducted to select the optimal parameters. The value we have obtained corresponds to the distribution of the test F1 for the corresponding set of model and optimizer hyperparameters.

## 5   Federated Approach

The goal of the series of experiments was to train the classical BiLSTM-CNNs-CRF and its modifications in a federated approach, to identify and find a solution to the problems encountered. Different metrics can serve as optimization functions in the implementation of the federated approach. In examples of model training on mobile devices, model weight and computational complexity are crucial. There are cases when the server cannot regularly receive model updates or send out a new version. Then it is critical to understand the minimum number of clients or their share required for training convergence. The experience of using federated training in the NER task speaks more about other problems.

Examples of applications are: training on customer medical records, medical records, payment documents, customer service calls. In all such cases, the distributed training will be conducted by the software provider or an independent participant for the common good of all participants in the training. This is why we do not consider improper motives or playing by the participants. Also, there is no question of communication, as there is constant communication between participants in such cases. The main issue in our research was cost reduction, which is primarily related to the learning time of the distributed model. For these reasons and for narrowing the space of the experiment, the number of users and tuned number of local epochs were fixed.

*Model Capacity.* Federated learning is often used when data is private. But a company may have some data for different reasons: open data, a special agreement with some customers to collect data, a way to ensure that no private data is collected. Suppose we have this situation and we have dataset. Before implementing the federated approach, it is necessary to check its appropriateness. We should proceed from the assumption that the quality of the model under the federated approach tends to the quality of the model trained on the whole dataset, which is now distributed. It is suggested that a simple sketch should be made to find out whether adding data will improve the quality of the current model. For this purpose, we will break down the data into a small number of identical parts. Then, we will train the model in 1 part and check the quality in the deferred sample. Then, we will train the model in 2 parts and evaluate it again. And so on, before we train and validate the model on 100% of available data.

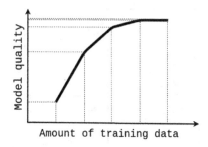 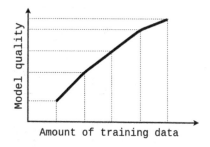

**Fig. 2.** Results of the capacity test of two abstract models. On the left we see the model with low capacity, on the right with high capacity. Before training model in federated approach we should estimate model capacity for expediency of realisation.

Let us look at possible results obtained for two different models in such an experiment in the Fig. 2. In the left side we see that adding data will most likely not lead to an increase in quality. Therefore, for such a model, it makes no sense to build federated training on clients. On the right picture, we can see that the quality of the model grows with the data quantity. Therefore, adding even more data may be reasonable. Such a model should be trained in a federated approach, because adding data would rather improve its quality.

There are two things we should note. First, the experiment should be conducted for each new model or architecture configuration, since each model has its own capacity. If the model has a large capacity, it needs more data to achieve a plateau of quality on validation. Models with low capacity learn fewer patterns from data, so they need less data. Second, by extrapolating model quality growth on more data, we involuntarily assume that data will come from a similar distribution, which of course is not always done. This point should be taken into account and have at least an approximate estimate of the nature of data on distributed clients.

*Distributed Data Preparation.* In the process of teaching one model on the whole dataset, most preprocessing steps are simple and even routine. With a federated approach, training becomes somewhat more difficult. Obviously, the initial data on all clients should be processed in the same way, i.e. we need the same preprocessing, creation of fiche, pretrained embedding if necessary. When creating NLP models, we often use dictionaries. Such dictionaries help to map tensor representation in the model (embedding) and real token or symbol. In the baseline model we also used such dictionaries for tokens and symbols. To create such a dictionary, we have collected tokens in the dataset and assigned them unique numbers corresponding to their embedding in the model.

Now, as part of a federated approach, we can not collect tokens from all clients, so it would violate the conditions of privacy. To create a single dictionary with a privacy condition, we can implement the client-central server interaction scheme shown in this Fig. 3. Every user after this procedure will be able to transform raw data to the model correct input.

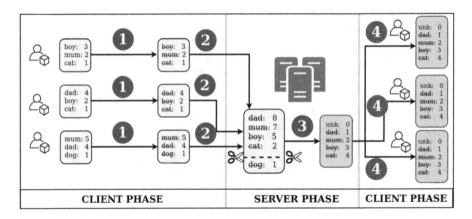

**Fig. 3.** Remote secure mapping. ① Each client preprocesses data and makes a frequency dictionary of tokens. ② Client hashes the keys in such dictionary and saves the pairs {*token* : *hash*} with itself. Then all clients send hashed dictionaries to the server. ③ On the central server we sum up all the dictionaries and create a single mapping by selecting the desired minimum frequency of token. Also, we add to the mapping special tokens, for example *<unk>*. ④ Send the mapping to all clients. Each client now has two dictionaries: its own {*token* : *hash*} and common {*hash* : *token_id*}. Applying one by one dictionary, each user will be able to get the correct input for the distributed model even for those tokens that were not in his original sample.

## 5.1   Experiments

**Federated Model Training.** In all our experiments with federated approach we will compare learning process and final model performance with baseline. We will estimate overall time to reach baseline model performance, number of global communication rounds, local training epochs in process. Also, we will check influence of different factors to model training, for example, usage of pretrained word vectors, neural model blocks.

During the experiments on federated model training for the NER, the quality of the baseline model was achieved. Obviously, this was longer than usual model training. The training time of the federated model has exceeded 2.58 times the training time of a single instance. The cost of model training is linearly related to the duration of training, so we measured and compared the duration of training in all experiments with the conventional and federated approach. It should be understood that with distributed parallel learning, the real learning speed is certainly higher. It is proportional to the number of users without taking into account the time spent. But to calculate the cost of training, you must use the total time spent on all the clients, which we have received.

*Learning Process and Results.* As already mentioned, the number of local epochs and users has been fixed in order to reduce the space of experiments. We take 10 for both parameters. The data divided equally between users. The division is fixed in all experiments. We conducted several experiments on initializing local optimizer. It was decided to initialize the local optimizer according to the number of passed global training rounds. The scheme of updating the model weights was taken from the article [10]. This scheme consists in simple taking the arithmetic mean of weights of all models that have locally learned the same number of epochs on the given clients. After averaging, we distribute the averaged model to the participants again. So the learning process lasts to the required quality. On the Fig. 4 we can see the average value of loss for clients during the training. We can see that every tenth local epoch of loss increases because the local model is replaced by a global model. Every round of local training epochs we overfit on local data, but in general we reduce losses as we learn, reaching the right quality.

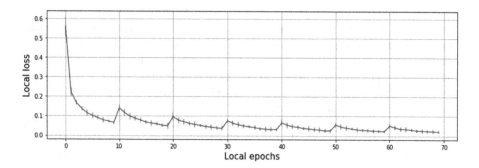

**Fig. 4.** Local losses. Mean value and 3 sigma interval for user distribution.

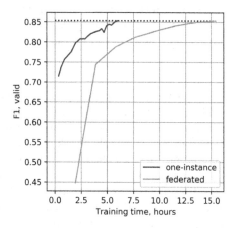

**Fig. 5.** Baseline and federated learning process.

The received training process is shown in Fig. 5. The training time of the federated model has exceeded 2.58 times the one-instance training time. It should be understood that with distributed parallel learning, the real learning speed is certainly higher. It is proportional to the number of users without taking into account the time spent. But to calculate the cost of training, we must use the total time spent on all the clients, which we have received.

**Pretrained Word Embeddings.** It is well known that the addition of pre-trained embeddings reduces the training duration of models, and often increases the final quality. Nowadays, most modern models for NER are obtained using BERT, ELMO, their modifications as well as large language models. The addition of pretrained embeddings to BiLSTM-CNNs-CRF also leads to an increase in quality. As you can see from the previous section, the implementation of training in a federated approach leads to a multiple increase in training duration. Use of pretrained vectors could speed up the learning process. Two experiments were conducted to test this hypothesis. BiLSTM-CNNs-CRF was taken as a baseline, word embedding was initialized by pretrained GloVe-100 vectors. Word embeddings were left as a variable parameter.

The results of the experiment are shown in Table 1 and the training graph in Fig. 6a. The learning time has diminished 2.19 times for one-instance and increased 1.64 times for federated. This is due to the fact that the model was better trained to quality, and recent epochs took a long time. For an honest assessment, it is necessary to compare the learning time at a certain quality level. This is done in Table 1. In third line there we can see how various model modifications achieve the quality of the initial level (85.33%) by learning the federated approach. We see that the use of pretrained vectors and convolutional blocks allows us to significantly increase the learning speed. Modification BiLSTM-CNNs-CRF is the fastest. The final quality increased by 4.67%. This allows us to speak about the high efficiency of the use of pretrained embeddings for federated learning.

**Char Embeddings.** The next experiment was not followed by a preliminary hypothesis or an expected result. The aim of the experiment was to compare the behaviour of convolutional and recurrent architectures for processing symbolic embedding in a federated approach. It was the block of symbolic embedding processing in the initial architecture that we changed during the experiment. RNN at its time was a breakthrough in sequence processing. But subsequent results of CNN application in the text allowed us to consider convolutional layers as an alternative. First of all, it is related to the speed of learning. In our experiments, learning speed is almost the most important criterion.

BiLSTM-CRF with pretrained embedding were used, also, CNN or RNN was used as a character processing block. Use of recurrent block to process symbolic text representation allowed to increase quality by 0.32% in valid Fig. 6b. In the course of this experiment, we did not get significant differences in the quality of the model, which cannot be said about the learning rate. The use of CNN allowed

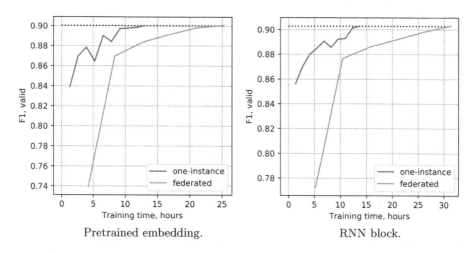

Pretrained embedding.                    RNN block.

**Fig. 6.** Impact to learning speed.

us to increase the training speed by 4% and 19% respectively for single-instance and federated approaches.

## 5.2   Results

In Table 1, the main results are the last 3 rows. For ease of comparison, we have scaled all experiments against the baseline experiment. The first line shows the training time for different modifications of the model. In the second line, we see how much longer training time increases when you move from one-instance to federated learning. This data is not so representative, because we need to remember that the final quality of models with pretrained embedding is higher, so they learn longer.

In order to estimate how quickly the models are trained to a specific level, the third line were created. In this line, we see how long various modifications of the models achieve the quality of baseline (85.33%) by learning the federated approach. We see that the use of pretrained vectors and convolutional blocks allows us to significantly increase the learning speed. At the same time, we should understand that the final quality of modification with convolutional is little worse to that with recurrent blocks.

Line 4 in Table 1 shows how time costs increase when moving from baseline one-instance to federated learning. The last line shows the $F1$ value on the validation dataset. If we speak about the final quality, the models with the use of pretrained vectors have bypassed those trained from scratch by 5%.

**Table 1.** Comparison of learning time. Symbol $x$ means 5.97 h on GCP Instance (8 CPU, 64 GB)

|  | CNN, scratch | CNN, GloVe | LSTM, GloVe |
|---|---|---|---|
| Baseline | x | 0.44x | 0.46x |
| Federated | 2.58x | 4.23x | 5.23x |
| Federated, 85% | x | 1.4x | 1.75x |
| Time increases | 2.58 | 3.18 | 3.8 |
| F1, valid | 85.33 | 90.00 | 90.32 |

## 6   Conclusion

In this work, a full process of preparation and training in a federated approach was carried out. A solution was found to the problems encountered. Such conclusions can be drawn for practical use.

- Before implementing a federated approach, it is necessary to conduct tests for the capacity module and, if possible, find out the nature of data distribution among the participants of training.
- For the model's auxiliary dictionaries, you can use the scheme described in this article before you start training.
- It is better to use pretrained embeddings and convolutional blocks to increase the learning speed. In case of federated learning is even more important than in case of single machine learning.

**Acknowledgments.** The research was supported by the Russian Science Foundation grant 19-11-00281.

## References

1. Li, J., Sun, A., Han, J., Li, C.: A survey on deep learning for named entity recognition. IEEE Trans. Knowl. Data Eng. 1 (2020). https://doi.org/10.1109/TKDE.2020.2981314
2. Ma, X., Hovy, E.: End-to-end sequence labeling via bi-directional LSTM-CNNs-CRF. arXiv preprint arXiv:1603.01354 (2016)
3. Baevski, A., Edunov, S., Liu, Y., Zettlemoyer, L., Auli, M.: Cloze-driven pretraining of self-attention networks. arXiv preprint arXiv:1903.07785 (2019)
4. Li, X., Feng, J., Meng, Y., Han, Q., Wu, F., Li, J.: A unified MRC framework for named entity recognition. arXiv preprint arXiv:1910.11476 (2019)
5. Voigt, P., Von dem Bussche, A.: The EU general data protection regulation (GDPR). A Practical Guide, 1st edn. Springer, Cham (2017). https://doi.org/10.1007/978-3-319-57959-7
6. Chik, W.B.: The Singapore Personal Data Protection Act and an assessment of future trends in data privacy reform. Comput. Law Secur. Rev. **29**(5), 554–575 (2013)

7. Pardau, S.L.: The California consumer privacy act: towards a European-style privacy regime in the United States. J. Tech. L. & Pol'y **23**, 68 (2018)
8. Sheth, A.P., Larson, J.A.: Federated database systems for managing distributed, heterogeneous, and autonomous databases. ACM Comput. Surv. (CSUR) **22**(3), 183–236 (1990)
9. Kurze, T., Klems, M., Bermbach, D., Lenk, A., Tai, S., Kunze, M.: Cloud federation. In: Cloud Computing 2011, pp. 32–38 (2011)
10. McMahan, H.B., Moore, E., Ramage, D., Hampson, S., et al.: Communication-efficient learning of deep networks from decentralized data. arXiv preprint arXiv:1602.05629 (2016)
11. Mohri, M., Sivek, G., Suresh, A.T.: Agnostic federated learning. arXiv preprint arXiv:1902.00146 (2019)
12. Li, T., Sahu, A.K., Zaheer, M., Sanjabi, M., Talwalkar, A., Smith, V.: Federated optimization in heterogeneous networks. arXiv preprint arXiv:1812.06127 (2018)
13. Konečný, J., McMahan, H.B., Yu, F.X., Richtárik, P., Suresh, A.T., Bacon, D.: Federated learning: strategies for improving communication efficiency. arXiv preprint arXiv:1610.05492 (2016)
14. Zhu, H., Jin, Y.: Multi-objective evolutionary federated learning. IEEE Trans. Neural Netw. Learn. Syst. **31**(4), 1310–1322 (2020)
15. Ge, S., Wu, F., Wu, C., Qi, T., Huang, Y., Xie, X.: FedNER: medical named entity recognition with federated learning. arXiv preprint arXiv:2003.09288 (2020)
16. Sang, E.F., De Meulder, F.: Introduction to the CoNLL-2003 shared task: language-independent named entity recognition. arXiv preprint arXiv:cs/0306050 (2003)
17. Reimers, N., Gurevych, I.: Optimal hyperparameters for deep LSTM-networks for sequence labeling tasks. arXiv preprint arXiv:1707.06799 (2017)

# E-hypertext Media Topic Model with Automatic Label Assignment

Olga Mitrofanova[1]([✉]), Anna Kriukova[2], Valery Shulginov[3], and Vadim Shulginov[4]

[1] Saint-Petersburg State University, Universitetskaya emb., 11, 199034 Saint-Petersburg,
Russian Federation
o.mitrofanova@spbu.ru
[2] Charles University, Ovocný trh 5, Prague 1 116 36, Czech Republic
krukova.ann@gmail.com
[3] Higher School of Economics National Research University, Myasnitskaya Street, 20,
101000 Moscow, Russian Federation
shulginov.val@yandex.ru
[4] Rostelecom, Svetlanskaya Street, 57, 690091 Vladivostok, Russian Federation
vadim.shulginov@yandex.ru

**Abstract.** This article deals with the principles of automatic label assignment for
e-hypertext markup. We've identified 40 topics that are characteristic of hypertext
media, after that, we used an ensemble of two graph-based methods using outer
sources for candidate labels generation: candidate labels extraction from Yandex
search engine (Labels-Yandex); candidate labels extraction from Wikipedia by
operations on word vector representations in Explicit Semantic Analysis (ESA).
The results of the algorithms are label's triplets for each topic, after which we
carried out a two-step evaluation procedure of the algorithms' results: at the first
stage, two experts assessed the triplet's relevance to the topic on a 3-value scale
(non-conformity to the topic/partial compliance to the topic/full compliance to
the topic), second, we carried out evaluation of single labels by 10 assessors who
were asked to mark each label by weights «0» – a label doesn't match a topic; «1»
– a label matches a topic. Our experiments show that in most cases Labels-Yandex
algorithm predicts correct labels but frequently relates the topic to a label that is
relevant to the current moment, but not to a set of keywords, while Labels-ESA
works out labels with generalized content. Thus, a combination of these methods
will make it possible to markup e-hypertext topics and create a semantic network
theory of e-hypertext.

**Keywords:** E-hypertext · Topic modelling · Label assignment · Media discourse

## 1 Introduction

The invention of e-hypertext is determined by the cultural context of the 20th century:
Theodor Nelson, who coined this term, defined it as an implementation of the post-
modernist concept in which the text exists only in relation to reading. He defined the
hypertext as «a series of text chunks connected by links, which offer a reader different

© Springer Nature Switzerland AG 2021
W. M. P. van der Aalst et al. (Eds.): AIST 2020, CCIS 1357, pp. 102–114, 2021.
https://doi.org/10.1007/978-3-030-71214-3_9

pathway» [1]. By having a direct connection from one position in a text to another, the reader is actively involved in the construction of the hypertext, which is very different from reading traditional linear text.

The structure of non-linear interconnections in Internet communications also modifies the author's interaction with e-hypertext. Following the principle of cooperation, the author creates a hypertext's structure according to two potential strategies of the reader: «selecting the text semantically related to the previously read section (coherence strategy) or choosing the most interesting text, delaying reading of less interesting sections (interest strategy)» [2]. So, every link becomes the «ephemeral contract» [3] between author and reader, which suggests that the semantics of hypertext transition is clear to each of the communication participants.

Another important factor determining the topic structure of e-hypertext is the type of discourse in which hypertext is created. We study the functioning of e-hypertext in media discourse, which provides communication among groups so large, heterogeneous, and widely dispersed that they could never interact face-to-face or through any other but mass-produced and technologically mediated message systems [4]. This defines with the standardization of e-hypertext connections between source and target text, which are used to extend the information and confirm the identity of the source text. Thus, the topical structure of e-hypertext becomes a lens for studying linguistic (discursive characteristics) and anthropological (users strategies) features of Internet communication.

This research describes the principle of automatic label assignment of e-hypertext components: source and target texts, which is a necessary stage in describing the topic structure of e-hypertext and simplifies the linguistic interpretation of hypertext transition semantics. Labels are considered as general terms or multiword expressions exhibiting the content of a topic in a precise and brief manner. As a result, this approach allows you to automate the creation of hypertext media systems through topic clustering of texts. This research is based on the e-hypertext media corpus, which includes two databases: the texts of a media discourse with paratextual elements (title, announcement, date, author), and a database of links' nominations taking into account the context of use, POS-tagging and vector representation.

## 2   E-hypertext Media Topic Model

The e-hypertext media corpus involved in our experiments includes texts and hypertext elements extracted from the Russian e-media «Kommersant», «Izvestia», «RBC», «Novaya Gazeta», «TASS», «Dozhd», «Vedomosti», «Interfax», etc. Alongside with the text of the article the corpus includes such metadata as media title, article title, subtitle, name of author, tags, date of publication. Experiments were carried out in the test set of 53 000 articles (total size 12 million tokens). The texts are linked by 70 000 tripartite hypertext elements. Development of the e-hypertext media corpus was performed by means of Python libraries BeautifulSoup [5], NLTK [6] and re [7]. The process of corpus development is discussed in detail in [4]. Corpus preprocessing implied tokenization, normalization, lowercase transformation, stop-word and punctuation deletion. Further we performed collocation analysis by means of phrases module in gensim library for Python [8]). Collocations were extracted and ranked according to the TF × IDF scheme.

Collocation extraction was necessary as it allowed to preserve multiword expressions in the topic model of the corpus. Topic modelling was fulfilled by the ensemble of algorithms including multiple learning method t-Distributed Stochastic Neighbor Embedding [9], DBSCAN clustering [10] and non-negative matrix factorization [11], these algorithms being applicable in processing high-dimensional datasets and clustering textual data as regards topic content. Thus, we generated a topic model for the e-hypertext media corpus which contained 40 topics of 10 topic words. In choosing such parameters of topic modelling we followed [12] where the given topic number and size were proposed for evaluation procedure.

A fragment of the output is given below:

Topic 2: *следствие, адвокат, уголовный, дело, обвинять, следователь, скр, арест, УК, фигурант, обвинение* (*investigation, lawyer, criminal, case, accuse, investigator, investigation committee, arrest, criminal code, person of interest, prosecution*).

Topic 4: *производство, предприятие, продукция, сеть, завод, технология, оператор, производитель, продукт, оборудование* (*manufacture, enterprise, production, network, plant, technology, operator, manufacturer, product, equipment*).

Topic 9: *матч, команда, сборный, чемпионат, клуб, футболист, игра, сборная, турнир, тренер* (*match, group, team (attr.), championship, club, football player, game, national team, tournament, coach*).

The next step of our research deals with topic interpretation in course of label assignment.

## 3    Approaches to Topic Labelling

As a rule, topic modelling does not imply assignment of topic labels as an obligatory intrinsic operation. However, labelling may evidently enhance linguistic interpretability of generated topics. There are various solutions to the problem of label assignment. Topic labelling requires automated or manual choice of candidate labels which are expected to be unigrams or multiword expressions with general meaning correlating with the meanings of topic words. The most evident ways of label assignment are the choice of the first word (or *n* first words) of a topic as a label or manual label assignment (e.g., [12, 13]). Those methods can hardly be used to form a solid baseline procedure for the following reasons. The first word (or *n* first words) of a topic may be inconsistent as topic labels: the ordering of words within a topic does not necessarily reflect hierarchical structure of the lexicon. On the one hand, topic words are expected to establish numerous paradigmatic (synonymic, hyponymic and similar relations, e.g., *завод – предприятие* (*plant – enterprise*), *команда – сборная* (*group – national team*), *матч – чемпионат – турнир* (*match – championship – tournament*), *топливо – нефть, бензин* (*fuel – oil, gasoline*), etc.), syntagmatic (e.g. *ядерный – комплекс, испытание* (*nuclear – complex, test*), *цена – вырасти* (*price – grow*), etc.), epidigmatic (derivational) relations (e.g., *обвинять – обвинение* (*accuse – prosecution*), *следствие – следователь* (*investigation – investigator*), *производство – производитель* (*manufacture – manufacturer*), *нефть – нефтяной* (*oil – oil (attr.)*)), etc.). On the other hand, their ordering within a topic can not be strictly defined in terms of abstractness-concreteness or taxonomic relations (hypernym vs. Hyponyms, holonym vs. Meronyms, etc.). Moreover, the first words in a topic may be polysemous (e.g., *акция, следствие* (*campain, investigation*) etc.), that depreciates them as possible topic labels. As for manual label assignment, this procedure relies upon on the intuition of researchers and annotators, it requires involvement of experts as annotators in case of domain-specific corpora processing (e.g. medical, legislative, etc. texts) thus it may bring subjectivity into empirical results. At the same time, human expertise remains a significant step in evaluation of topic labelling results.

Computational linguistics provides several automatic techniques of label assignment, these techniques are distributed into classes as regards:

1) the sources of labels: inner sources (the labels are extracted from the corpus involved in topic modelling) or outer sources (the labels are extracted from reference corpora, output of search engines, knowledge databases (Wikipedia, WordNet, etc.);
2) type of algorithms involved: supervised or unsupervised;
3) type of labels: single words, phrases or both.

The works on label assignment using inner sources describe algorithms based on a) definition of Kullbach-Leibler distance between word distributions and maximization of mutual information between candidate labels and topics [14]; b) proper reranking of topical words as regards their attraction to the topic [12]; c) ranking candidate labels by means of summarization algorithms [15]; d) candidate label collocations extraction from documents most relevant to topics, mapping candidates to word vectors and letter trigrams, ranking candidates according to similarity between topics and label vectors [16]; e) detection of documents closest to topics, extracting single terms and multiword expressions and ranking them by information measures [17], etc.

There are various approaches to label assignment using outer sources, the most significant are as follows: a) using terms from Google Directory (gDir) hierarchy [18]; b) extracting article titles from Wikipedia or DBpedia and further ranking them as candidate labels [19, 20]; c) using web as a corpus for extracting candidate labels by searching Google and ranking candidates by PageRank [13]; d) using Wikipedia titles as candidate labels, ranking candidates by operations with neural embeddings for words and documents [21], e) incorporating a formal ontology into topic model for knowledge extraction (KB LDA) [22], f) using $k$-nearest neighbor clustering and similarity-preserved hashing for fast assignment of labels for newly emerging topics [23], etc.

In the study we proceed from two assumptions. First, topics consisting of sentences are more accessible for interpretation than topics consisting of 2–3 keywords (this is where we see the disadvantage of method b)).

Secondly, the source of topics interpretation should have a genre similar to the dataset. Some of these methods are based on hard-to-reach data: a) Google Directory is no longer an actual source and e) shows stable results compared to the usual LDA but requires ontology. As a result, methods c) and d) are close to our approach and are analogous to English.

# 4 Our Approach to Topic Labelling

Procedures of automatic label assignment are widely discussed and evaluated for English corpora and knowledge resources, the Russian data being poorly represented in research projects. Our recent investigations are aimed to fill in the gap. In this study we used an ensemble of two graph-based methods using outer sources for candidate labels generation [24]: a) candidate labels extraction from Yandex search engine with their further ranking by TextRank (this is a graph-based model that takes into account the value of the each graph's vertex depending on how many links it forms) [25, 26]; b) candidate labels extraction from Wikipedia by operations on word vector representations in Explicit Semantic Analysis (ESA) model [27, 28]. These procedures comprise two stages: candidate extraction and ranking. Both methods are compatible with any topic models and were tested on LDA from scikit-learn [29] for the corpus of Russian encyclopedic texts on linguistics. Experimental results and their evaluation proved that both methods are applicable. Candidate labels extraction from Yandex provides more consistent labels, although this algorithm turned out to be time-consuming (search engine imposes restrictions to the number of queries from an IP per minute) and hard to reproduce (the output may be adjusted to informational preferences of particular users). Losses in the quality of labels generated by Yandex-based algorithm are explained by the unstability of the external source. Candidate labels provided by ESA seem to be less general (in most cases they correspond to hyponyms of topic words), however the given approach is not time-consuming and provides reproducible results due to the stability of Wikipedia dump. Losses in the quality of labels in case of ESA labelling procedure deal with the specific content of the external source. Thus, the ensemble of two labelling methods could counterbalance and mitigate their drawbacks.

## 4.1 Topic Labelling Using Yandex

The algorithm for candidate labels extraction from Yandex search engine (Labels-Yandex) [25, 26] is an elaboration of the procedure originally designed for visual information processing [13]. At the stage of candidate extraction the first 10 topical words for each topic form a separate query to Yandex. The output of the query consists of top 30 titles for documents retrieved by the search engine. The list of titles is transformed into a continuous text which is subjected to preprocessing, namely, stop-words removal and lemmatization which is performed by pymorphy2 [30]. At the stage of candidate ranking a text is converted to an oriented graph $G = \{V, E\}$, where $V$ is a set of nodes corresponding to lemmata, and $E$ is a set of weighted edges marked by values of co-occurrence frequency within the text. Two nodes are connected by an edge if they occur together in a context window $[-1, +1]$. Further TextRank values are calculated for all nodes of the graph. Lemmata obtaining higher scores are considered as more significant, while edges getting larger weights imply strong semantic relations in word pairs. The algorithm was adjusted so that it could generate not only single word labels, but also multiword expressions. We introduced morphosyntactic patterns $Adj+N$, $N+N$, $N+prep+N$, etc., which were used to extract key phrases, each of them was assigned a TextRank value calculated as a sum of weights for each component.

## 4.2 Topic Labelling by Means of ESA

Explicit Semantic Analysis (ESA) is a variety of distributional semantic models projecting words from Wikipedia dump to a high-dimensional vector space. The original ESA algorithm was developed to improve monolingual and crosslingual search [31], the paper [24] discusses the first experience in using it for label assignment, the papers [27, 28] show its applications in detecting Russian text similarity/relatedness. In ESA model each article is treated as a separate «concept» represented by a vector generalizing all words co-occurring in an article. Wikipedia is converted to a term-document matrix, its cells containing TF-IDF values showing association strength in word – concept pairs. For each word ESA creates an inverted index showing its relations with the concepts (i.e. articles in which a word occurs). Concepts with low weights for a given word are removed from the model, this allows to reduce irrelevant links in ESA.

The algorithm for candidate labels extraction from Wikipedia by operations on word vector representations in ESA model (Labels-ESA) is based on the assumption that candidate labels considered as Wikipedia article titles should have vectors close to the topic words vectors. At the stage of candidate extraction the first 10 topical words for each topic form an averaged query vector, the output of ESA contains lists of article titles ordered according to cosine values indicating closeness of concept vectors and a query vector. The step of candidate ranking is performed in the same way as for topic labelling using Yandex.

## 5 Experiments on Topic Labelling for the E-hypertext Media Corpus

### 5.1 Experimental Procedure

Each topic was submitted as an input to the algorithms of topic labelling Labels-Yandex and Labels-ESA. The first three candidate labels ranked by the algorithms were chosen for further analysis (cf. Table 1).

### 5.2 Evaluation of Results

Experimental results require verification. As regards topic labelling, verification poses certain problems, the main obstacle being the absence of the golden standard for this procedure for any language (none for English, not to mention Russian). In our study we carried out a two-step evaluation procedure which is based on [13, 21].

First, we performed annotation of label triplets generated by Yandex-based algorithm and ESA. The annotation was based on a questionnaire modified from [24, 26]. Two assessors were asked to mark label triplets according to 3-value scale of weights:

«0» – a triplet doesn't cover the content of a topic and can't be used as a topic label;
«1» – a triplet somehow reflects the content of a topic and to more or less extent can be used as a topic label;
«2» – a triplet covers the content of a topic and can be used as a topic label.

**Table 1.** Examples of candidate labels generated by Labels-Yandex and Labels-ESA.

| Labels-Yandex | | | Labels-ESA | | |
|---|---|---|---|---|---|
| Topic 2: *следствие, адвокат, уголовный, дело, обвинять, следователь, СКР, арест, УК, фигурант, обвинение* (investigation, lawyer, criminal, case, accuse, investigator, investigation committee, arrest, criminal code, person of interest, prosecution) | | | | | |
| дело в отношении адвокатов РФ (*case against lawyers of the Russian Federation*) | дело в отношении (*case against*) | дело о взяточничестве (*case of bribery*) | уголовный кодекс (*criminal code*) | уголовное обвинение (*criminal charge*) | проект уголовного уложения (*criminal code draft*) |
| Topic 21: *акция, митинг, задержать, Навальный, полиция, мероприятие, проведение, оппозиционер, площадь, участник* (campaign, rally, detain, Navalny, police, event, holding, oppositionist, square, participant) | | | | | |
| задержание на акциях в поддержку (*detention at support events*) | акции в поддержку (*actions in support*) | организаторы несогласованных митингов в Москве (*organizers of uncoordinated rallies in Moscow*) | список маршей (*list of marches*) | полиция (*police*) | сплит акций (*stock split*) |

We calculated averaged weights for label triplets: Yandex label triplets get high weight 1,4 and ESA triplets get medium weight 0,98 (maximum threshold 2), preliminary results being satisfactory.

Inter-annotator agreement was calculated by Kendall's coefficient of concordance (Kendall's $W$), its values being very high both for Yandex label triplets ($W = 0,75$) and for ESA label triplets ($W = 0,87$). The agreement between the raters proves to be quite strong, that testifies in favour of data and algorithms consistency.

Second, we carried out evaluation of single labels by 10 independent assessors who were asked to mark each label by weights «0» – a label doesn't match a topic; «1» – a label matches a topic, cf. Table 2 (the main purpose of the second stage is to verify the received data, so we reduced the choice to two options). In this case we used a more general questionnaire as we addressed the intuition of the native speakers of Russian, not necessarily specialists in NLP and Data Science. The assessors were unaware of the label sources and were supposed to evaluate 6 labels per topic.

We calculated averaged weights for Labels-Yandex and for Labels-ESA: $w$(Labels-Yandex) = 0,68, $w$(Labels-ESA) = 0,39, medium weight 0,54 (cf. in experiments on the corpus of Russian encyclopedic texts on linguistics [24] $w$(Labels-Yandex) = 0,54, $w$(Labels-ESA) = 0,47, medium weight 0,51). Labels-Yandex performs better than Labels-ESA as it addresses to a wide range of web-documents of various subject areas, styles and genres, while ESA uses Wikipedia texts which reveal the traits of scientific style. On the one hand, in previous tests on linguistic texts the performance of two algorithms in question was almost equal due to the choice of the subject area which is well-represented both in the web and in wiki-articles. On the other hand, in experiments with the hypercorpus which contains news articles Labels-Yandex algorithm provides much more precise labels than Labels-ESA, this sharp difference is explained by semantic and stylistic homogeneity of the corpus and search engine output and proves the divergence of news texts and encyclopedic articles.

**Table 2.** Examples of weighting candidate labels generated by Labels-Yandex and Labels-ESA.

| | Labels-Yandex | | | Labels-ESA | | |
|---|---|---|---|---|---|---|
| Topic 22: *цена, нефть, рост, топливо, баррель, нефтяной, бензин, вырасти, снижение, фас (price, oil, growth, fuel, barrel, oil (attr.), gasoline, grow, decline, FAS)* | | | | | | |
| | **цена на бензин и нефть** (***price of gasoline and oil***) | **бензин и нефть** (***gasoline and oil***) | цена на фьючерс на нефть (*price of oil futures*) | **нефть** (*oil*) | страны по добыче (*countries in mining*) | лицензия на добычу (*production licence*) |
| Assessor 1 | 1 | 1 | 1 | 1 | 1 | 1 |
| Assessor 2 | 1 | 1 | 0 | 1 | 0 | 0 |
| Assessor 3 | 1 | 1 | 1 | 1 | 0 | 0 |
| Assessor 4 | 1 | 1 | 1 | 1 | 1 | 1 |
| Assessor 5 | 1 | 1 | 1 | 1 | 1 | 1 |
| Assessor 6 | 1 | 1 | 1 | 1 | 0 | 0 |
| Assessor 7 | 1 | 1 | 1 | 1 | 0 | 0 |
| Assessor 8 | 1 | 1 | 1 | 1 | 0 | 0 |
| Assessor 9 | 1 | 1 | 0 | 0 | 0 | 0 |
| Assessor10 | 1 | 1 | 1 | 1 | 1 | 0 |
| Total weight | **10** | **10** | 8 | 9 | 4 | 3 |
| | Labels-Yandex | | | Labels-ESA | | |
| Topic 38: *Роснефть, нефть, месторождение, нефтяной, иск, соглашение, добыча, акция, дивиденд, сделка (Rosneft, oil, field, oil (attr.), lawsuit, agreement, mining, share, dividend, deal)* | | | | | | |
| | **нефть в России** (***oil in Russia***) | тонны в год (*tons per year*) | Тоталь в России (*Total in Russia*) | **месторождение (field)** | лицензия на добычу (*production licence*) | **добыча нефти** (***oil production***) |
| Assessor 1 | 1 | 1 | 0 | 1 | 1 | 1 |
| Assessor 2 | 1 | 0 | 0 | 0 | 0 | 1 |
| Assessor 3 | 1 | 1 | 1 | 1 | 1 | 1 |
| Assessor 4 | 1 | 1 | 1 | 1 | 1 | 1 |
| Assessor 5 | 1 | 0 | 1 | 1 | 1 | 1 |
| Assessor 6 | 1 | 0 | 0 | 1 | 1 | 1 |
| Assessor 7 | 1 | 0 | 0 | 1 | 1 | 1 |
| Assessor 8 | 1 | 0 | 0 | 1 | 0 | 1 |
| Assessor 9 | 1 | 0 | 0 | 0 | 0 | 1 |
| Assessor10 | 1 | 0 | 0 | 1 | 0 | 1 |
| Total weight | **10** | 3 | 3 | **8** | 6 | **10** |

## 6  Discussion

We performed the analysis of raters' label evaluation and tried to choose triplets of the best labels. To do this, we ranged total weights of labels as shown in Table 2.

Of 40 label triplets 20 contained 2 Yandex labels and 1 ESA label. In 13 triplets Yandex label was the best (e.g., Topic 16), in 6 triplets ESA labels got the highest scores (e.g., Topic 4), in 1 triplet Yandex and ESA labels got equal weights (Topic 27).

Topic 4: *производство, предприятие, продукция, сеть, завод, технология, оператор, производитель, продукт, оборудование* (*manufacture, enterprise, production, network, plant, technology, operator, manufacturer, product, equipment*).

Label triplet: *заводы* (*plants*) (10, ESA), *заводы и станки* (*plants and machines*) (9, Yandex), *завод пищевого оборудования* (*food equipment factory*) (6, Yandex).

Topic 16: *встреча, лидер, переговоры, саммит, КНДР, визит, состояться, обсудить двусторонний МИД* (*meeting, leader, talks, summit, North Korea, visit, take place, discuss, bilateral, Foreign Ministry*).

Label triplet: *встреча лидеров России и КНДР* (*meeting of Russia and North Korea leaders*) (9, Yandex); *Россия и КНДР* (*Russia and North Korea*) (8, Yandex), *Восточная Азия* (*East Asia*) (4, ESA).

Topic 27: *фильм, режиссер, картина, кино, кинотеатр, актер, театр, продюсер, фестиваль, зритель* (*film, director, picture, movie, cinema, actor, theatre, producer, festival, audience*).

Label triplet: *театр в кино* (*theatre in cinema*) (5, Yandex), *кино в театре* (*cinema in theatre*) (5, Yandex), *советское кино* (*soviet cinema*) (5, ESA).

Further, 8 of 40 label triplets contained 1 Yandex label and 2 ESA labels. In 3 triplets Yandex label was evaluated higher (e.g., Topic 17), in 5 triplets ESA labels won (e.g. Topic 2).

Topic 2: *следствие, адвокат, уголовный, дело, обвинять, следователь, СКР, арест, УК, фигурант, обвинение* (*investigation, lawyer, criminal, case, accuse, investigator, investigation committee, arrest, criminal code, person of interest, prosecution*).

Label triplet: *уголовное обвинение* (*criminal charge*) (9, ESA) *уголовный кодекс* (*criminal code*) (8, ESA), *дело в отношении адвокатов РФ* (*case against lawyers of the Russian Federation*) (6, Yandex).

Topic 17: *британский, Великобритания, Солсбери, Скрипаль, Лондон, отравление, дипломат, вещество, новичок, март* (*british, Great Britain, Salisbury, Skripal, London, poisoning, diplomat, novichok, substance, March*).

Label triplet: *Скрипали в Великобритании* (*the Skripals in Great Britain*) (10, Yandex), *новичок* (*novichok*) (6, ESA), *Лондон* (*London*) (4, ESA).

The rest 12 triplets contained 3 Yandex labels (e.g., Topic 13), there were no triplets including ESA label only.

Topic 13: *налог бюджет фонд доход налоговый Минфин триллион ставка НДС федеральный* (*tax, budget, fund, income, tax (attr.), Ministry of Finance, trillion, rate, VAT, federal*).

Label triplet: *налоги в бюджет* (*taxes to the budget*) (10, Yandex), *поступления в бюджет* (budget receipts) (10, Yandex), *рост доходов бюджета от повышения* (*growth of budget revenues from the increase*) (7, Yandex).

In general, we got 28 triplets with both Yandex and ESA labels, where Yandex got the upper hand in 16 cases, ESA won in 11 cases, not to mention a single draw. That provides evidence in favour of Labels-Yandex and Labels-ESA ensemble.

Linguistic data analysis shows that Labels-Yandex algorithm gives quite concrete labels reflecting the agenda of the moment, while Labels-ESA works out labels with generalized content. Our experiments show that in certain cases Labels-Yandex algorithm fails to predict correct labels. Let's consider Topic 10 and its labels:

Topic 10: *губернатор отставка пост должность господин заместитель область администрация назначить совет (Governor, resignation, post, position, mister, deputy director, region, authority, appoint, board).*

Label triplet: *губернаторы (governors) (10, ESA); государс твенная администрация (state administration) (8, ESA), правительство в отставке (government in retirement) (6, Yandex).*

In this case Labels-Yandex generated a label *правительство в отставке (government in retirement)* which reflected the news of the moment (January 15, 2020) [32], but could hardly correspond to a news cluster related to the topic from the hypercorpus. At the same time, Labels-ESA provided rather consistent labels like *государственная администрация (state administration)* or *губернаторы (governors)*. So, news dynamics may be responsible for the shifts in the content of labels extracted from the search engine, while Wikipedia texts remain stable.

That's why it is reasonable to use both Labels-Yandex and Labels-ESA as an ensemble of algorithms in automatic topic label assignment: a) internal ranking provided by both algorithms allows to choose 3 reliable labels from a mixed set of candidates, b) both algorithms provide multiword expressions alongside with single terms as reliable candidate for topic labels, c) the procedure is suitable for topic models of any type, irrespective of the topic modelling scheme and type of models (unigram or n-gram models), d) working together, two algorithms reduce losses in the quality of labels caused by the specificity of external sources and amend the total result.

## 7  Summary

The automatic topic label assignment is an important stage in the development of the semantic network theory of e-hypertext. This approach makes it possible to mark out multiple nodes of hypertext, rubricating them by topic, which will be the basis for studying the semantics of hypertext transitions and, as a result, it will be possible to determine the potential of a text to reveal its hypertextuality in a specific type of discourse.

Our experiments with the ensemble of topic labelling algorithms Labels-Yandex and Labels-ESA represent the first full-scale research performed on a large Russian e-hypertext corpus. We are the first to perform the procedure of topic label assignment working with the Russian e-media texts.

The choice of topic labelling algorithms Labels-Yandex and Labels-ESA which use external sources of labels (web-documents produced by Yandex search engine and the Russian Wikipedia dump) is justified by the fact that the Russian e-hypertext corpus is constantly extending and requires synchronization of its topics with the data from the web.

Results achieved in our experiments may serve as a starting point for further elaboration of automatic topic labelling techniques applied to Russian corpora as we provide both datasets, evaluation procedure and baseline scores. Thus, our research fills in the gap in contemporary Russian NLP research and may draw attention of computational linguists towards generalization and linguistic interpretation of topic models.

**Acknowledgements.** The reported study was funded by RFBR according to the research project № 18-312-00010. The authors express their deep gratitude to Aliia Erofeeva (CCG.ai, Cambridge, UK) and Kirill Sukharev (ETU «LETI», Saint-Petersburg, Russia) for their help in the development of topic labelling software.

# References

1. Nelson, T.: Literary Machines. Mindful Press, Sausalito (1993)
2. Salmerón, L., Kintsch, W., Cañas, J.J.: Reading strategies and prior knowledge in learning from hypertext. Mem. Cognit. **34**, 1157–1171 (2006)
3. Vandendorpe, C.: From Papyrus to Hypertext: Toward the Universal Digital Library (Topics in the Digital Humanities). University of Illinois Press (2009)
4. Shulginov, V.A., Shulginov, V.A., Mitrofanova, O.A.: Topic organization of e-hypertext media: corpus driven research. In: R. Piotrowski's Readings in Language Engineering and Applied Linguistics (PRLEAL 2019), CEUR Workshop Proceedings, vol. 2552, pp. 299–312 (2019)
5. BeautifulSoup. https://pypi.org/project/beautifulsoup4/. Accessed 08 July 2020
6. NLTK. https://www.nltk.org/. Accessed 08 July 2020
7. re. https://github.com/python/cpython/blob/3.8/Lib/re.py. Accessed 08 July 2020
8. genism. https://radimrehurek.com/gensim. Accessed 08 July 2020
9. t-Distributed Stochastic Neighbor Embedding. https://scikit-learn.org/stable/modules/generated/sklearn.manifold.TSNE. Accessed 08 July 2020
10. DBSCAN clustering. https://scikit-learn.org/stable/modules/generated/sklearn.cluster.DBSCAN.html. Accessed 08 July 2020
11. Non-negative matrix factorization. https://radimrehurek.com/gensim/models/nmf.html. Accessed 08 July 2020
12. Lau, J.H., Newman, D., Karimi, S., Baldwin, T.: Best topic word selection for topic labelling. In: COLING'10 Proceedings of the 23rd International Conference on Computational Linguistics, Stroudsburg, PA, Association for Computational Linguistics, pp. 605–613 (2010)
13. Aletras, N., Stevenson, M., Court, R.: Labelling topics using unsupervised graph-based methods. In: Proceedings of the 52nd Annual Meeting of the Association for Computational Linguistics (Volume 2: Short Papers), Baltimore, Maryland, ACL, pp. 631–636 (2014)
14. Mei, Q., Shen, X., Zhai, C.: Automatic labeling of multinomial topic models. In: Proceedings of the 13th ACM SIGKDD International Conference on Knowledge Discovery and Data Mining KDD 2007. pp. 490–499. ACM Press, New York (2007)
15. Cano Basave, A.E., He, Y., Xu, R.: Automatic labelling of topic models learned from twitter by summarisation. In: Proceedings of the 52nd Annual Meeting of the Association for Computational Linguistics (Volume 2: Short Papers), Stroudsburg, PA, USA, Association for Computational Linguistics, pp. 618–624 (2014)

16. Kou, W., Li, F., Baldwin, T.: Automatic labelling of topic models using word vectors and letter trigram vectors. In: Zuccon, G., Geva, S., Joho, H., Scholer, F., Sun, A., Zhang, P. (eds.) AIRS 2015. LNCS, vol. 9460, pp. 253–264. Springer, Cham (2015). https://doi.org/10.1007/978-3-319-28940-3_20

17. Nolasco, D., Oliveira, J.: Detecting knowledge innovation through automatic topic labeling on scholar data. In: 49th Hawaii International Conference on System Sciences (HICSS), Koloa, HI, pp. 358–367. IEEE Computer Society (2016)

18. Magatti, D., Calegari, S., Ciucci, D., Stella, F.: Automatic labeling of topics. In: ISDA 2009 9th International Conference on Intelligent Systems Design and Applications, Pisa, pp. 1227–1232. IEEE (2009)

19. Lau, J.H., Grieser, K., Newman, D., Baldwin, T.: Automatic labelling of topic models. In: Proceedings of the 49th Annual Meeting of the Association for Computational Linguistics: Human Language Technologies, Stroudsburg, PA, vol. 1, pp. 1536–1545. Association for Computational Linguistics (2011)

20. Hulpus, I., Hayes, C., Karnstedt, M., Greene, D.: Unsupervised graph-based topic labelling using DBpedia. In: Proceedings of the Sixth ACM International Conference on Web Search and Data Mining WSDM 2013, pp. 465–474 (2013)

21. Bhatia, S., Lau, J.H., Baldwin, T.: Automatic labelling of topics with neural embeddings. In: 26th COLING International Conference on Computational Linguistics, 2016, pp. 953–963 (2016)

22. Allahyari, M., Pouriyeh, S., Kochut, K., Arabnia, H.R.: A knowledge-based topic modeling approach for automatic topic labeling. Int. J. Adv. Comput. Sci. Appl. **8**(9), 335–349 (2017)

23. Mao, X., Hao, Y.-J., Zhou, Q., Yuan, W., Yang, L., Huang, H.: A novel fast framework for topic labeling based on similarity-preserved hashing. In: COLING 2016, pp. 3339–3348 (2016)

24. Kriukova, A., Erofeeva, A., Mitrofanova, O., Sukharev, K.: Explicit semantic analysis as a means for topic labelling. In: Ustalov, D., Filchenkov, A., Pivovarova, L., Žižka, J. (eds.) AINL 2018. CCIS, vol. 930, pp. 110–116. Springer, Cham (2018). https://doi.org/10.1007/978-3-030-01204-5_11

25. Mirzagitova, A., Mitrofanova, O.: Automatic assignment of labels in topic modelling for Russian corpora. In: Proceedings of 7th Tutorial and Research Workshop on Experimental Linguistics, ExLing 2016/A. Botinis, ed. Saint Petersburg: International Speech Communication Association, 2016, pp. 115–118 (2016)

26. Erofeeva, A., Mitrofanova, O.: Automatic Topic label assignment in topic models for russian text corpora. In: Structural and Applied Linguistics, Saint-Petersburg, vol. 12, pp. 122−147 (2019). (in Russian)

27. Kriukova, A., Mitrofanova, O., Sukharev, K.: Measuring semantic relatedness of russian texts by means of explicit semantic analysis. In: Kalinichenko, L., Manolopoulos, Y., Stupnikov, S., Skvortsov, N., Sukhomlin, V. (eds.) Data Analytics and Management in Data Intensive Domains: XX International Conference DAMDID/RCDL'2018 (October 9–12, 2018, Moscow, Russia): Conference Proceedings /, pp. 284–288. FRC CSC RAS, Moscow (2018)

28. Kriukova, A., Mitrofanova, O., Sukharev, K., Roschina, N.: Using explicit semantic analysis and Word2Vec in measuring semantic relatedness of russian paraphrases. In: Alexandrov, D.A., Boukhanovsky, A.V., Chugunov, A.V., Kabanov, Y., Koltsova, O. (eds.) DTGS 2018. CCIS, vol. 859, pp. 350–360. Springer, Cham (2018). https://doi.org/10.1007/978-3-030-02846-6_28

29. Scikit-learn. https://scikit-learn.org/stable/. Accessed 08 July 2020

30. Korobov, M.: Morphological analyzer and generator for Russian and Ukrainian languages. In: Khachay, M.Y., Konstantinova, N., Panchenko, A., Ignatov, D.I., Labunets, V.G. (eds.) AIST 2015. CCIS, vol. 542, pp. 320–332. Springer, Cham (2015). https://doi.org/10.1007/978-3-319-26123-2_31

31. Gabrilovich, E., Markovitch, S.: Computing semantic relatedness using Wikipedia-based explicit semantic analysis. In: Proceedings of the 20th International Joint Conference on Artificial Intelligence (IJCAI), pp. 1606–1611 (2007)
32. RIA News. https://ria.ru/20200115/1563456719.html. Accessed 08 July 2020

# Convolutional Variational Autoencoders for Spectrogram Compression in Automatic Speech Recognition

Olga Yakovenko[1(✉)] and Ivan Bondarenko[2]

[1] Center of Financial Technologies, Novosibirsk, Russia
olya.yakovenko@bk.ru
[2] Novosibirsk State University, Novosibirsk, Russia
i.yu.bondarenko@gmail.com

**Abstract.** For many Automatic Speech Recognition (ASR) tasks audio features as spectrograms show better results than Mel-frequency Cepstral Coefficients (MFCC), but in practice they are hard to use due to a complex dimensionality of a feature space. The following paper presents an alternative approach towards generating compressed spectrogram representation, based on Convolutional Variational Autoencoders (VAE). A Convolutional VAE model was trained on a subsample of the LibriSpeech dataset to reconstruct short fragments of audio spectrograms (25 ms) from a 13-dimensional embedding. The trained model for a 40-dimensional (300 ms) embedding was used to generate features for corpus of spoken commands on the GoogleSpeechCommands dataset. Using the generated features an ASR system was built and compared to the model with MFCC features.

**Keywords:** Variational autoencoder · Speech recognition · Audio feature representation

## 1 Introduction

Automatic recognition of the spoken language has already became a part of a daily life for many people in the modern world. Although algorithms for automatic speech recognition have progressed greatly throughout the last years, most of the applications still utilize a basic set of features – Mel-frequency cepstral coefficients. These kind of features are processed rapidly and can produce good results, but recent research [1] has proven that raw FFT spectrograms give better accuracy on ASR tasks when combined with Deep Neural Networks (DNN) for feature-extracting.

The drawback of using spectrograms as the inputs to a recognition system is that a lot of computational power has to be used to process a single audio. With the increase of length of the audio increases the time spent computing an optimal path from first to last sound. Embeddings, generated by VAE, can

© Springer Nature Switzerland AG 2021
W. M. P. van der Aalst et al. (Eds.): AIST 2020, CCIS 1357, pp. 115–126, 2021.
https://doi.org/10.1007/978-3-030-71214-3_10

be used as a compressed version of the general FFT spectrogram features for both traditional Gaussian Mixture Models and modern DNNs. Moreover, this approach can be used for dimensionality reduction of input data [2] that can subsequently be used for visualization of audio data or for reduction of occupied space on a hard drive.

Traditionally VAE is used as a generative model, similar to Generative Adversarial Networks (GAN). For the past several years there has been going a lot of research on the topic of audio generation with VAEs, including the fascinating DeepMind's VQ-VAE [3], which is a successor of WaveNet [4]. The main features of those two systems is that having the train samples of some audio clips the system learns the probability distributions over observed features, making it possible to generate new samples by selecting other values from these distributions.

VAE was found out to be a good tool for dimensionality reduction using the bottleneck features of the hidden layer. For example, authors of [2] prove in their paper that VAE is a natural evolution of the robust PCA algorithm. Basic Autoencoders (AE) are also known to be able to reduce the dimensionality of the data, but the resulting embedding does not serve as a good generalization, because the hidden representations of the intermediate layer are poorly structured due to non-probabilistic nature of AE.

Over the past few years, there has also been attempts to perform construction of audio embeddings. The main goal of this approach is to be able to compare many utterances to some etalon utterance or to compare different utterances against each other and present all of them in some feature space. This approach could be found to be used in the spoken term recognition by example [5], for speaker identification or verification using Siamese Neural Networks [6] and for semantic embedding construction based on context based on Recurrent Neural Networks (RNN) and skipgram/cbow architectures [7].

It is important to notice, that most of the presented approaches use recurrent structures for analyzing variable-length audio data. Recurrent structures enable good quality of automatically determining features, relevant to the task, having approximately same length. The drawback here is that if the system was trained on samples of length from 4 s to 10 s, it will produce unexpected results for samples much longer than 10 s. Researcher is forced to either adapt his data to fit the model, or to build his own model that would fit his data. These drawbacks can be avoided by using fixed-length features, such as MFCC or VAE-based features, presented in the following paper.

Thus, in this paper an approach towards feature generation for audio data is presented. The first part is dedicated to the feature generation itself and the ability to reconstruct the signal from the resulting embedding, while the second part describes the experiments with the embeddings for an ASR task.

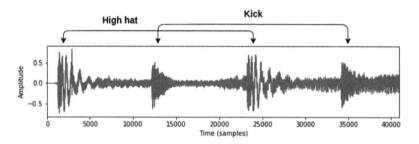

**Fig. 1.** Amplitude audio representation.

## 2 Traditional Audio Features for ASR/TTS Tasks

Audio wave is a sequence of positive and negative values, the values represent the amplitudes of the fluctuation of a sound device. The frequency and the amount of the fluctuation defines different sounds.

### 2.1 Raw Wave

Sometimes raw signal is used for speech recognition or generation tasks. The only preprocessing that the sound may undergo is discretization of an analog sound when it is being recorded. It is possible to analyse a discrete signal, if the task itself is not too complex (e.g. recognition of loud noises, of sound patterns) but also if the system consists of an end-to-end structure that besides solving the main tasks is also capable to detect frequencies in a signal. Some of those end-to-end architectures include 1D Convolutions or PixelCNNs (Fig. 1 and 2).

### 2.2 Spectrogram

A traditional approach towards deep analysis of speech and sounds is mostly realized through spectrograms using a Short-Time Fourier Transform (STFT). Spectrograms are quite useful for manual analysis: for example, for detection of formants in human's speech. Generally, spectrograms are very good means of learning information from audio data, since sounds, that are produced by the vocal cord and, more importantly, speech, are characterized by cooccurrence of different harmonics and their change through time. STFT is applied to a discrete signal, so the data is broken up into overlapping chunks. Each chunk is Fourier transformed, and the result is added to a matrix, which records magnitude and phase for each point in time and frequency. This can be expressed as:

$$\mathbf{STFT}\{x[n]\}(m,\omega) \equiv X(m,\omega) = \sum_{n=-\infty}^{\infty} x[n]w[n-m]e^{-j\omega n} \qquad (1)$$

with signal $x[n]$ and window $w[n]$.

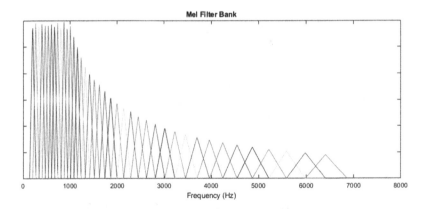

**Fig. 2.** Mel-scaled filterbanks.

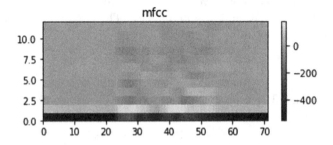

**Fig. 3.** MFCC of the LibriSpeech sample 1779-142733-0000.wav.

The magnitude squared of the STFT yields the spectrogram representation of the Power Spectral Density of the function:

$$spectrogram\{x[n]\}(m, \omega) \equiv |X(m, \omega)|^2 \tag{2}$$

The described spectrogram was generated for the LibriSpeech dataset during the experiments with VAE.

One of the commonly used modifications of STFT spectrograms are mel-spectrograms. During this modification each of the time frames in a spectrogram are passed through series of triangular filters (mel-scaled filters). Mel filters are based on a hand-crafted mel-scale, which represents the extent to which humans can distinguish sounds of different frequencies.

All of the aforementioned approaches have led us to the traditional approach for creation of audio features: Mel-Frequency Cepstral Coefficients (MFCC).

### 2.3   MFCC

Once we have the mel-spectrograms, we can then proceed to the computation of the Mel-Frequency Cepstral Coefficient (MFCC). It consists of three simple steps:

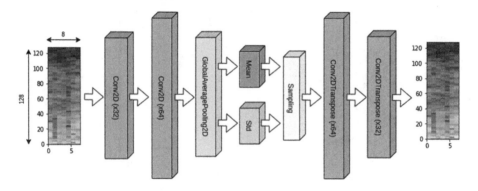

**Fig. 4.** VAE architecture.

1. Compute elementwise square of each value in the mel-spectrogram;
2. Compute elementwise logarithm of each element;
3. Apply discrete cosine transform to the list of mel log powers, as if it were a signal.

The final result will also be a matrix, quite unreadable by a human eye, but surprisingly effective when used for both Speech to Text (STT) and Text to Speech (TTS). To this day MFCC features are used mostly in all speech-related technologies.

Mel-spectrogram and MFCC are means towards compressing audio data without erasing the information relevant to speech, since these features are further used in applications, connected to speech.

Here we determine the goal of this study: we believe that it is possible to compress audio in analogous way, but with the help of neural network, specifically, a Variational Autoencoder. We are going to cover the architecture in the next chapter.

## 3    Audio VAE

As was mentioned in the introduction, Variational Autoencoders can be successfully used not just as a generative network, but also as a dimensionality reduction tool. This is the one of the main reasons for selecting this architecture. The other reason for selecting a VAE for audio compression is the assumption that a VAE that is trained on a big clean corpus of speech will then be able to extract features from a spectrogram that are relevant only to the speech. Therefore, the encoder of the resulting VAE will be able to conduct denoising, selection of relevant frequencies and compression simultaneously. As an interesting additional feature, this kind of network may be used for generating sounds that sound close to the true signal, that way acting as a option for augmentation.

It is also important to specify that encoder and decoder of the VAE include convolutional layers instead of fully-connected layers to evade a big amount

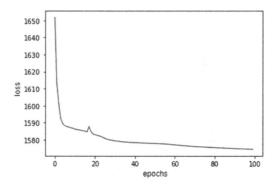

**Fig. 5.** Train loss dynamics during training for 100 epochs.

of parameters to optimize and improve the recognition of formants of various frequencies and scales.

### 3.1 Architecture

VAE architecture was inspired by traditional VAE for image generation and segmentation. The VAE neural network is symmetric, where encoder and decoder parts include 2 convolution layers with kernel sizes 8 and stride 2. The encoder and decoder part are connected by a bottleneck which consists of 2D global average pooling layer, then followed by mean and variance layers of the VAE and finally the sampling layer (Fig. 4).

As for any other VAE, special loss function had to be defined that consists of two parts: reconstruction loss and regularisation. In our case reconstruction loss is Mean Squared Error (MSE) between the true values and predicted values and regularisation is estimated by Kullback–Leibler divergence between the distribution that is present in the data and the distribution that has been modeled by our encoder.

$$\mathcal{L}(\phi, \theta, \mathbf{x}, \mathbf{x}') = \|\mathbf{x} - \mathbf{x}'\|^2 + -0.0005 * D_{\mathrm{KL}}(q_\phi(\mathbf{h}|\mathbf{x})\|p_\theta(\mathbf{h})) \tag{3}$$

where $p_\theta(\mathbf{x}|\mathbf{h})$ is the directed graphical model that we want to approximate, encoder is learning an approximation of this model $q_\phi(\mathbf{h}|\mathbf{x})$ to the Gaussian multivatiate (in our case) distribution $p_\theta(\mathbf{h}|\mathbf{x})$ where $\phi$ and $\theta$ denote the parameters of the encoder (recognition model) and decoder (generative model) respectively.

### 3.2 Dataset

Training of the system was carried out on a subset of the LibriSpeech dataset [8]. LibriSpeech flac files were transformed to mono-channel wav, sampling rate 16 kHz, signed PCM little endian with bit rate 256 kB/s and bit depth 16. The subset was formed as follows:

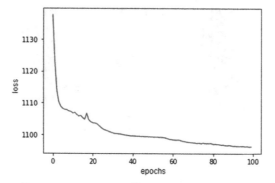

**Fig. 6.** Test loss dynamics during training for 100 epochs.

- Dataset was split into train and test with 2097 audiofiles representing train and 87 representing test.
- Train set included utterances from 1031 speakers, test set – 40 speakers.
- Spectrograms were calculated from the audio data.
- Each utterance was split into fragments of 0.025 s with stride of 0.01 s.

That left us with 1047736 samples for training and 30548 samples for testing. One input sample had the shape of 8 timestamps by 128 wave frequency stamps.

The choice for making such a small window (a small 25 ms window) was determined by different ASR systems, which use MFCC as their features. Later we want to compare the performance of an ASR system with VAE features and MFCC features, thus it is important to maintain the same conditions for feature generation for clear comparison.

The dynamics for the training of the VAE audio features are shown on the Fig. 5 for the train set and Fig. 6 for the test set.

### 3.3   Reconstruction

The system described earlier was used for the encoding and reconstruction of the audio. Once we have trained the VAE, we can compress and decompress some samples from the dataset, some examples of the original and reconstructed samples can be seen in the Fig. 7.

We can see from the pictures, that despite a big compression rate from 1024 elements of the spectrogram to just 13 elements of the latent vector, the information about the energised parts of the spectrum is still preserved. Moreover, the system seems to have learned the common localisation of the human speech (lower frequencies) and the way it can change over time.

The next step would be to visually compare the VAE features that we have achieved with MFCC features. MFCC for a file from the LibriSpeech dataset we have calculated earlier, see Fig. 3, and now we can see the analoguous features for the VAE generated encoding on the Fig. 8. It is evident from the pictures,

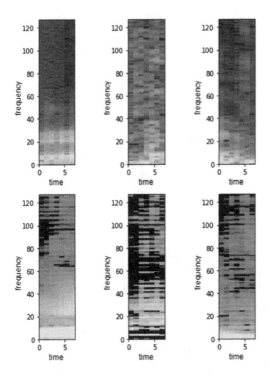

**Fig. 7.** Examples of reconstruction (lower) of the samples from the 13-dimensional hidden representation of the real sample (upper). The dark purple segments in the reconstructed pictures are the elements of matrix, that are equal to 0. (Color figure online)

that VAE has a better ability to code necessary segments of audio where human speech takes place, and seems to ignore the any kind of other sounds, that may sound on the background. On the other hand, MFCC features seem to detect some of the activity in the background, which may not reflect on ASR systems too well.

### 3.4   Generation

Since the trained encoder of the VAE is an approximation to a Gaussian multivariate distribution and the decoder is a directed graphical model, we can sample examples of spectrograms from the distribution and visualize them using our decoder.

For every component in the latent representation 4 evenly spaced numbers are chosen in the interval $[-1, 1]$. Then the resulting 4 vectors that differ by one value are transformed into spectrograms. Examples for some components are presented in Fig. 9.

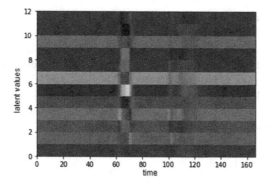

**Fig. 8.** Example of VAE encoding of the LibriSpeech sample 1779-142733-0000.wav.

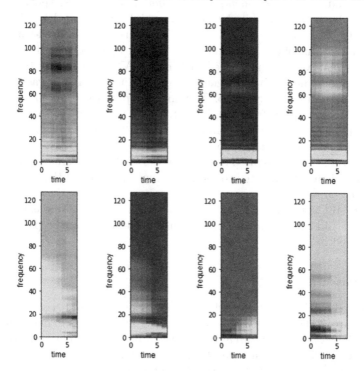

**Fig. 9.** Samples from the multivariate (13 variables) Gaussian distribution, modeled by VAE, alternation of the 2-nd (top) and 5-th (bottom) variable, while the other are kept as mean values for the distribution.

It can be seen from the figures, that the VAE has learned how human speech may sound and has successfully encapsulated that knowledge in a multivariate Gaussian distribution.

## 4    Experiments

It is essential to compare developed approach with the existing. Therefore, experiments were carried out to evaluate the performance of resulting embeddings in a speech recognition task in comparison to MFCC features. Although the previous part was dedicated to reconstructing 25 ms feature frames, in this part 300 ms feature frames are regarded. The experiment with 25 ms frames and comparison with MFCC features in a Kaldi environment is planned for the nearest future.

### 4.1    Dataset

The GoogleSpeechCommands [9] dataset was used for training and testing. Dataset includes audio fragments of 30 different commands, spoken in noisy conditions. The choice of this dataset was mainly determined by the relative simplicity to test it for both VAE and MFCC features: it is fixed-length audio while containing one of the 30 commands in a single audio. One of the other main reasons was that we have mentioned earlier, that VAE has a great potential for smoothing audio and performing general noise reduction, and it was interesting to test the system on noisy audio. Training was performed on 46619 samples, testing on 11635 samples from the dataset. Audio data had the same format as LibriSpeech: mono-channel wav, sampling rate 16 kHz, signed PCM little endian with bit rate 256 kB/s and bit depth 16.

VAE features were generated in a window 0.3 s with 0.1 overlap. The resulting vectors were then concatenated to serve as a representation of the feature vector of a spoken command.

For the MFCC features an analysis window of length 0.025 s was used with a step of 0.01 s and 26 filterbanks. For one window 13-dimensional cepstrum vector was generated.

### 4.2    Architecture

With this type of dataset it is possible to solve a simple multiclass task, since all the audios are of the same length and one audio only belongs to one class simultaneously. The ASR system architecture was a basic Multi-Layered Perceptron (MLP) of the following structure: 2 fully-connected layers of 100 hidden units with 0.2 dropout rate and ReLU activation followed by a softmax layer (Fig. 10).

Although we are solving such a simple ASR task (command recognition), that is quite far away from many practical ASR systems, still solving this task may give us the understanding the potential of the VAE features in comparison to MFCC.

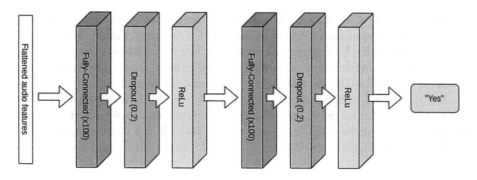

**Fig. 10.** ASR Neural Network structure.

### 4.3   Results

The results of the experiments are shown in the Table 1 below.

**Table 1.** Speech recognition results

| Results | VAE | MFCC |
|---|---|---|
| Train accuracy | **0.45** | 0.41 |
| Test accuracy | **0.49** | **0.49** |
| MLP training time | **42 s** | 63 s |
| Best epoch | **12** | 17 |
| Train size | **204 MB** | 668 MB |

What we can see from the results in the table is that VAE does not concede to the MFCC features, while generally taking up less space. That means that being 3× smaller, VAE features carry approximately the same amount of acoustic information as the MFCC features. This can be very useful in the cases when the audio information has to be stored somewhere (e.g. etalon voice samples, database of speech segments). Furthermore, when the input to the neural network is smaller, as in the case of VAE features in comparison to MFCC, then the amount of parameters in a network also becomes smaller and this leads to faster convergence of the neural network (as it is also highlighted in the results table).

## 5   Conclusions

This research has shown that convolutional VAEs are capable of reconstructing audio data in the form of spectrograms. They succeed at representing fixed-length audio data fragments for the ASR tasks. As a result, convolutional VAEs feature extracting have shown relatively good results in a recognition task on a noisy dataset, in comparison to traditional MFCC features. Finally, the embeddings, generated by VAE occupy 3 times less space in comparison to MFCC, despite being as informative. The resulting algorithms, scripts and models are available on GitHub for any researcher that would be interested [10].

The further work that has to be done is to test the 13-dimensional latent features of the VAE as an input for a proper ASR system (e.g. Kaldi, DeepSpeech). It is planned to carry out the experiment using Kaldi in the nearest future. This will be done for English and Russian languages, for the Voxforge dataset.

# References

1. Amodei, D., et al.: Deep speech 2: end-to-end speech recognition in English and Mandarin. In: ICML (2016)
2. Dai, B., et al.: Hidden talents of the variational autoencoder (2017)
3. van den Oord, A., et al.: Neural discrete representation learning. In: NIPS (2017)
4. van den Oord, A., et al.: WaveNet: a generative model for raw audio. In: SSW (2016)
5. Zhu, Z., et al.: Siamese recurrent auto-encoder representation for query-by-example spoken term detection. In: Interspeech (2018)
6. Milde, B., Biemann, C.: Unspeech: unsupervised speech context embeddings. In: Interspeech (2018)
7. Chung, Y.-A., Glass, J.R.: Speech2Vec: a sequence-to-sequence framework for learning word embeddings from speech. In: Interspeech (2018)
8. LibriSpeech. http://www.openslr.org/12/
9. Google Speech Commands. https://ai.googleblog.com/2017/08/launching-speech-commands-dataset.html
10. Audio_vae. https://github.com/nsu-ai-team/audio_vae

# Computer Vision

# Unsupervised Training of Denoising Networks

Alexey Kovalenko$^{(\boxtimes)}$ (ID) and Yana Demyanenko$^{(\boxtimes)}$ (ID)

Southern Federal University, Rostov-on-Don, Russia
alexey.s.russ@mail.ru, demyanam@gmail.com
https://www.sfedu.ru/

**Abstract.** This work explore approach for image denoising of received by CMOS sensor. Proposed pipeline solves the problem of unsupervised training neural network architectures for image denoising which uses datasets without clean data. This approach bases on theoretical background about image restoration proposed by Nvidia researchers. We implemented custom denoising neural network architectures using specifics of noise distribution. Networks are trained on custom images collection.

**Keywords:** Image denoising · Unsupervised learning · Neural networks · Learning image denoising

## 1 Introduction

Noise reduction is common problem in computer vision. Any image captured by CMOS sensor contains noise. This noise appears in useful signal due to errors in reception of optical radiation by sensor. Clear image signal will be denoted by $I$, and noise component denoted by $\alpha$. Assuming that the process of noise occurrence is an absolutely random process from the distribution of $P$. Then the matrix of the final image can be denoted by formula (1).

$$\tilde{I} = I + \alpha, \ \alpha \sim P \tag{1}$$

Since the error in receiving optical signal depends on physical device of the CMOS sensor, that for each model of the sensor there will be a unique distribution of $P$, which generates the noise component of the signal.

The goal of this work is approximation of the function $\phi : R^n \longrightarrow R^n$ by neural network, which has the following property:

$$\forall \tilde{I} \Longrightarrow \phi(\tilde{I}) = I \tag{2}$$

For build of approximation of the mapping $\phi$, the neural network $f$ will be trained to solve the following optimization problem:

$$\min_{w} \|f(\tilde{I}, w) - I\|_{L_2}, \tag{3}$$

where $w$ are the neural network weights $f$, $L_2$ is Euclidean norm.

© Springer Nature Switzerland AG 2021
W. M. P. van der Aalst et al. (Eds.): AIST 2020, CCIS 1357, pp. 129–139, 2021.
https://doi.org/10.1007/978-3-030-71214-3_11

## 2   Related Works

Significant contribution to the denoising problem by neural networks is made by the work of Nvidia researchers. This work is titled Noise2 Noise: Learning Image Restoration without Clean Data [1]. The main idea of this work is using representation of noise as a composition of a clear signal and unique noise received at different points in time to train a neural network to restore a clean image signal. A noisy image with the noise component $\alpha_1$ is forward to neural network and the network is required to predict the same image, but with the noise component $\alpha_2$. Assuming that the noise was obtained randomly, the neural network is not able to predict it, and therefore, during training, the neural network seeks to reconstruct the image with some losses. The disadvantage of this approach is the use of only a pair of images of one scene during training, which can result in large losses of the useful signal during network operation.

There is also a lot of researches touch upon problem of training noise reduction networks on datasets containing images without a noise component. An example of this approach is the work of engineers from Google which titled Unprocessing Images for Learned Raw Denoising [2]. In this work, the authors will train the network using the standard error function $L_1$. And tested this network on a Darmstadt Noise Dataset [3].

Darmstadt Noise Dataset datasets contains pairs of images. Each pair consists of an image taken with correctly selected camera parameters for shooting and an image having noises arising from incorrect parameters. Networks trained on such data sets do not solve the problem of suppressing noise arising from the CMOS sensor of a camera even in the most correctly selected shooting parameters.

Thus, for training a network focused on denoising from a specific CMOS sensor, it becomes necessary to collect data and develop a method for training a network on them.

Together with Noise2Noise work [1], there are state-of-the-art approaches such as Noise2Void [4], Noise2Self [5] and Deep Image Prior [6], contributing to the field of unsupervised learning of denoising neural networks.

## 3   Dataset

### 3.1   Collecting Dataset

Images for training dataset were obtained using a certain device, *Apple iPhone X*, which has a camera consisting of two sensors, with the characteristics given in the following source [7].

From this device a series of RAW images of seven scenes with different lighting and color component ware obtained. A scene made by shooting an static picture of the real world, getting the matrix (1). To do this, the device was fixed on a tripod in a stationary state and the shooting process was started from a wireless device, thereby obtaining a set of images $\{\tilde{I}_k^q\}_{k=1}^N$, where $q$ is the sequence number of the scene being shot. For each series ISO, focus and color temperature values were fixed when shooting.

Each scene contains about 14 photos. Total number of frames from all sets is 95 images.

The maximum number of images in one series is limited to 20 frames because CMOS sensor heats with long time the shooting process continues and additional signal distortions appear due to thermal effects. Due to this effect of which the pixel-by-pixel correspondence of frames in the series to each other is lost.

The result is a dataset consisting of 125 images taken on one sensor.

Also, to compare the results, 6 series were made, shot by next web camera in the resolution of $1920 \times 1080$. Short videos were shot and divided into frames. Total number of frames in these series was 811.

## 3.2  Analysing Dataset

During the shooting process, slight distortion of the frame may occur when the photosensor is heated. Or imperceptible displacements of the device during shooting, as well as a change in external conditions, such as lighting or movement of objects in the frame, are also possible. For an collected set of images, a shift value of more than one pixel between frames of the same series is already critical.

To analyze the quality of the obtained series of images, the distributions of $e^q$ (5) deviations of each image from the average over all images from the series $\hat{I}^q$ by the Euclidean metric.

$$\hat{I}^q_{i,j} = \frac{\sum_k \tilde{I}^q_{k\,i,j}}{N}, \text{ where } N \text{ is the number of frames in series } q \qquad (4)$$

$$e^q_k = \|\hat{I}^q - \tilde{I}^q_k\|_{L_2} \qquad (5)$$

Additionally, the series is normalized by metric $L_1$:

$$e^q_k = \frac{e^q_k}{\sum_i (e^q_i)}$$

Then, using the values of the sample $e^q$ (5), we plot the density curves of the normal distribution $\mathcal{N}(\mu, \sigma)$ with the parameters:

$$\mu^q = \mu(e^q), \ \sigma^q = \sigma(e^q) \qquad (6)$$

Thus, the mathematical expectation of $\mu^q$ (6) shows how near the average image of $\hat{I}^q$ (4) to the image theoretically obtained from a pure signal $I^q$:

$$\hat{I}^q \xrightarrow[\mu^q \to 0]{} I^q \qquad (7)$$

And also, the near to 0 the value of the standard deviation of the sample $\sigma^q$ (6), the more images from the series there is a signal of static scene, fixed at time $t = t_0$, that is, how much the scene is unchanged between frames.

**Fig. 1.** Series of images with color histograms, the parameter $\sigma$ of this series is 0.00608

For example, the deviation parameter $\sigma$ (6 may be quite large with different illumination of the same scene between frames (see Fig. 1).

For demonstrate the various values of the deviation parameter, a graph is constructed (see Fig. 2).

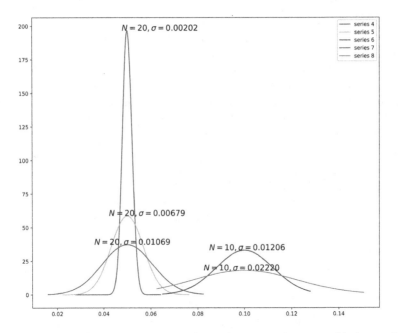

**Fig. 2.** Normal distribution density plots for series shot under natural light conditions

When choosing data for training neural networks, the condition is set that the parameter $\sigma$ (6) should not exceed the value of 0.007.

## 4    Method

In this work we produce a novel method for training denoising neural network architectures. This approach combine traditional supervised method for training autoencoder neural network architectures and approach produced by authors of Noise2Noise work [1].

When we have image series of one scene obtained in different time (see Sect. 3), we can average this image series and made approximation of expected clear image. Obtained mean image can be used for train the denoising neural network as the expected output signal $y'$ in solving of loss $E$ minimization problem (9). An input signal $x$ can be taken random image from series under consideration. But in this case value $y'$ consists noise component from input signal $x$. If input image has been excluded from series before averaging step, then we made approximation of clear image without noise component of input signal and this setting is consistent with theory of Noise2Noise work [1]. This ground truth signal $y'$ are constructed using the formula (8). Thereby, when we train denoising network to predict signal $y'$, it train to extract additional features from input signal by Noise2Noise approach logic for achieving more denoising quality. Scheme of this approach also is shown (see Fig. 3).

$$X_i = \tilde{I}_i, \ y = \frac{\sum_{k=1, k \neq i}^{N} \tilde{I}_k}{N-1} \tag{8}$$

$$\min_{w} E(f(x), y') \tag{9}$$

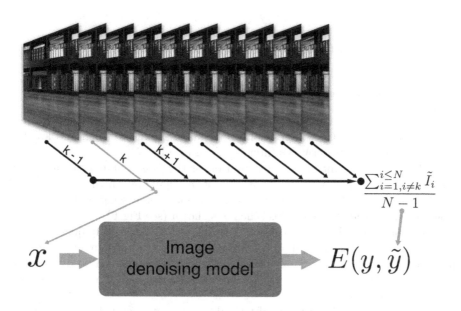

**Fig. 3.** Scheme of training pipeline

# 5   Architecture

## 5.1   Base Architecture Block

To build architectures in this work for test proposed approach, the blocks from the ResNet [8] architecture were taken and modified. From ResNet blocks batch normalization layers has been excluded. The scheme of these blocks is shown (see Fig. 4).

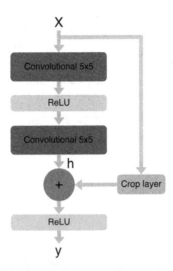

**Fig. 4.** Detailed scheme of modified ResNet block

## 5.2   Architecture with Frequency Components Splitting

The noise in question (1) belongs to the high frequency component of the image signal. An example of decomposition of an image signal into high and low frequencies using the Butterworth filter is shown (see Fig. 5).

Taking into account such a distribution of the noise component, an architecture was developed with the separation of the signal into high-frequency and low-frequency signals and their further independent processing by the convolutional layers of the network. This scheme is shown (see Fig. 6).

Implemented architecture (see Fig. 6) consists of a layer of frequency signal decomposition and sequential processing of parts of the input signal using convolutional blocks. Then the signals from blocks with are concatenated by channels and processed by a 3D convolutional layer to extract high-level features with interchannel dependence.

This architecture has 511390 of trainable parameters and the weight of this model is 2 megabytes when using single precision FP32 numbers to store the parameters.

**Fig. 5.** From left to right: original part of the image, 2D Fourier spectrum for the grayscale image, rendering of the Butterworth filter, low-frequency component of the image, high-frequency component of the image

## 6   Training

In the experiment we train architecture with frequency components splitting (see Sect. 5.2) by proposal approach (see Sect. 4) on our dataset (see Sect. 3).

To use all frames from a series of images, one frame from a series was use to the network input, and a series averaging without an input frame was used as the expected output signal (see Sect. 4). Based on the sample $\{\tilde{I}_k\}_{k=1}^N$, the input $X$ and the expected at the output of the network $y$ are constructed using the formula (8).

Denoising model was trained on random cropped images and using the following augmentation: rotations by 90, 180, 270° and flips.

For denoising models training was implemented pipeline based on PyTorch framework. As optimizer for was models training used radam optimizer [9].

## 7   Inference

On inference input image unfolds on grid of crop. Network forwards on each crop with test-time data augmentation with transforms from training [10]. Result image computed by formula (10), where $T_i$ - augmentation transform, $T_i^{-1}$ - inverse transform for correspondent transform $T_i$.

$$I_{result} = \frac{\sum_{n=1}^{8} T_i^{-1}(f(T_i(x)))}{8} \tag{10}$$

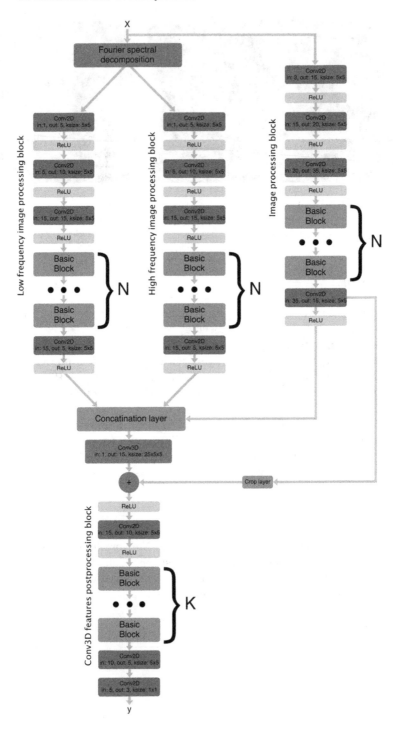

**Fig. 6.** Detailed architecture diagram of denoising network with frequency decomposition of input image

# 8 Results

To test the quality of noise reduction, the PSNR (Peak Signal-to-Noise Ratio) metric is usually used. But this metric is incorrect for a datasets where there is no clean samples. PSNR between output of the architecture which not completely reduced noise and average series value can be more then PSNR between output of the architecture which completely reduce noise component and the average series. This is due to the fact that a noise component remains in the average value of the series. In this case PSNR measure incorrect, but we have not another way to comparison our solution with state-of-the-art denoising approaches.

We compare our method with state-of-the-art approaches for training unsupervised image denoising neural networks. Before comparison neural network of each work we train by authors pipeline on our dataset. In addition for comparison taken an unsupervised image denoising algorithm from the OpenCV library: Non-local means denoising [11]. Table 1 gives comparison results. Furthermore visual comparison of inference all approaches is shown (see Fig. 7). Additional shown comparison with Non-local means denoising [11] (see Fig. 8).

**Table 1.** Comparison proposal approach with related works

| Approach | PSNR |
|---|---|
| Noise2Noise | 34.6 |
| Noise2Void | 30.53 |
| Noise2Self | 33.7 |
| Non-local means denoising | 36.04 |
| Deep image prior | 36.34 |
| Our | **37.1** |

These models work only with frames received from the CMOS sensor from which the dataset was obtained for their training. An example of work on frames from another sensor is shown (see Fig. 9). Poor prediction quality is noticeably.

# 9 Summary

As a result pipeline was implemented for training the noise reduction model. It does not depend on datasets with clean sample images. A neural network architecture was implemented and tested to effectively solve the noise reduction problem without clean data.

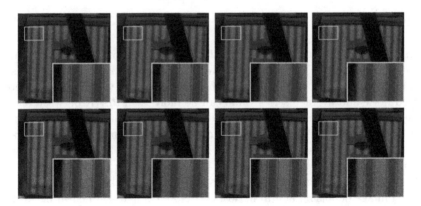

**Fig. 7.** From left to right on first line: image obtained by averaging all frames in series, original image, result of Noise2Noise, result of Noise2Self, from left to right of second line: result of Noise2Void, result of the non-local means denoising, result of Deep Image Prior, result of the architecture with frequency decomposition of input signal

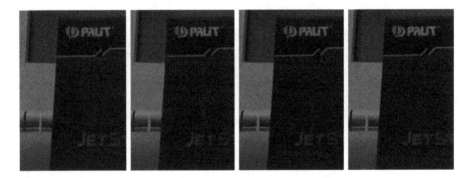

**Fig. 8.** From left to right: original image, image obtained by averaging all frames in series, result of the architecture with frequency decomposition of input signal, result of the non-local means denoising

According to the results, proposed model solves the optimization problem (3) quite efficiently. All experiments were performed on the dataset collected for this work.

The source code is available at the following link:
https://github.com/AlexeySrus/image_denoising.

**Fig. 9.** From left to right: original image, image obtained by averaging all frames in series, result of the architecture with frequency decomposition of input signal, result of the non-local means denoising

# References

1. Lehtinen, J., Munkberg, J., Laine, S., Karras, T., Aittala, M., Aila, T.: Noise2Noise: learning image restoration without clean data. CoRR abs/1803.04189 (2018). http://arxiv.org/abs/1803.04189
2. Brooks, T., Mildenhall, B., Xue, T., Chen, J., Sharlet, D., Barron, J.: Unprocessing images for learned raw denoising. CoRR abs/1811.11127 (2018). http://arxiv.org/abs/1811.11127
3. Plotz, T., Roth, S.: Benchmarking denoising algorithms with real photographs. CoRR abs/1707.0131 (2017). http://arxiv.org/abs/1707.01313
4. Krull, A., Buchholz, T.-O., Jug, F.: Noise2Void - learning denoising from single noisy images. CoRR abs/1811.10980 (2018). http://arxiv.org/abs/1811.10980
5. Batson, J., Royer, L.: Noise2Self: blind denoising by self-supervision. CoRR abs/1901.11365 (2019). https://arxiv.org/abs/1901.11365
6. Ulyanov, D., Vedaldi, A., Lempitsky, V.: Deep image prior. CoRR abs/1711.10925 (2017). https://arxiv.org/abs/1711.10925
7. GSMARENA. https://www.gsmarena.com/apple_iphone_x-8858.php. Accessed 20 Jun 2020
8. He, K., Zhang, X., Ren, S., Sun, J.: Deep residual learning for image recognition. CoRR abs/1512.03385 (2015). http://arxiv.org/abs/1512.03385
9. Liyuan, L., et al.: On the variance of the adaptive learning rate and beyond. CoRR abs/1908.03265 (2019). http://arxiv.org/abs/1908.03265
10. Ayhan, M., Berens, P.: Test-time data augmentation for estimation of heteroscedastic aleatoric uncertainty in deep neural networks. In: MIDL 2018 Conference (2018). https://openreview.net/forum?id=rJZz-knjz
11. Buades, A., Coll, B., Morel, J.-M.: Median filtering: a new insight. Image Process. On Line (2011). https://doi.org/10.5201/ipol.2011.bcm_nlm

# Efficient Group-Based Cohesion Prediction in Images Using Facial Descriptors

Ilya Gavrikov[✉] and Andrey V. Savchenko

HSE University, Laboratory of Algorithms and Technologies for Network Analysis,
Nizhny Novgorod, Russia
ilsgavrikov@gmail.com, avsavchenko@hse.ru

**Abstract.** In this paper we study the problem of predicting the cohesiveness and emotion of a group of people in photo. We proposed a fast approach, consisting of face detection by using MTCNN, aggregation of facial features (age, gender and embeddings) extracted by multi-task MobileNet, prediction of group cohesion and classification of emotional background using multi-output convolution neural network. Experimental study on the Group Affect Dataset from EmotiW 2019 challenge demonstrated that our approach allows to achieve an improvement of quality and even to reduce the running time of an algorithm's work when compared to known solutions. As a result, we obtained mean squared error 0.63 for cohesion prediction, which is 0.21 lower when compared to baseline CapsNet.

**Keywords:** Group cohesiveness prediction · Group emotion recognition · Facial analysis · Deep Convolutional Neural Network (CNN)

## 1 Introduction

Prediction of group-level emotion [1] and cohesiveness is very useful for various companies in order to analyze employee's emotional state throughout the day and build a relationship between their emotional state and group cohesion [2]. As the cohesiveness of a group is a crucial indicator of success of a group of people, the problem of predicting the perceived cohesiveness of a group of people in image becomes one of the main tasks in the EmotiW (Emotion Recognition in the Wild) 2019 challenge [3]. The usage of the CapsNet (Capsule Network) fitted for emotion recognition with seven labels, made it possible to obtain the baseline with MSE (mean squared error) equal to 0.84 [3] by feeding aggregated emotions into regression CNN (Convolutional Neural Network). The fusion of three models with face detection and feature extraction for support vector regression and aggregation of predictions for all faces lead to MSE 0.66 on validation set [4]. An ensemble of three branches that process global image, poses based on skeleton

© Springer Nature Switzerland AG 2021
W. M. P. van der Aalst et al. (Eds.): AIST 2020, CCIS 1357, pp. 140–148, 2021.
https://doi.org/10.1007/978-3-030-71214-3_12

(a) Strongly disagree     (b) Disagree     (c) Agree     (d) Strongly agree

**Fig. 1.** Sample images from Group Affect Dataset with various cohesiveness

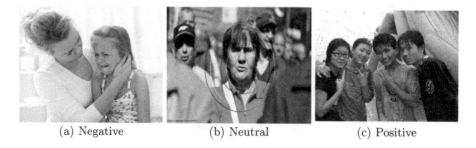

(a) Negative     (b) Neutral     (c) Positive

**Fig. 2.** Sample images from Group Affect Dataset with various group emotions

images and faces reached MSE 0.6493 on the validation set [5]. The first place in this challenge (MSE 0.52 on validation set) was obtained by a hybrid CNN with analysis of scene, faces, skeletons and UV coordinates [6]. Analysis of faces, bodies and the whole image lead to MSE 0.56 [7].

Unfortunately, all these techniques are very slow and cannot be used in embedded solutions for video analytics. Hence, in this paper we propose a lightweight solution based on facial feature extraction with MobileNet v1 [8] that simultaneously extract facial embeddings [9] and predict its gender and age [10]. We propose to extend this network by computing the average of its outputs for all detected faces and feed the overall descriptor of a group of people into a simple multi-task neural network for group-level emotion recognition and cohesiveness prediction. The main contribution of this paper is to experimentally demonstrate that information about facial features from very large external dataset of celebrities [11] may be used to train emotion classifier using rather small Group Affect Dataset [3]. As a result, in contrast to existing studies [5], we obtain rather accurate group-based cohesion prediction technique based on processing of facial features only.

The rest of the paper is organized as follows. In Sect. 2 we introduce the proposed algorithm. Experimental results for the dataset from EmotiW 2019 are presented in Sect. 3. Concluding comments are given in Sect. 4.

## 2   Proposed Approach

The task of this paper may be formulated as follows. Given an input image of a group of people, it is necessary to predict their cohesiveness, i.e. a measure

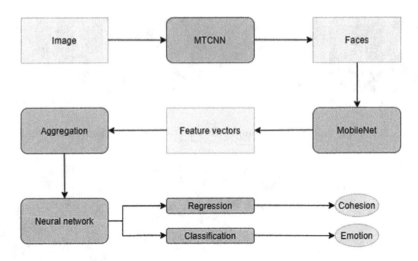

**Fig. 3.** Proposed pipeline

of bonding between group members [3]. It is an ordinal regression task with 4 variants of bonding (very weak, weak, strong and very strong, see Fig. 1). In addition, it is necessary to classify emotion of a group of people. We will use three classes from Group Affect Dataset: positive, neutral and negative (Fig. 2) [2].

The complete pipeline of the proposed Algorithm 1 is shown in Fig. 3. In this paper we test the hypothesis that all necessary information to solve this problem is reflected in the faces of persons for a group. Hence, at first, it is necessary to detect $R$ faces in a given photo. We will use MTCNN (multi-task CNN) facial detector [12] which is fast and accurate for rather large faces, though it does not obtain the state-of-the-art results for more complex photos [13]. After that, the features are extracted from every $r$-th facial region ($r = 1, 2, ..., R$). As the faces are observed in unconstrained conditions, modern transfer learning and domain adaptation techniques can be used. According to these methods, the large external dataset of celebrities, e.g., VGGFace2 [14], is used to train a deep CNN in order to let the neural network to learn reliable facial features. The outputs of one of the last layers of this CNN form $D$-dimensional features (embeddings/descriptor) $\mathbf{x}_r$ of the $r$-th face. It is typical to use embeddings from pre-trained CNNs, e.g., VGGFace-16 [15], ResNet-50 [11], ArcFace [16], etc.

In order to speed-up the decision process, we propose to use lightweight multi-task MobileNet [10]. In addition to facial embeddings, it can simultaneously predict age and gender of a person, so that we decided to concatenate two values (estimate of posterior probability for a male gender and predicted age) and extracted facial features. Next, it is necessary to aggregate $R$ facial features into a single descriptor of a photo. We will compute the mean of $L_2$-normalized features of individual faces [17]. We propose to fed the resulting mean descriptor into the multi-task neural network for predicting group cohesion and emotional background. Our neural net consists of the fully connected (dense) layer with

---

**Algorithm 1.** Proposed procedure of video-based cohesion prediction.

---

**Require:** Video frames or images $\{X(t)\}, t = 1, 2, ..., T$
**Ensure:** Cohesion and Emotion label of the given video
 1: **for** each frame $t = 1, 2, ..., T$ **do**
 2:    Obtain $R \geq 0$ facial regions using, e.g., MTCNN face detector
 3:    **for** each facial area $r = 1, 2, ..., R$ **do**
 4:       Extract embeddings and simultaneously predict age and gender using the multi-output MobileNet [10]
 5:       Concatenate embeddings and predicted age and estimate of male gender posterior probability into a single descriptor $\mathbf{x}_r(t)$
 6:    **end for**
 7:    Compute the frame feature vector $\mathbf{x}(t)$ as an average of embeddings $\{\mathbf{x}_r(t)\}, r = 1, 2, ..., R$ for all facial regions and normalize it
 8:    Feed the features into the multi-output neural network
 9:    Assign the vector of scores $s_{c;cohesion}(t)$ and $s_{c;emotion}(t)$ from the output of regression and classification layers for cohesion and emotion prediction, respectively
10: **end for**
11: Compute the cohesion scores $s_{c;cohesion}$ as an average of scores $\{s_{c;cohesion}(t)\}, t = 1, 2, ..., T$ for all frames
12: Compute the group-level emotion scores $s_{c;emotion}$ as an average of scores $\{s_{c;emotion}(t)\}, t = 1, 2, ..., T$ for all frames
13: Return the cohesion and group emotion categories with the maximal scores $\mathrm{argmax} s_{c;cohesion}$ and $\mathrm{argmax} s_{c;emotion}$.

---

600 neurons and ReLU activation, conv1d, MaxPooling1d, flattening, dense with 200 neurons and two output layers, namely, a layer with softmax activation and 3 outputs for emotion recognition and one linear output for group cohesiveness prediction.

Our approach (Fig. 3) has been implemented in a simple demo application[1], which can predict group cohesiveness and emotions given an input photo and video from web camera. Sample screen shots of our application for high and low predicted cohesiveness are shown in Fig. 4.

## 3   Experiments

In the experimental study we used the Group Affect Dataset from EmotiW 2019 [3] with 9,815 images. The training set contains 3100 images for each class of emotions and 1141 pictures for strongly disagree (ground-truth label '0'), 1561 for disagree (label '1'), 4601 for agree ('2') and 1997 for strongly agree ('3') levels of group cohesion. The averaged facial features extracted by a CNN are fed into two fully connected layers to predict group emotions and cohesiveness (Fig. 3). The output of the latter linear layer is in the range between 0 and 3.

---

[1] https://github.com/gbcodes/GroupCohesiveness.

(a) (b)

**Fig. 4.** Sample results of group-level cohesion prediction

**Table 1.** Results for group cohesion prediction, VGGFace2 facial features

| Supervised learning methods | MSE |
|---|---|
| Ordinal Ridge [16] | 1.01 |
| Ordinal Ridge Regularized [16] | 0.92 |
| Logistic All Threshold [16] | 1.07 |
| Logistic All Threshold Regularized [16] | 0.87 |
| Logistic Immediate Threshold [16] | 1.20 |
| Logistic Immediate Threshold Regularized [16] | 0.97 |
| Logistic SE (Squared Error) [16] | 1.06 |
| Logistic Squared Error Regularized [16] | 0.85 |
| Multi-class Logistic Regression | 1.03 |
| Least Absolute Deviation | 1.05 |
| Catboost [17] | 0.96 |
| Catboost-based [17] Ordinal Classifier | 0.87 |

The predicted cohesiveness is rounded to the nearest number (0, 1, 2, 3) and the MSE with the ground-truth cohesiveness level is calculated.

In the first experiment we used conventional ResNet-50 [11] trained on VGGFace2 dataset [14] in order to extract $D = 2048$ facial features. We used ordinal regression methods from mord package [18] and Catboost [19]. Their hyper-parameters (regularization coefficient, learning rate, etc.) have been tuned on the 20% of training set using cross validation (GridSearchCV). The results on validation set are shown in Table 1.

Here the lowest MSE is obtained for LogisticSE ordinal regression [18] with regularization, which will be used in the next experiment to choose the best facial descriptor. Logistic SE obtained not the best result, but regularization greatly improves the quality of prediction. The text on page 4 is about Logistic

**Table 2.** Results for facial descriptors for group cohesion prediction, LogisticSE

| Facial Descriptor | Feature extraction time, ms | | |
|---|---|---|---|
| | MSE | CPU | GPU |
| VGGFace (VGG-16) [15] | 1.12 | 109.74 | 8.61 |
| VGGFace2 (ResNet-50) [14] | 0.85 | 57.14 | 11.54 |
| Multi-output MobileNet [10] | 0.80 | 19.94 | 4.76 |

SE with regularization. It has already been mentioned. We compared three pre-trained CNNs for feature extraction, namely, 1) VGGFace-16 [15] that extracts $D = 4096$ features; 2) VGGFace2 (Res-Net-50) [11] ($D = 2048$ features); and 3) multi-output MobileNet [10] trained on the same VGGFace-2 dataset [14]. The dimensionality of the feature vector extracted by the latter network ($D = 1024$) is the lowest one. Keras framework with TensorFlow 1.15 backend were used in all experiments. The results are presented in Table 2.

Here we additionally measure an average inference time to extract features of one face by using CPU (12 core AMD Ryzen Threadripper) or GPU (Nvidia GTX1080TI) on a special server. It is surprisingly that the lightweight descriptor (MobileNet that simultaneously extracts facial features and predicts age and gender) outperforms the deeper CNNs including VGGFace2 that achieves the state-of-the-art results in face recognition [14].

In order to speed up decision-making, we examined the usage of principal component analysis (PCA) and processed only small number of first principal components. The dependence of MSE on the number of components for facial features from MobileNet [10] is shown in Fig. 5. Here we compared our simple multi-task neural network with one convolutional layer with a similarly engineered networks with 2 and 3 sequentially connected conv1d layers. These models were fitted with batch size 32 during 100 epochs with early stop.

Here the lowest MSEs are achieved by a network with one convolutional layer. Unfortunately, regression for small number of principal components causes significant (up to 0.3) increase in MSE. However, usage of all principal components in the proposed approach leads to the lowest MSE for cohesion prediction (0.63). It is 0.03 and 0.02 lower when compared to ensembles of several complex models [4,5]. Moreover, our approach much better than the baseline multi-task CNN of the challenge's organizers [3].

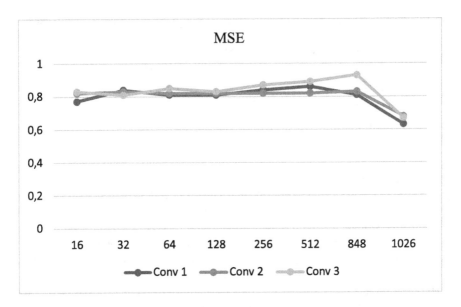

**Fig. 5.** Dependence of MSE of cohesiveness prediction on the number of principal components

## 4    Conclusion

In this paper we proposed an efficient approach (Fig. 3) to process the images of a group of people, for which MSE of cohesiveness prediction on a validation set is 0.21 lower when compared to MSE (0.84) of the baseline from the EmotiW 2019 [3]. The proposed approach is implemented in a publicly-available demo application (Fig. 4). In this demo, we predict age and gender of each person and predict the cohesiveness and emotion of the whole group. It is rather fast (+10 FPS for at most 16 persons in a group using Nvidia GTX1080 Ti GPU) due to the usage of MobileNet. Our preliminary results demonstrated that our model can be used even at Android mobile device with 5 FPS for a small group of 3 persons.

Our approach is not obviously the best one, as our MSE on validation set is 0.11 and 0.07 greater than MSEs of the first [6] and second [7] places in the EmotiW 2019 challenge. However, their running time is much worth: slower than 0.35 and 0.2 FPS for ensembles from [6] and [7], respectively. Another advantage of our model is the possibility to simultaneously recognize gender and age of each person in a photo/video, which can be important for potential usage of cohesiveness prediction in industrial applications. Finally, our results prove that the claim from [5] that "faces features may not reflect the group cohesiveness as well as other features" is incorrect. Indeed, it is necessary to use the pre-trained facial descriptors in order to obtain the high-quality group-based cohesion prediction. However, it is important to emphasize that in this task our MobileNet [10] obtains lower MSE than much more powerful ResNet-50 that is characterized by the state-of-the-art results in face identification [14]. It seems that the lightweight CNN with low number of parameters is more appropriate for such tasks with rather small number of training examples.

Unfortunately, our group-level emotion recognition accuracy is rather low (0.69), so that it is necessary to further improve our model. Moreover, in future, it is important to extract the faces from a group photo, which significantly influence the overall cohesiveness score. In this case it will be possible to find persons who, e.g., are less concentrated on the task than the majority of other persons in a photo. Finally, it is necessary to examine the state-of-the-art face detectors, e.g., RetinaFace [13], instead of MTCNN in order to locate more faces accurately. However, it is still possible that better facial detectors won't lead to the better quality of cohesiveness prediction, because very small faces do not have robust facial features. For example, experiments from paper [17] clearly demonstrate that usage of TinyFaces detector provide lower accuracy of group emotion recognition when compared to Viola-Jones detector, if VGGFace features are used [15]. In such case, it is possible that we should use other facial descriptors trained on facial images with small resolution [1].

**Acknowledgements.** The work of A.V. Savchenko is supported by RSF (Russian Science Foundation) grant 20-71-10010.

# References

1. Tarasov, A.V., Savchenko, A.V.: Emotion recognition of a group of people in video analytics using deep off-the-shelf image embeddings. In: van der Aalst, W.M.P., et al. (eds.) AIST 2018. LNCS, vol. 11179, pp. 191–198. Springer, Cham (2018). https://doi.org/10.1007/978-3-030-11027-7_19
2. Dhall, A., Goecke, R., Ghosh, S., Joshi, J., Hoey, J., Gedeon, T.: From individual to group-level emotion recognition: EmotiW 5.0. In: Proceedings of the 19th International Conference on Multimodal Interaction (ICMI), pp. 524–528. ACM (2017)
3. Ghosh, S., Dhall, A., Sebe, N., Gedeon, T.: Predicting group cohesiveness in images. In: Proceedings of the International Joint Conference on Neural Networks (IJCNN), pp. 1–8. IEEE (2019)
4. Zhu, B., Guo, X., Barner, K., Boncelet, C.: Automatic group cohesiveness detection with multi-modal features. In: Proceedings of the International Conference on Multimodal Interaction (ICMI), pp. 577–581. ACM (2019)
5. Zou, B., Lin, Z., Wang, H., Wang, Y., Lyu, X., Xie, H.: Joint prediction of group-level emotion and cohesiveness with multi-task loss. In: Proceedings of the 5th International Conference on Mathematics and Artificial Intelligence, pp. 24–28 (2020)
6. Xuan Dang, T., Kim, S.H., Yang, H.J., Lee, G.S., Vo, T.H.: Group-level cohesion prediction using deep learning models with a multi-stream hybrid network. In: Proceedings of the International Conference on Multimodal Interaction (ICMI), pp. 572–576. ACM (2019)

7. Guo, D., Wang, K., Yang, J., Zhang, K., Peng, X., Qiao, Y.: Exploring regularizations with face, body and image cues for group cohesion prediction. In: Proceedings of the International Conference on Multimodal Interaction (ICMI), pp. 557–561. ACM (2019)
8. Howard, A.G., et al.: MobileNets: efficient convolutional neural networks for mobile vision applications. arXiv preprint arXiv:1704.04861 (2017)
9. Savchenko, A.: Efficient statistical face recognition using trigonometric series and CNN features. In: Proceedings of the 24th International Conference on Pattern Recognition (ICPR), pp. 3262–3267. IEEE (2018)
10. Savchenko, A.V.: Efficient facial representations for age, gender and identity recognition in organizing photo albums using multi-output convNet. PeerJ Comput. Sci. **5**, e197 (2019)
11. He, K., Zhang, X., Ren, S., Sun, J.: Deep residual learning for image recognition. In: Proceedings of International Conference on Computer Vision and Pattern Recognition (CVPR), pp. 770–778. IEEE (2016)
12. Zhang, K., Zhang, Z., Li, Z., Qiao, Y.: Joint face detection and alignment using multitask cascaded convolutional networks. IEEE Sig. Process. Lett. **23**(10), 1499–1503 (2016)
13. Deng, J., Guo, J., Ververas, E., Kotsia, I., Zafeiriou, S.: RetinaFace: single-shot multi-level face localisation in the wild. In: Proceedings of the International Conference on Computer Vision and Pattern Recognition (CVPR), pp. 5203–5212. IEEE (2020)
14. Cao, Q., Shen, L., Xie, W., Parkhi, O.M., Zisserman, A.: VGGFace2: a dataset for recognising faces across pose and age. In: Proceedings of 13th International Conference on Automatic Face & Gesture Recognition (FG 2018), pp. 67–74. IEEE (2018)
15. Parkhi, O.M., Vedaldi, A., Zisserman, A.: Deep face recognition (2015)
16. Deng, J., Guo, J., Xue, N., Zafeiriou, S.: ArcFace: additive angular margin loss for deep face recognition. In: Proceedings of International Conference on Computer Vision and Pattern Recognition (CVPR), pp. 4690–4699. IEEE (2019)
17. Rassadin, A., Gruzdev, A., Savchenko, A.: Group-level emotion recognition using transfer learning from face identification. In: Proceedings of the 19th International Conference on Multimodal Interaction (ICMI), pp. 544–548. ACM (2017)
18. Pedregosa-Izquierdo, F.: Feature extraction and supervised learning on fMRI: from practice to theory. Ph.D. thesis (2015)
19. Prokhorenkova, L., Gusev, G., Vorobev, A., Dorogush, A.V., Gulin, A.: CatBoost: unbiased boosting with categorical features. In: Advances in Neural Information Processing Systems (NIPS), pp. 6638–6648 (2018)

# Automatic Grading of Knee Osteoarthritis from Plain Radiographs Using Densely Connected Convolutional Networks

Alexey Mikhaylichenko$^{(\boxtimes)}$ and Yana Demyanenko

Institute of Mathematics, Mechanics and Computer Science,
Southern Federal University, Rostov-on-Don, Russia
alexey.a.mikh@gmail.com, demyanam@gmail.com

**Abstract.** In this paper, we consider densely connected convolutional networks and their applicability to the problem of assessment of knee osteoarthritis (OA) severity in the five-point Kellgren-Lawrence scale. First, we use trained from scratch Single Shot Detector (SSD) to localize knee joint areas in radiographs. Then, we apply DenseNets to quantify OA stages in the images of detected knee joints. We consider networks of different depths, trained both from scratch and pre-trained on the ImageNet dataset and fine-tuned in the images from Osteoarthritis Initiative dataset (OAI). Also, different loss functions are examined to understand which one gives the best training results. In the knee joint localization task, we obtain an accuracy of 94.03% under the Jaccard index threshold of 0.75. Also, our classifier outperforms the current state-of-the-art with accuracy of 71% in the classification task.

**Keywords:** Knee osteoarthritis · Kellgren and Lawrence grading · Convolutional neural network · Classification

## 1 Introduction

Osteoarthritis (OA) is one of the most common diseases of the musculoskeletal system. Such diseases reduce the patient's working capacity and lead to large socio-economic losses. A significant part of the costs appears due to the late diagnostics of disease. Currently, there is no effective treatment for osteoarthritis, except for the replacement of the entire joint with an artificial one after its complete destruction. Therefore, the only way to reduce the costs is the early diagnosis when it is still possible to slow down the process of joint destruction.

The most common method of knee osteoarthritis diagnostics relies on X-rays images analysis. The main symptoms of OA are the degeneration and wear of the articular cartilage. However, cartilage tissue is not directly visible in radiographs. Therefore, with using X-rays, the progression of OA is assessed mainly by indirect signs—the narrowing of joint space or the appearance of osteophytes.

© Springer Nature Switzerland AG 2021
W. M. P. van der Aalst et al. (Eds.): AIST 2020, CCIS 1357, pp. 149–161, 2021.
https://doi.org/10.1007/978-3-030-71214-3_13

Osteoarthritis is a phased disease. Despite the existence of various systems for assessing the stages of osteoarthritis progression, the diagnosis is highly dependent on the subjectivity and experience of the expert. In addition, the widely used Kellgren-Lawrence grading scale [1] is very ambiguous [5,6]. The use of automated analysis could reduce the influence of the subjectivity factor and increase the accuracy and the reliability of the diagnosis.

## 2   Related Works

In the paper [7] OA stage classification is examined using Decision Trees, Naive Bayes classifier, Bayesian (Probabilistic) networks, and logistic regression by various texture characteristics (histogram features, Haralick features, etc.). There are 130 X-ray images applied in this study.

Minciullo et al. [8] present an original approach to classify OA using shape information from lateral knee radiographs. In this work, the authors use a statistical shape model for key points of knee contour detection, which are then utilized to the OA classification with random forests. The achieved performance for OA classification to 5 grades is 47,9% in the set of 300 images. The authors declare that it is comparable with similar techniques applied to the frontal view.

Antony et al. [9] suggest simple, trained from scratch convolutional neural network (CNN) which consists of 5 convolutional layers and a fully-connected layer. They optimize a weighted combination of cross-entropy loss and mean squared error, and achieve classification accuracy at about 60%. In the study [10], the authors use pre-trained CNNs with various architectures for classifying the OA grade, which are fine-tuned in the OAI dataset.

Tiulpin et al. [14] present a new CNN model not only for OA classification but also for the assessment of the joint space narrowing and the osteophytes presence. Their approach is based on the ensembling of two neural networks, each of which also consists of two parts: the pre-trained on the ImageNet convolutional layers, and 7 independent fully-connected layers. To connect these convolutional and fully connected layers, they utilize an average pooling layer after the convolutional block. They evaluate various network backbones from Resnet family with squeeze-excitation (SE) blocks [13]. The reached average multi-class accuracy is 66.68% for the OA grading task. In [15] the Deep Siamese CNNs are examined. They use the symmetry in the image, and their network consists of two branches, each of which works with different parts of the knee joint. The classification accuracy is 67.49%.

Wahyuningrum et al. [18] suggest a three-step method: preprocessing, features extraction using simple VGG-based CNN, and classification of the knee OA severity using LSTM. For the analysis, the work uses a 400 × 100 images of the knee joint area. The LSTM network is trained by optimizing Stochastic Gradient Descent (SGD). The average classification accuracy is 75.28%, but the accuracy of KL grade 1, presented in the article, is very different from the corresponding accuracy of all previous works and results of our experiments and needs to be rechecked.

Liu et al. [21] use the Faster R-CNN [20] for the simultaneous location of the knee joint area and classification OA severity grade for this area. The focal loss function is used for training to address the class imbalance. The own dataset containing 1385 X-ray images is used as a dataset for the analysis instead of the OAI dataset. The achieved classification accuracy is 74,3%.

Pingjun et al. [11] propose an adjustable ordinal loss instead of cross-entropy loss, to assign a higher penalty to misclassification with a larger distance between the real and the predicted KL grades. Several networks of well-known architectures, such as ResNet, VGG, etc., were trained in this work. The best result (69.7%) was obtained on the fine-tuned VGG-19 model with the proposed ordinal loss.

In this paper, we study the applicability of DenseNets to the problem of assessment knee osteoarthritis severity. We consider networks of different depths, both trained from scratch and pre-trained on the ImageNet, and evaluate different loss functions to improve the training process. Different ensembling methods are used for evaluation.

DenseNets is also used in [19], but for the case of the abbreviated Kellgren-Lawrence scale—KL grades of 0–1 were grouped since the clinical response for these two grades are usually similar. U-Net network is utilized for knee joints localization. Also, the authors considered only ensembles of different model checkpoints and only cross-entropy loss function for training.

## 3    Materials and Methods

### 3.1    Osteoarthritis Grades

There are various systems for assessing the progression of knee osteoarthritis. One of the standards is the Kellgren-Lawrence (KL) grading scale [1,2]. There are 5 grades of OA severity in this scale. A brief description of each stage is given in the Table 1, samples of knee joint images for each grade are shown in Fig. 1.

**Table 1.** The description of Kellgren-Lawrence (KL) grading system through JSN designated joint space narrowing.

| Grade | Verbal description |
|---|---|
| Grade 0 | None: no JSN or reactive changes |
| Grade 1 | Doubtful: doubtful JSN, possible osteophytic lipping |
| Grade 2 | Minimal: definite osteophytes, possible JSN |
| Grade 3 | Moderate: moderate osteophytes, definite JSN, some sclerosis, possible bone-end deformity |
| Grade 4 | Severe: large osteophytes, marked JSN, severe sclerosis, definite bone ends deformity |

The Kellgren-Lawrence scale has been criticized for the emphasis on the osteophytes, the overall grading of OA from normal to severe, and insensitivity to changes [3]. Osteoarthrosis of the minimum severity is diagnosed in grade 2 [4], therefore, in some studies, stage 0 and 1 are combined into one (normal).

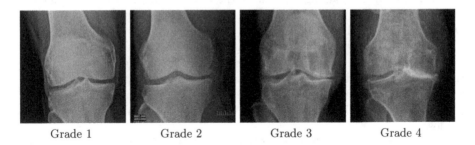

| Grade 1 | Grade 2 | Grade 3 | Grade 4 |

**Fig. 1.** Knee joint samples of significant KL grades

## 3.2  Data

The Osteoarthritis Initiative (OAI) dataset is a standard dataset for research on knee osteoarthritis using radiographs. It contains an archive of clinical data from 4,796 patients aged 45 to 79 years. In addition, each radiograph contains information about the measurements, osteophytes, joint space narrowing, the age of the patient, the stage of OA (including Kellgren-Lawrence grades) taken from few experts, etc.

In this study, we used bilateral posterior-anterior (PA) fixed-flexion knee radiographs for our experiments. Since the OAI dataset does not contain ground truth for knee joint detection, we manually annotate knee joints on all images.

After filtering, when low-quality images (very blurry, overexposed, etc.) were removed from the dataset, 4130 X-ray images with 8260 knee joints were remained. We randomly split all radiographs into train, validation, and test sets with a ratio of 7:1:2. The distribution of the knee joint images with OA on the Kellgren-Lawrence scale is given in Table 2:

**Table 2.** The distribution of images by Kellgren-Lawrence grades, used in this research. The numbers in the table indicate the number of knees images, available in each group.

| Group | KL-0 | KL-1 | KL-2 | KL-3 | KL-4 | Images |
|---|---|---|---|---|---|---|
| Train | 2295 | 1051 | 1504 | 752 | 175 | 6604 |
| Validation | 319 | 148 | 223 | 111 | 25 | 826 |
| Test | 639 | 296 | 447 | 223 | 51 | 1656 |

The OAI dataset is a highly unbalanced dataset, therefore we utilized data augmentations during models training. We applied random horizontal flip, random rotation and scale, saturation, and brightness. After all of these augmentations, we were rescaling the images to 224 × 224. All transformations were performed in random order on the fly.

### 3.3   Knee Joints Localization

Mainly, there are two steps in assessing knee osteoarthritis severity from raw X-ray images: knee joint localization and classifying the detected knee joint into OA grades. Several methods have been developed for knee joint detection in modern works.

One of the simplest methods is given in [17]—localization of the knee joint area there is performed by calculating the center of mass of the histogram, calculated by the sum of intensities of image rows. Similar histograms are also used in [12]. In [10], it is proposed to use the SVM classifier to search for ROI. In paper [16] for knee joint detection template matching is used. Solutions using deep learning have also been proposed: trained from scratch fully convolutional network (FCN) [9], and customized YOLOv2 network [11].

In this paper, we consider Single Shot Detector (SSD) for knee joint detection. This architecture has higher accuracy and performance with a comparison with the previous approaches like YOLO, R-CNN, etc. [22]. This network composes of two parts—features maps extraction from the input image, and applying convolution filter to detect objects. For features extraction mainly used pre-trained well-know architectures—VGG-16, Resnet-101, Inception [23] etc. We utilized MobileNet [24], pre-trained on COCO (Common Objects in Context) dataset. This architecture allows achieving high accuracy comparable with the state-of-the-art approaches while having significantly less number of parameters and high speed.

### 3.4   Quantifying Osteoarthritis Severity

In this study, we investigate the use of networks from DenseNet family [25] for assessing the severity of knee OA through classification. We were interested in the claim that dense connections have a regularizing effect, which reduces overfitting on tasks with the smaller training sets [25]. It means, that DenseNets work well on small datasets, which is important in our case. We consider networks of various depth, different loss functions, and different training processes.

As our initial approach, we trained from scratch DenseNet-121 and DenseNet-161, where the top fully connected layer was replaced with the layer with 5 outputs of the Kellgren-Lawrence grades.

Next, we examined the use of pre-trained on the ImageNet networks to solve the problem. In various options, we used both the training only the last fully-convolutional layer (freezing weights on convolutional layers), and fine-tuning the entire network.

Then we considered the effectiveness of applying various loss functions to train networks in both cases. We used the cross-entropy loss as the main loss function. We also experimented with adjustable ordinal loss [11].

In addition to the experiments, we also investigated using the convolution layer with window size equal to the layer size between convolutional and fully-connected layers instead of a simple average pooling. We assumed that some information is lost in this transition, and the result can be improved by changing the transition layer. However, this modification did not lead to significant improvements. Therefore, we present only the results with the average pooling in the paper.

### 3.5 Models Ensemble

A neural network ensemble combines the outputs of multiple individually trained neural networks in order to improve generalization ability and reduce dispersion in model predictions. In most cases, the best results of large machine learning competitions are achieved by an ensemble of models, rather than by a single model.

There are many different types of ensembles. One of them is stacking, which involves training a new learner to combine the predictions of several models. We used the most common ensemble approach (the simplest form of stacking)—unweighted averaging [26]. Subsequently, we picked the models, summed their predictions, and propagated them through the softmax layer. Eventually, the class probability of the KL grade $k$ for given image $x$ was inferred as follows:

$$P(y = k|x) = \frac{\exp\left[\sum_{m=1}^{M} P_m(y = k|x)\right]}{\sum_{j=1}^{K} \exp\left[\sum_{m=1}^{M} P_m(y = j|x)\right]},$$

where $M = 3$ is the number of models in the ensemble, $K = 5$ is number of classes and $P_m(y = k|x)$ an individual network output before the softmax layer (unnormalised probability distribution). We considered different sets of models as parts of ensembles: of same network with different training checkpoints, of the same network with different random seeds (21, 42, and 84), and an ensemble of networks with different structures (a combination of DenseNet-121 and DenseNet-161). In the last two cases, we chose models that the best performed on the validation set.

## 4    Experiments and Results

### 4.1    Knee Joints Detection

A comparison of the accuracy of the proposed method and methods from previously published works is given in Table 3. We use the Intersection over Union (IoU), also known as the Jaccard index, between manual annotations and detection results to evaluate localization accuracy:

$$IoU(A, B) = \frac{|A \cap B|}{|A \cup B|},$$

where A and D are the manually annotated and the automatically detected bounding boxes of knee joint, respectively. The numbers in the table indicate the percentage of correctly detected knee joints, where the Jaccard index values

**Table 3.** A comparison of the accuracy of the proposed method and methods from previously published works, based on the Jaccard index (J). YOLOv2 and SSD based networks are fine-tuned models.

| Methods | $J \geq 0.25$ | $J \geq 0.5$ | $J \geq 0.75$ | Mean |
|---|---|---|---|---|
| Linear SVM + Sobel [10] | ~81.8% | 38.6% | – | 0.36 |
| Region proposal [12] | – | – | – | 0.8399 |
| FCN [9] | 100% | 99.9% | 89.2% | 0.83 |
| YOLOv2 [11] | – | – | 92.2% | **0.858** |
| SSD, our method | 100% | 100% | **94.03%** | 0.844 |

are greater than 0.25, 0.5 and 0.75, and the mean value of Jaccard index is given for all samples. A dash means that the corresponding data is not available.

A comparison of the results shows that methods knee joint localization based on convolutional neural networks reach higher detection accuracy than other approaches. Moreover, simpler methods do not have any significant advantages over neural networks. Besides, the experiments have shown that using a pre-trained network in the localization problem allows greater accuracy compared to a network trained from scratch.

### 4.2 The Knee Osteoarthritis Classification on the Kellgren-Lawrence Scale

**Training Process.** All the models were trained using the adaptive moment estimation (Adam) optimizer [27] with a learning rate of $1e-3$ and a weight decay (L2-norm regularization) of $1e-4$. During the training, the learning rate was decayed by 5% every 5 epochs. The batch size was fixed to 64. In total, we trained every model for 75 epochs.

First, we re-implemented the model described in the paper [9]. This network produces two outputs—one for regression and the other one for classification. In our implementation, we used only classification output and $224 \times 224$ pixel images as the network input (instead of $300 \times 200$ from original work).

Secondly, we trained DenseNet-121 and DenseNet-161 from scratch on the OAI dataset. Next, we performed a fine-tuning of DenseNets that were pre-trained on the ImageNet dataset. We executed different experiments: both, we trained only the FC layer with a frozen convolution part, and we unfroze the convolutional layers and trained the full network. We found that training the full model gives a much better result than the training of the network with frozen convolution layers. Besides, the network with frozen convolution layers is very quickly overfitted. The results of testing only the fully trained models are described below. We used this procedure three times with different random seeds (21, 42, and 84) and selected model snapshots with the best accuracy.

We investigated training our DenseNets with two loss functions: cross-entropy loss and adjustable ordinal loss [11]. We implemented an adjustable ordinal matrix, described in the original paper, to denote the penalty weights between the predicted and the real grade, and use it for the training process. However, validation performance in the multi-class average accuracy was comparable with the case using cross-entropy loss.

Finally, we used our model snapshots to make ensembles of these models and check them on the test set. The final results are presented in Table 4.

In the case of unbalanced datasets, a common practice is using various weighted data sampling strategies like under-sampling (removing samples from the majority class) or over-sampling (adding more examples from the minority class). We tested these strategies, however, they did not lead to an improvement in the final scores.

**Evaluation.** In our experiments, we got multi-class accuracy 61.78% for Antony's CNN [9], which we trained only for classification using stochastic gradient descent and Nesterov momentum to reproduce the original results (60.3%). The improved accuracy may be due to the fact that we applied the L2-norm weight regularization for all layers, not just for the last two convolutional layers and the fully connected layer, as in [9].

**Table 4.** Comparison of different classifiers on the Kellgren-Lawrence grading task. We trained multiple networks and selected the best snapshots for each model. The main variants of DenseNet are compared with both cross entropy loss (CE) and ordinal loss (Ordinal) from [11]

| Method | Accuracy | Precision | Recall | $F_1$ |
|---|---|---|---|---|
| Antony et al. [9] | 61.78% | 0.58 | 0.62 | 0.55 |
| Tiulpin et al. [14] | 66.68% | – | – | – |
| Siamese CNNs [15] | 67.49% | – | – | – |
| VGG-19 [11] | 69.70% | – | – | – |
| CNN-LSTM [18] | 75.28% | – | – | – |
| Faster R-CNN [20] | 74.30% | – | – | – |
| DenseNet-121-CE | 68.98% | 0.66 | 0.69 | 0.66 |
| DenseNet-121-Ordinal | 67.39% | 0.67 | 0.67 | 0.67 |
| DenseNet-161-CE | 67.69% | 0.67 | 0.68 | 0.66 |
| DenseNet-161-Ordinal | 67.94% | 0.66 | 0.68 | 0.66 |
| 3× DenseNet-121-DRS | **71.08%** | 0.68 | **0.71** | 0.68 |
| 3× DenseNet-161-DRS | 70.17% | **0.69** | 0.70 | **0.69** |
| 3× DenseNet-121-DCH | 69.69% | 0.66 | 0.70 | 0.66 |
| 3× DenseNet-161-DCH | 70.10% | 0.68 | 0.70 | 0.68 |
| 4× DenseNet-121/161 | 70.47% | 0.68 | 0.70 | 0.67 |

We trained DenseNet-121 and DenseNet-161 with cross-entropy and adjustable ordinal losses (DenseNet-[121/161]-CE, DenseNet-[121/161]-Ordinal). The ordinal loss showed better result than cross-entropy only for DenseNet-161. Despite this, the enhancement of this approach, probably, may provide better results than using cross-entropy loss.

Training such complex networks from scratch, as expected, did not bring any improvements and the accuracy of the classification was significantly less than in the case of pre-trained models, therefore, these results are not presented in the table above.

As we mentioned before, we experimented with different sets of models for ensembles: ensemble from the same model with different random seeds (3× DenseNet-[121/161]-DRS), ensemble the same model from defferent checkpoints (3× DenseNet-[121/161]-DCH) and ensemble from combination of two DenseNet-121 and two DenseNet-161 (4× DenseNet-121/161). The best result—average multiple-class accuracy of **71.08%**—was achieved by the ensemble of three DenseNets, trained with different random seeds (21, 42, and 84). Classification metrics and the confusion matrix with ROC curves for the best ensemble are presented in Table 5 and Fig. 2 respectively.

**Table 5.** Classification metrics achieved by the ensemble of three models, trained with different random seeds (3× DenseNet-121-DRS)

| KL Grade | Accuracy | Precision | Recall | $F_1$-score | AUC |
|---|---|---|---|---|---|
| 0 | 92.02% | 0.70 | 0.92 | 0.80 | 0.92 |
| 1 | 16.22% | 0.44 | 0.16 | 0.24 | 0.78 |
| 2 | 72.26% | 0.72 | 0.72 | 0.72 | 0.91 |
| 3 | 83.41% | 0.82 | 0.83 | 0.82 | 0.98 |
| 4 | 62.75% | 0.97 | 0.63 | 0.76 | 0.99 |
| Mean | 71.08% | 0.68 | 0.71 | 0.68 | – |

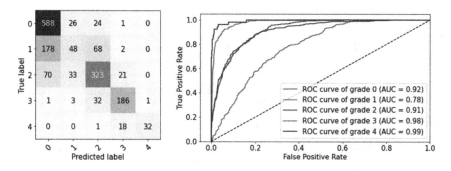

**Fig. 2.** Confusion matrix and ROC curves of KL grading for the best ensemble of 3× pre-trained DenseNet-121 with different random seeds. Average multi-class accuracy is 71.08%

The classification accuracy reported in [20] and [18] is higher than in our research, however, it is incorrect to compare our results with these cases directly.

To train and test the model in [20], a completely different dataset was used, several times smaller in volume (1385 base images) and less balanced (especially for grade 1, which gives the biggest error) than the OAI dataset.

The results of classification metrics for grade 1 in [18] differ fundamentally from the observations in both our and previous studies—the accuracy for grade 1 is comparable to the accuracy of other grades, while in all other studies this is not the case (see Table 5 for details). Therefore, the claimed results need to be rechecked and confirmed.

## 5   Discussion

Analysis of the confusion matrix and classification metrics for each class shows that the main misclassification occurs between neighboring classes. Moreover, the biggest error is noted for KL grade 1, due to the small variations between grade 0 and grade 1. This fact was also noted in previous works.

The analysis of the saliency map, calculated by the Grad-CAM algorithm [28] for the final DenseNet-121, shows that classification does not depend much on the accuracy of the joint area localization—the main pixels that make the greatest contribution to decision making are found directly in the area of osteophytes and in the joint space (Fig. 3). This hypothesis is supported by experiments that are used true knee joint images, shifted by a random number of pixels vertically and horizontally—the final accuracy was comparable to the results obtained over true areas.

Due to the high requirements for accuracy in the healthcare sector, the apply of fully automatic classification methods can be dangerous at this stage. However, the described method can be used to help the expert make a decision by providing additional information about the radiograph, such as the probability distribution of OA grades, the attention map, etc. (Fig. 3).

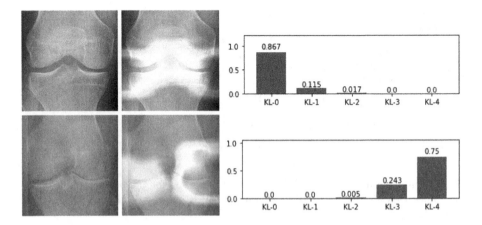

**Fig. 3.** Example of applying the Grad-CAM algorithm to the best DenseNet-121 and probability distribution for the presence of OA by the Kellgren-Lawrence scale.

There are several ways to improve the approach described in the work. First, it is to explore more complicated ensemble methods (e.g., super learner, stochastic weight averaging, etc.). Based on the results obtained, it can be concluded that the ensembles can significantly increase the final accuracy. Secondly, the study of training methods, which will reduce the classification error for KL grade 1, is seen as a perspective. And finally, it is important to decrease the false-negative rate of the network, which is especially important in the field of diagnostics of diseases.

# 6  Conclusion

In this paper, we applied the SSD model for knee joint localization and investigated the applicability of DenseNets and their ensembles to quantifying knee osteoarthritis from plain radiographs.

In the detection task the SSD model achieved the results comparable with the current state-of-the-art. The mean Jaccard index is 0.844, and the percentage of the images with $J > 0.75$ is 94.03%, which is better than with previous methods.

In the grading knee OA task DenseNets also outperformed previous approaches with average multiple-class accuracy 71.08% in an ensemble of three DenseNet-121, trained with different random seeds. DenseNet-161 as a whole did not show better results than DenseNet-121, which may be due to insufficient data for training. Also, the usage of ordinal loss did not lead to a significant increase in classification accuracy. However, after the improvements aimed at reducing the classification errors of neighboring grades, or errors for grade 1 (doubtful in KL scale), the prediction accuracy might significantly increase.

We also provide trained models and source codes for all analyzed models. All of this is publicly available on Github: https://github.com/almikh/grading-knee-oa-using-densenets.

# References

1. Kellgren, J., Lawrence, J.: Radiological assessment of osteo-arthrosis. Ann. Rheum. Dis. **16**, 494–502 (1957)
2. Schiphof, D., Boers, M., Bierma-Zeinstra, S.: Differences in descriptions of Kellgren and Lawrence grades of knee osteoarthritis. Ann. Rheum. Dis. **67**, 1034–1036 (2008)
3. Altman, R., Gold, G.: Atlas of individual radiographic features in osteoarthritis, revised. Osteoarthr. Cartil. **15**(Suppl A), A1–A56 (2007)
4. Altman, R., Asch, E., Bloch, D., Bole, G., Borenstein, D., Brandt, K., et al.: Development of criteria for the classification and reporting of osteoarthritis: classification of osteoarthritis of the knee. Arthritis Rheum. **29**, 1039–1049 (1986)
5. Culvenor, A., Engen, C., Engebretsen, L., Risberg, M.: Defining the presence of radiographic knee osteoarthritis: a comparison between the Kellgren and Lawrence system and OARSI atlas criteria. Osteoarthr. Cartil. (22), 265 (2014)
6. Sheehy, L.: Validity and sensitivity to change of three scales for the radiographic assessment of knee osteoarthritis using images from the multicenter osteoarthritis study (MOST). Osteoarthr. Cartil. **23**, 1491–1498 (2015)

7. Chan, S., Dittakan, K.: Osteoarthritis stages classification to human joint imagery using texture analysis: a comparative study on ten texture descriptors. In: Santosh, K.C., Hegadi, R.S. (eds.) RTIP2R 2018. CCIS, vol. 1036, pp. 209–225. Springer, Singapore (2019). https://doi.org/10.1007/978-981-13-9184-2_19

8. Minciullo, L., Cootes, T.: Fully automated shape analysis for detection of Osteoarthritis from lateral knee radiographs. In: 23rd International Conference on Pattern Recognition (ICPR), pp. 3787–3791 (2016)

9. Antony, J., McGuinness, K., Moran, K., O'Connor, N.: Automatic detection of knee joints and quantification of knee osteoarthritis severity using convolutional neural networks. arXiv:1703.09856 [cs.CV], 29 March 2017

10. Antony, J., McGuinness, K., Moran, K., O'Connor, N.: Quantifying radiographic knee osteoarthritis severity using deep convolutional neural networks. arXiv:1609.02469 [cs.CV], 8 September 2016

11. Pingjun, C., Linlin, G., Xiaoshuang, S., Kyle, A., Lin, Y.: Fully automatic knee osteoarthritis severity grading using deep neural networks with a novel ordinal loss. Comput. Med. Imaging Graph. **75**, 84–92 (2019). https://doi.org/10.1016/j.compmedimag.2019.06.002

12. Tiulpin, A., Thevenot, J., Rahtu, E., Saarakkala, S.: A novel method for automatic localization of joint area on knee plain radiographs. In: Sharma, P., Bianchi, F.M. (eds.) SCIA 2017. LNCS, vol. 10270, pp. 290–301. Springer, Cham (2017). https://doi.org/10.1007/978-3-319-59129-2_25

13. Hu, J., Shen, L., Sun, G.: Squeeze-and-excitation networks. In: 2018 IEEE/CVF Conference on Computer Vision and Pattern Recognition, pp. 7132–7141 (2018)

14. Tiulpin, A., Saarakkala, S.: Automatic grading of individual knee osteoarthritis features in plain radiographs using deep convolutional neural networks. arXiv:1907.08020, Image and Video Processing (2019)

15. Tiulpin, A., Thevenot, J., Rahtu, E., Lehenkari, P., Saarakkala, S.: Automatic knee osteoarthritis diagnosis from plain radiographs: a deep learning-based approach. arXiv:1710.10589 [cs.CV], 29 October 2017

16. Norman, B.D., Pedoia, V., Noworolski, A., Link, T.M., Majumdar, S.: Automatic knee Kellgren Lawrence grading with artificial intelligence. Osteoarthr. Cartil. **26**, S436–S437 (2018)

17. Anifah, L., Purnama, I.K., Hariadi, M., Purnomo, M.H.: Osteoarthritis classification using self organizing map based on Gabor kernel and contrast-limited adaptive histogram equalization. Open Biomed. Eng. J. **7**, 18–28 (2013). https://doi.org/10.2174/1874120701307010018

18. Wahyuningrum, R.T., Anifah, L., Purnama, I.K., Purnomo, M.H.: A new approach to classify knee osteoarthritis severity from radiographic images based on CNN-LSTM method. In: 2019 IEEE 10th International Conference on Awareness Science and Technology (iCAST), pp. 1–6 (2019). https://doi.org/10.1109/ICAwST.2019.8923284

19. Norman, B., Pedoia, V., Noworolski, A., et al.: Applying densely connected convolutional neural networks for staging osteoarthritis severity from plain radiographs. J. Digit. Imaging **32**, 471–477 (2019). https://doi.org/10.1007/s10278-018-0098-3

20. Ren, S., He, K., Girshick, R., Sun, J.: Faster R-CNN: towards real-time object detection with region proposal networks. IEEE Trans. Pattern Anal. Mach. Intell. **39**, 1137–1149 (2017). https://doi.org/10.1109/TPAMI.2016.2577031

21. Liu, B., Luo, J., Huang, H.: Toward automatic quantification of knee osteoarthritis severity using improved Faster R-CNN. Int. J. Comput. Assist. Radiol. Surg. **15**, 457–466 (2020). https://doi.org/10.1007/s11548-019-02096-9

22. Liu, W., et al.: SSD: single shot MultiBox detector. In: Leibe, B., Matas, J., Sebe, N., Welling, M. (eds.) ECCV 2016. LNCS, vol. 9905, pp. 21–37. Springer, Cham (2016). https://doi.org/10.1007/978-3-319-46448-0_2

23. Szegedy, C., Liu, W., Jia, Y., Sermanet, P., et al.: Going deeper with convolutions. arXiv:1409.4842 [cs.CV], 17 September 2014

24. Howard, A.G., Zhu, M., Chen, B., Kalenichenko, D., et al.: MobileNets: efficient convolutional neural networks for mobile vision applications. arXiv:1704.04861, Computer Vision and Pattern Recognition (2017)

25. Huang, G., Liu, Z., Weinberger, K.Q.: Densely connected convolutional networks. arXiv:1608.06993 [cs.CV], 28 January 2018

26. Cheng, J., Aurélien, B., Mark, L.: The relative performance of ensemble methods with deep convolutional neural networks for image classification. J. Appl. Stat. **45**, 2800–2818 (2018). https://doi.org/10.1080/02664763.2018.1441383

27. Kingma, D.P., Ba, J.: Adam: a method for stochastic optimization. arXiv:1412.6980 [cs.LG] (2014)

28. Selvaraju, R.R., Cogswell, M., Das, A., et al.: Grad-CAM: visual explanations from deep networks via gradient-based localization. In: 2017 IEEE International Conference on Computer Vision (ICCV), pp. 618–626 (2017). https://doi.org/10.1109/ICCV.2017.74

# A Novel Approach to Measurement of the Transverse Velocity of the Large-Scale Objects

Ivan Goncharov[1], Alexey Mikhaylichenko[1(✉)], and Anatoly Kleschenkov[2,3]

[1] Southern Federal University, Institute of Mathematics,
Mechanics and Computer Science, Rostov-on-Don, Russia
`ivan.goncharov6@gmail.com`, `alexey.a.mikh@gmail.com`
[2] Southern Federal University, Physics Faculty, Rostov-on-Don, Russia
`aktech@inbox.ru`
[3] ISOSCAN, Ltd., Rostov-on-Don, Russia

**Abstract.** This paper presents a novel approach for measuring the transverse velocity of large-scale objects based on stereo vision. The suggested approach uses a high-speed stereo camera located perpendicular to the traffic lane, and consists in matching frames from the left and right cameras. Compared to methods based on monocular vision, this approach solves the problem of dependence of object speed on distance. The proposed algorithm is part of a system for non-contact measurement of large-sized objects and has a calculated measurement error that does not exceed 1.5% of the measured speed up to 30 km/h while having low computational complexity.

**Keywords:** Computer vision · Speed measurement · Stereo vision

## 1 Introduction

Systems for vehicle detection and speed measurement have become an integral part of our daily life. This is evidenced by the growing interest of researchers in this area as well as the density of implementation of these systems in the environment. As digital cameras are becoming cheaper and able to produce images with higher quality, video-based systems are becoming increasingly popular for speed measurement tasks. Most of the published works are devoted to the problem of traffic flow mean speed estimation, which is necessary when building traffic management or smart city systems. The current paper considers the speed estimation approach for use in a different application problem.

A suggested approach was developed as part of the system for non-contact measurement of large-sized objects using laser triangulation [7]. The specificity of this task is the need to get speed values with the highest possible frequency. This is necessary to minimize the error in measuring the length of the object. To ensure the required frequency of obtaining the speed values, the proposed approach uses a stereo vision system running at 250 frames per second.

© Springer Nature Switzerland AG 2021
W. M. P. van der Aalst et al. (Eds.): AIST 2020, CCIS 1357, pp. 162–169, 2021.
https://doi.org/10.1007/978-3-030-71214-3_14

In this paper, we introduce the pipeline for the transverse velocity measurement of large-scale objects with a measurement error not exceeding 2%. In Sect. 3 we describe computer vision techniques and stereo vision algorithms used in the suggested approach. In Sect. 4 we show the measurement results obtained from applying the described method to real-world data. Finally, in Sect. 5, we discuss results and potential improvements.

## 2    Related Works

Currently, many approaches to speed measurement have been proposed using both classical computer vision algorithms and machine learning approaches. However, most of them have a similar sequence of actions required to calculate the speed.

The first step is to detect a moving object in the frame. In the second step, the object's motion vectors are calculated. And, in general, algorithms differ in methods for solving the problems of the above stages. For example, the authors of [4] suggest a deep convolutional neural network to detect a moving object. This approach shows stable results and is less dependent on the equipment used. Also, simpler algorithms for detecting a moving object are presented—background subtraction [8] and neighboring frames difference histogram filtering [1,2,6]. These are conventional algorithms with low computational complexity. But these methods are sensitive to conditions such as shadows and illumination variations.

Keypoints tracking [5,8] or optical flow computation [2,4] are often used for building motion vectors. Approaches based on optical flow computation demonstrate more accurate results. This is due to sensitivity of keypoint localization algorithms to lighting conditions—feature point position may differ in neighboring frames due to changing lighting.

As already mentioned in Sect. 1, a special feature of our task is the need to obtain the instantaneous speed of a moving object during the entire passage through the measuring system. In addition, the moving object is located in close proximity to the measuring system. These features make it impossible to use existing methods of speed measurement with the required accuracy.

## 3    Materials and Methods

In this section, we introduce our method for measurement the transverse velocity of the large-scale objects. The proposed approach is based on classical computer vision algorithms and consists of several important steps. We will discuss, in turn, our strategy for improving the results obtained and reducing the computational complexity.

In speed measurement approaches based on monocular vision, there is a problem with evaluating a perspective conversion. To avoid this, we suggest using the following property of the stereo camera.

Consider $X$—some point in space, $(x, y)$—coordinates of the corresponding point on the left camera image plane. By shifting the point $X$ in a direction strictly perpendicular to the optical axes of the cameras, we get the point $X'$ with coordinates $(x, y)$ of the corresponding point on the **right** camera image plane. In this case, the distance between the points $X$ and $X'$ will be equal to the stereo camera baseline (Fig. 1). It is important to note that this fact is true for all points and does not depend on the coordinates of their projections on the image plane, nor the distance to the stereo camera.

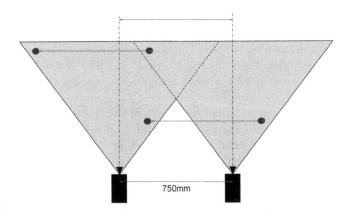

**Fig. 1.** Theoretical basis

If optical axes of the cameras are not parallel, the distance between points will be equal to the length of baseline projection on the plane of points $X, X'$. This fact allows us to determine the speed of a moving object using the conventional formula: $v = \frac{S}{t}$, by defining the unknown values as follows:

1. find a pair of frames with the same pixel coordinates of the object (according to property described above, the distance will match camera baseline);
2. extract the timestamps of the receiving frames, thus obtaining time.

Thus, speed estimation task is reduced to searching for a pair of frames $(frame_l, frame_r)$, where the object has the same pixel coordinates. Or, rephrasing—find frame $frame_l$ from the first camera $C_l$ on moving direction for frame $frame_r$ from the second camera $C_r$ on moving direction, where the pixel coordinates of the object will match up to some infinitesimal $\theta$—measurement error.

An important requirement is that the object's motion vector is perpendicular to the optical axes of the cameras. In some applications, this cannot be guaranteed but in our case, this is due to the operating conditions. The proposed solution is used to measure the speed of vehicles, so the stereo camera module is located on the side of the roadway. This ensures that the above requirement is met.

## 3.1   Input Data

The suggested approach uses a stereo camera consisting of two high-speed cameras with external shutters synchronization. The cameras are placed exactly parallel at a distance of 750 mm. The stereo pair is located perpendicular to the lane at a height of about 3 m. The working frame rate is 250 frames per second. Figure 2 shows a sample of input data.

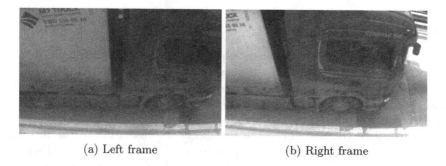

(a) Left frame                          (b) Right frame

**Fig. 2.** Input data sample (before rectification procedure)

## 3.2   Matching Function

Let's consider several features of the input data:

1. the camera angles of the subject are completely the same for left and right cameras, due to their parallel arrangement in stereo pair;
2. position of the object can only be changed along the coordinate axis X of the image plane due to perpendicular arrangement of stereo pair relative to the lane

Because of this, the image search function with matching pixel coordinates of the object has been replaced with the search function for the most similar image. To determine the similarity of images, the difference module function is used:

$$S(F_r) = min\{\sum |F_r - F_{l_i}|, i \in Q\}, \tag{1}$$

where $F_r$—image from the right camera; $F_l$—image from the left camera; $Q$—set of previous frames from the left camera. The minimum of this function will correspond to the most similar image in search neighborhood.

## 3.3   Camera Calibration

Due to the extreme naivety, the image comparison function is sensitive to any outliers in the data. Such outliers, for example, are lens distortion and mismatch of image lines in the left and right stereo image frames. Camera calibration [9], stereo calibration, and application of rectification transform [3] significantly minimize the number of false positives of the image comparison function.

An important addition to the conventional camera calibration algorithm is filtering input patterns by distribution on the frame. The main motivation is to exclude patterns with a potentially high re-projection error from the calibration procedure. It is enough to leave the minimum possible number of normally distributed patterns on the frame. This approach has improved the quality of camera calibration by 20% according to the RMSE metric.

## 3.4   Increasing Resolution

The accuracy of speed measurement is affected not only by the quality of the target search function but also by the resolution of the proposed approach.

In the proposed method (Sect. 3), the speed is determined by the time the pixel coordinates of the object match on the left and right frames of the stereo camera. Since the frame rate is not infinite, the time obtained during stereo frame matching will be discrete. This directly affects the resolution of the method. Obviously, the resolution is inversely proportional to the discreteness. Figure 3 demonstrates a dependence graph of discreteness/FPS in the range of the measured speed.

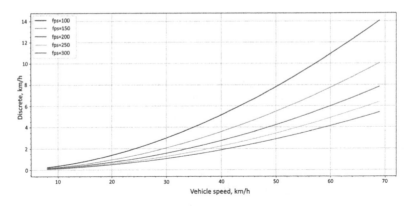

**Fig. 3.** Dependence of speed measurement discreteness on frame rate

You can see that as the speed increases, the discreteness of measurements increases, and therefore the resolution decreases. This is clearly noticeable in Fig. 4 – as the speed increases, the decrease in resolution becomes more and more obvious. This fact clearly demonstrates the need to increase the frame rate.

Decimation and binning operations make it possible to increase the FPS in most machine vision cameras. However, when they are applied, the frame size changes, and therefore requires updating the matrices of the camera model [3]. Calibrating the camera on a smaller frame means a deliberate increase of the re-projection error, and therefore a decrease in the quality of the rectification transform. Therefore, we propose to change the matrices as follows:

$$K_i^{'} = SK_i$$
$$S = \begin{bmatrix} k_h & 1 & k_h \\ 1 & k_v & k_v \\ 1 & 1 & 1 \end{bmatrix}, \tag{2}$$

where $K_i$—camera calibration matrix, $k_h$—horizontal scaling factor; $k_v$—vertical scaling factor.

$$H_i^{'} = SH_iS^{-1}$$
$$S = \begin{bmatrix} k_h & 0 & 0 \\ 0 & k_v & 0 \\ 0 & 0 & 1 \end{bmatrix}, \tag{3}$$

where $H_i$—homography matrix of $i$-th camera in stereo pair.

### 3.5 Reducing the Search Space Dimension

Target search function of the proposed approach searches for a match in some search buffer. In general, the search buffer is a set of all frames received after the previous one found. However, as FPS increases, the buffer size also increases and the calculation of the search function for buffer can't be completed in real time.

We propose to use a heuristic which physical value can be represented as follows: given that the speed values are obtained with an interval of $\frac{1}{FPS}$ s, it can be assumed that the speed cannot change vastly between two neighboring measurements. Then the index of the correct solution in the buffer will not change or will change slightly. Therefore, we consider only a certain part (5–10 frames nearby with the index of the previous one found) of the entire buffer.

## 4    Results

In this section, we analyze the results obtained by applying proposed approach to real-world data. We first describe accuracy and computational complexity and then discuss the results of practical implementation.

Figure 4 shows several speed graphs samples of various vehicles. Each individual graph contains measurements of the speed during the entire passage through the measuring system. Such low measurement discreteness is provided by high frame rate—250 frames per second.

We did not use the proposed method as a separate solution. As mentioned above, the proposed method is part of the system for non-contact measurement of large-sized objects using laser triangulation. Therefore, the accuracy of the proposed method was checked by comparing the actual lengths of objects and the results obtained. Of course, this does not directly indicate the accuracy of

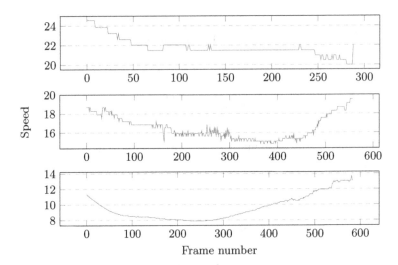

**Fig. 4.** Speed graphs samples of various objects for the entire time spent in the stereo camera frame

speed measurement, but it provides indirect confirmation of the validity of the results, since many experiments were performed with different length and nature of movement.

It is important to note that the proposed approach is a real-time solution. Which means that the algorithm's running time is less than $\frac{1}{250}$ s. This is provided by the simplicity of the search function and the efficiency of the heuristic.

## 5    Conclusion

In this paper, we proposed a novel approach for real-time speed estimation using side view stereo video images. Due to its simplicity, the resulting method was able to perform with a frequency of 250 frames per second which reduces the estimated error rate to 1.5% of the measured value. The approach was tested for almost half a year on real-world data. This solution is not suitable for standard traffic monitoring systems due to focusing on the presence of a single vehicle in the frame. However, it can be extended for this by adding the segmentation of vehicles in the frame.

We continue to work on improving the accuracy and stability of the proposed approach. As future work, we intend to reduce the effect of the image background on the target search function. A separate issue is to reduce the influence of shadows of scene objects on the search function.

## References

1. Dailey, D., Cathey, F., Pumrin, S.: An algorithm to estimate mean traffic speed using uncalibrated cameras. IEEE Trans. Intell. Transp. Syst. **1**, 98–107 (2000). https://doi.org/10.1109/6979.880967

2. Dogan, S., Temiz, M., Külür, S.: Real time speed estimation of moving vehicles from side view images from an uncalibrated video camera. Sensors (Basel, Switz.) **10**, 4805–4824 (2010). https://doi.org/10.3390/s100504805
3. Hartley, R., Zisserman, A.: Multiple view geometry in computer vision (2003)
4. Hua, S., Kapoor, M., Anastasiu, D.C.: Vehicle tracking and speed estimation from traffic videos. In: 2018 IEEE/CVF Conference on Computer Vision and Pattern Recognition Workshops (CVPRW), pp. 153–1537 (2018)
5. Kumar, T., Kushwaha, D.: An efficient approach for detection and speed estimation of moving vehicles. Procedia Comput. Sci. **89**, 726–731 (2016). https://doi.org/10.1016/j.procs.2016.06.045
6. Madasu, V.K., Hanmandlu, M.: Estimation of vehicle speed by motion tracking on image sequences. In: 2010 IEEE Intelligent Vehicles Symposium, pp. 185–190 (2010)
7. Mikhaylichenko, A.A., Kleshchenkov, A.B.: Approach to non-contact measurement of geometric parameters of large-sized objects. Program. Comput. Softw. **44**(4), 271–277 (2018). https://doi.org/10.1134/S0361768818040096
8. Yabo, A., Arroyo, S., Safar, F., Oliva, D.: Vehicle classification and speed estimation using computer vision techniques (2016)
9. Zhang, Z.: A flexible new technique for camera calibration. IEEE Trans. Pattern Anal. Mach. Intell. **22**(11), 1330–1334 (2000)

# Social Network Analysis

# How the Minimal Degree of a Social Graph Affects the Efficiency of an Organization

Ilya Samonenko[1]([envelope]), Tamara Voznesenskaya[1], and Rostislav Yavorskiy[2]

[1] Higher School of Economics, Moscow 101000, Russia
{isamonenko,tvoznesenskaya}@hse.ru
[2] Tomsk Polytechnic University, Tomsk 634050, Russia
ryavorsky@tpu.ru

**Abstract.** This paper continues our research on how the communication structure of an organization affects its efficacy. We assume that the organization consists of several types of professionals (researchers, engineers, testers etc.) and each project team must include at least one professional of each type. We study a dynamic simulation model, which defines a team creation process, based on the social graph of the organization. The greatest lower bound of the average utilization rate is found under (is based on) the assumption that each node in the social graph has a degree of at least $\delta$. This theoretical result is supported by a series of computer simulations.

**Keywords:** Professional network · Self-organizing teams · Team model · Social graph · Utilization rate · Simulation modelling

## 1 Introduction

The topic of team modelling and performance analysis generates regular interest among researchers, see [5]. Analysis and improvement of employee utilization are widely covered in business administration literature, but papers, which use mathematical modelling for that are quite rare. In our previous paper [8] a literature overview of team modelling approaches and the utilization rate analysis was presented. An important insight of the literature study is that a meaningful modelling approach should take into account the skill profiles of the employees, their roles and communications.

A simple example which we have in mind is a group of researchers and engineers working at a rather big academic research center, see [7]. Some of the group members know each other, and these connections are used to form project teams when information about a new grant, a call for papers or other opportunity occurs. We use a simulation modeling approach to study the process of information distribution and team forming in response to that new opportunity. We are especially interested in the efficiency of such organization as the average number of well-formed project teams per opportunity. The same model type

© Springer Nature Switzerland AG 2021
W. M. P. van der Aalst et al. (Eds.): AIST 2020, CCIS 1357, pp. 173–181, 2021.
https://doi.org/10.1007/978-3-030-71214-3_15

could also be applied to other cases, where the project teams are self-organized by the employees on the basis of existing social connections.

An *average utilization rate* of a group is the proportion of specialists involved in projects to the total number of specialists. We are interested in studying this important aspect of the organization performance. In particular, we want to analyze how the minimal degree of the social graph affects the utilization rate of an organization.

The horizontal connection between members of an organization is an important factor. Many researchers consider different characteristics of these connections in team modeling. For example, team sports performance analysis (see e.g. [4]) pay attention not only to social graph topology but also to the complexity of interpersonal interactions of players. In [2], a model of multi-agent team performance for the RoboCup Rescue domain is considered. The authors simulated the behavior of small teams of 2–3 robots, working on civilian rescue tasks to study the synergy among them. The different characteristics of intra-team interaction are considered in [3]. Computer simulation to support the systematic design of organization was also studied in [1].

The paper is organized as follows. In Sect. 2, we provide formal definitions, then different restrictions and conditions on the model are considered. The main part of the paper consists of theorems in Sect. 3, which formalize some properties of the model with respect to these conditions. Section 4 provides the simulation analysis and Sect. 4 concludes.

## 2   Definitions

We consider here a simplified version of a more general simulation model developed for analysis of team performance in a research centre [7,8]. Input parameters of the model are the following:

- $\mathbf{P} = A \sqcup B$ is a finite set of group members (employees, professionals) of two categories: type $A$ (researchers) and type $B$ (engineers), where $|A| = n_1$ and $|B| = n_2$;
- $G \subseteq \mathbf{P} \times \mathbf{P}$ is an undirected social graph of an organization, so $G(p_1, p_2)$ indicates that employees $p_1, p_2 \in \mathbf{P}$ have established relations enough to invite each other to a new project.

In general, we assume that for a successful project one needs a team, which includes $k_1$ researchers and $k_2$ engineers. The focus of this paper is where $k_1 = k_2 = 1$.

The simulation process is executed in the following way (it is described in detail in [6]):

- **Team initialization.** When an employee decides to initiate a new project, a new team is created, which consists of only the initiator.

- **Invitation process.** Each team member sends an invitation to join his project to each of his contacts in the social graph of the organization. An employee always accepts the invitation if their competencies are needed for this team and they have time available. Each employee can only join one project.
- **Team finalization:**
  - **Success.** A project team is formed where $k_1$ researchers and $k_2$ engineers have joined it.
  - **Failure.** It may be that the combined skills of members are not sufficient and there is no one else available with the required skill set. Then the team formation process is stopped and members who already joined become available again.

We assume that the receiver of an invitation makes the decision and responds immediately (from a practical point of view this means that all the communications are fast and decisions are instant in comparison to the duration of projects).

The simulation process sequentially runs the rules above for all the agents (employees) until the formation of new teams is impossible.

For a given graph $G$, let $w$ denote the number of employees who are finally involved in a project team. Then the utilization rate is defined as a fraction:

$$\rho(G) = \frac{w}{n_1 + n_2}.$$

Note that the team building is a non-deterministic process, so value $\rho(G)$ will differ for different simulations, see detailed [6].

## 3   Upper and Lower Bounds for the Utilization Rate

We assume that each project needs exactly one researcher and one engineer.

The upper bound for $\rho(G)$ is rather straightforward. Since any project needs exactly one researcher and one engineer, if $n_1 \neq n_2$ then at least $|n_1 - n_2|$ employees will always be idle. So,

$$\rho(G) \leq \frac{2\min(n_1, n_2)}{n_1 + n_2}.$$

This upper bound is the lowest, see [8] for the detailed proof.

Next we prove the lower bound for the utilization rate for the model defined above under the assumption that each node in the social graph has a degree greater than $\delta$. This generalizes our previous results, see Fig. 1 and [8], which correspond to case $\delta = 1$.

For $k_1 = k_2 = 1$ connections between employees of the same category do not affect the team formation. So, we assume that the social graph is bipartite.

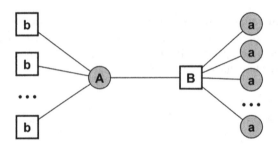

**Fig. 1.** Social graph with the worst utilization rate, see [8]

**Theorem 1 (lower bound for the utilization rate).** *Let $G = (A \sqcup B, E)$ denote a bipartite social graph of an organization, $|A| = n_1$, $|B| = n_2$. Assume that each node in $G$ has a degree equal to or greater than $\delta$, then:*

$$\rho(G) \geq \frac{2\delta}{n_1 + n_2} \qquad (1)$$

*and this lower bound is the greatest.*

*Proof.* Under the assumption of the two skills model, every project team consists of exactly two members, a researcher (from $A$) and an engineer (from $B$).

First, we explicitly construct a social graph and a simulation run for which

$$\rho(G) \geq \frac{2\delta}{n_1 + n_2}.$$

Then, it will be shown that for any graph and simulation, $\rho(G)$ cannot be lower.

Consider Fig. 2. It is a bipartite graph with $n_1$ nodes of type $A$ and $n_2$ nodes of type $B$. The nodes are divided into four groups:

- group I consists of $(n_2 - \delta)$ nodes of type B;
- group II consists of $\delta$ nodes of type A;
- group III consists of $\delta$ nodes of type B;
- group IV consists of $(n_1 - \delta)$ nodes of type A.

Each node from group I is linked to each node from group II. Similarly, each node from group II is linked to each node from group III, each node from group III is linked to each node from group IV. It is clear that the degree of nodes in groups I and IV are equal to $\delta$, and the degree of nodes in groups II and III are equal $n_2$ and $n_1$ correspondingly, which is greater that $\delta$.

For this structure of the social graph, it may happen that nodes from groups II and III form $\delta$ pairs and all the remaining nodes will be idle. For this run the utilization rate equals

$$\rho = \frac{2\delta}{n_1 + n_2}.$$

Let's now show that this is the strict lower bound and any simulation makes at least $\delta$ teams. This could be shown by induction on $\delta$. Case $\delta = 1$ is trivial,

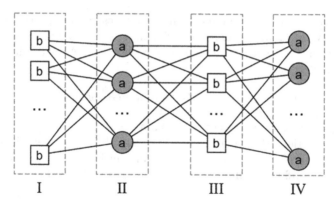

**Fig. 2.** The worst case social graph of a two-skilled organization: group I consists of $(n_2 - \delta)$ nodes, groups II and III contain $\delta$ nodes, group IV consists of $(n_1 - \delta)$ nodes

since each node has at least one connection, at least one pair will be formed. Consider any bipartite graph such that each degree of every node is equal or greater than $\delta$. Take an arbitrary node of any type, and any one from it's connections to form a team. In the subgraph, all the nodes have a degree $\delta - 1$ or greater, so at least $\delta - 1$ teams will be formed according to the induction hypothesis. □

From a practical point of view this result looks quite frustrating. For example, assume $n_1 = n_2 = 15$ and $\delta = 3$, this has a balanced group of 30 workers in which everyone is linked to at least three potential collaborators. According to the reason above it may happen that only 6 teams are formed and the remaining 18 members are forced to be idle. The group utilization rate for this example is $\rho = 40\%$, which is far from intuitively expected $\rho = 100\%$, which is potentially reachable when the social graph is more dense.

One of the useful implications is the fact that measures aimed at establishing more connections in a team do not always guarantee a substantial growth of the average utilization rate.

## 4    Simulation Analysis

In this section we describe in detail the simulation process because its parameters strongly influence the observed results.

We implement function $GetRandomGraph(n, \delta)$ that creates a random bipartite graph in the following way. First, for given $n$ and $\delta$ graph $G = (A \sqcup B, \emptyset)$ with $n$ vertices and no edges is created, where $A$ stands for a set of vertices corresponding to researchers, $B$ is the set of vertices corresponding to engineers, and $|A| = |B| = n/2$. Then, each vertex $a \in A$ is connected to $\delta$ different vertices from $B$ with equal probability, and vice versa. Each vertex $b \in B$ is connected to $\delta$ different vertices from $A$ with equal probability. It may happen that some edges are created twice, so we remove the duplicates. Such a procedure guarantees that every researcher is connected with no less than $\delta$ engineers, and every engineer is connected with no less than $\delta$ researchers. As a result we have a random bipartite graph

$$G = (A \sqcup B, E) = GetRandomGraph(n, \delta),$$

with the degree of each vertex not less than $\delta$.

The first series of experiments was aimed at finding dependencies between guaranteed minimal node degree $\delta$ and utilization rate $\rho(GetRandom Graph(n, \delta))$ for different values of $n = 20, 30, 40$ and $50$, see Fig. 3.

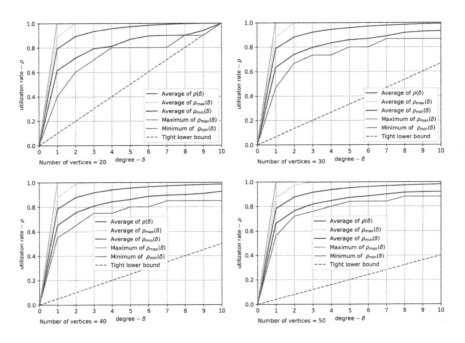

**Fig. 3.** Dependencies between $\delta$ and $\rho$ for $n = 20, 30, 40$ and $50$.

The second series of experiments was aimed at finding dependencies between the size of graph $n$ and the utilization rate $\rho(GetRandomGraph(n, \delta))$ for different values of $\delta = 1, 2, \ldots, 6$, see Fig. 4.

For each fixed value of $n$ and $\delta$, $T = 500$, graphs $G_i(n, \delta)$ where $i = 1, \ldots, T$, were generated randomly. Then, for each graph $G_i(n, \delta)$ we have run $N = 200$ simulations $\rho_{s_j}(G_i(n, \delta))$, $j = 1, \ldots, N$, and compute the following characteristics.

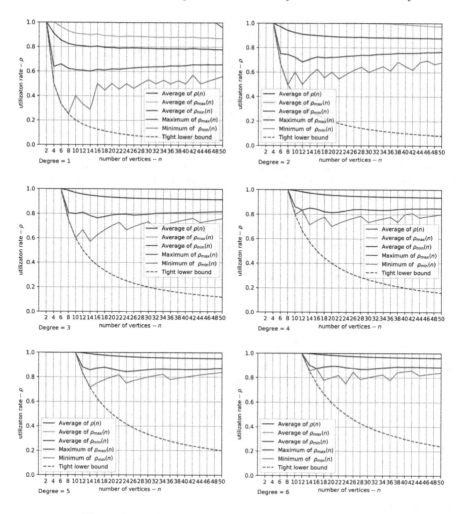

**Fig. 4.** Dependencies between $n$ and $\rho$ for $\delta = 1, 2, \ldots, 6$

$$Average\ of\ \rho(n, \delta) = \frac{1}{T}\sum_{i=1}^{T}\frac{1}{N}\sum_{j=1}^{N}\rho_{s_j}(G_i(n, \delta)),$$

$$Average\ of\ \rho_{max}(n, \delta) = \frac{1}{T}\sum_{i=1}^{T}\max_{j=1..N}\rho_{s_j}(G_i(n, \delta)),$$

$$Average\ of\ \rho_{min}(n, \delta) = \frac{1}{T}\sum_{i=1}^{T}\min_{j=1..N}\rho_{s_j}(G_i(n, \delta)),$$

$$Maximum\ of\ \rho_{max}(n, \delta) = \max_{i=1..T}\max_{j=1..N}\rho_{s_j}(G_i(n, \delta)),$$

$$Minimum\ of\ \rho_{min}(n, \delta) = \min_{i=1..T}\min_{j=1..N}\rho_{s_j}(G_i(n, \delta)).$$

As a result, we found the following.

1. Typically, random simulation results are far better than the worst possible bound established, according to our theorem. For $\delta = 1$, the average value of utilization rate $\rho(n, \delta)$ is around 80%, while the theoretical minimum is below 10% for $n > 20$, see blue and dashed graphs on Fig. 3 and Fig. 4.
2. Average $\rho(n, \delta)$ increases with $\delta$, which complies with the intuition that adding more connections increases utilization, see Fig. 3. The graphs show that for $\delta = 3$ (each employee knows at least three colleagues from the other category), the average utilization is almost certainly greater than 80%.
3. For a fixed $\delta \in \{3, 4, 5, 6\}$ average the utilization $\rho(n, \delta)$ is rather stable and does not depend on $n$, see Fig. 4.
4. At $\delta \geq 3$, the average value is about 90%, and with the increase of $\delta$, does not grow significantly. Therefore, for a high probability of getting on a team, it is enough for an engineer or researcher to have three contacts. This means a simple practical recommendation: build a team so that each specialist has at least three contacts with specialists with other skills.

## 5   Conclusion

In this paper, an analysis of the utilization rate for an organization with a guaranteed minimum degree of a bipartite social graph has been provided. We generalized our previous results [8] for the greatest lower bound of the utilization rate. Simulation analysis gave us some ideas and possible directions for future research. In particular, we plan to consider different distributions of node degrees for this graph generation algorithm. Besides, it is interesting to study the generalization of this model with a larger number of skills and the evolving topology of social graphs.

## References

1. Jin, Y., Levitt, R.E., Kunz, J.C., Christiansen, T.R.: The virtual design team: a computer simulation framework for studying organizational aspects of concurrent design. Simulation **64**(3), 160–174 (1995)
2. Liemhetcharat, S., Veloso, M.: Weighted synergy graphs for effective team formation with heterogeneous ad hoc agents. Artif. Intell. **208**, 41–65 (2014)
3. Peltomäki, M., Alava, M.: Correlations in bipartite collaboration networks. J. Stat. Mech. Theory Experiment 2006 (2005). https://doi.org/10.1088/1742-5468/2006/01/P01010
4. Ribeiro, J., Silva, P., Duarte, R., Davids, K., Garganta, J.: Team sports performance analysed through the lens of social network theory: implications for research and practice. Sports Med. **47**, 1–8 (2017)
5. Salas, E., Goodwin, G.F., Burke, C.S.: Team Effectiveness in Complex Organizations: Cross-disciplinary Perspectives and Approaches. Routledge, Milton Park (2008)

6. Samonenko, I., Voznesenskaya, T.: The influence of self-organizing teams on the structure of the social graph. In: TMPA: International Conference on Software Testing, Machine Learning and Complex Process Analysis. Springer (2019, to appear)
7. Voznesenskaya, T.V., Krasnov, F.V., Yavorsky, R.E., Chesnokova, P.V.: Modeling self-organizing teams in a research environment. Bus. Inform. **13**, 7–17 (2019). https://doi.org/10.17323/1998-0663.2019.2.7.17
8. Yavorskiy, R., Voznesenskaya, T., Samonenko, I.: Effect of social graph structure on the utilization rate in a flat organization. In: AIST 2019. CCIS, vol. 1086, pp. 214–224. Springer, Cham (2020). https://doi.org/10.1007/978-3-030-39575-9_22

# Data Analysis and Machine Learning

# Object-Attribute Biclustering
# for Elimination of Missing Genotypes
# in Ischemic Stroke Genome-Wide Data

Dmitry I. Ignatov[1]([✉])[iD], Gennady V. Khvorykh[2][iD], Andrey V. Khrunin[2][iD], Stefan Nikolić[1][iD], Makhmud Shaban[1][iD], Elizaveta A. Petrova[3], Evgeniya A. Koltsova[3], Fouzi Takelait[1][iD], and Dmitrii Egurnov[1][iD]

[1] National Research University Higher School of Economics, Moscow, Russia
dignatov@hse.ru
[2] Institute of Molecular Genetics of National Research Centre "Kurchatov Institute", Moscow, Russia
{khvorykh,khrunin}@img.ras.ru
[3] Pirogov Russian National Research Medical University, Moscow, Russia
http://www.hse.ru, http://img.ras.ru, http://rsmu.ru

**Abstract.** Missing genotypes can affect the efficacy of machine learning approaches to identify the risk genetic variants of common diseases and traits. The problem occurs when genotypic data are collected from different experiments with different DNA microarrays, each being characterised by its pattern of uncalled (missing) genotypes. This can prevent the machine learning classifier from assigning the classes correctly. To tackle this issue, we used well-developed notions of object-attribute biclusters and formal concepts that correspond to dense subrelations in the binary relation *patients* × *SNPs*. The paper contains experimental results on applying a biclustering algorithm to a large real-world dataset collected for studying the genetic bases of ischemic stroke. The algorithm could identify large dense biclusters in the genotypic matrix for further processing, which in return significantly improved the quality of machine learning classifiers. The proposed algorithm was also able to generate biclusters for the whole dataset without size constraints in comparison to the In-Close4 algorithm for generation of formal concepts.

**Keywords:** Formal concept analysis · Biclustering · Single nucleotide polymorphism · Missing genotypes · Data mining · Ischemic stroke

## 1 Introduction

The recent progress in studying different aspects of human health and diversity (e.g., genetics of common diseases and traits, human population structure, and relationships) is associated with the development of high-throughput genotyping technologies, particularly with massive parallel genotyping of Single Nucleotide Polymorphisms (SNPs) by DNA-microarrays [1]. They allowed the determination of hundreds of thousands and millions of SNPs in one experiment and were

© Springer Nature Switzerland AG 2021
W. M. P. van der Aalst et al. (Eds.): AIST 2020, CCIS 1357, pp. 185–204, 2021.
https://doi.org/10.1007/978-3-030-71214-3_16

the basis for conducting genome-wide association studies (GWAS). Although thousands of genetic loci have been revealed in GWAS, there are practical problems with replicating the associations identified in different studies. They seem to be due to both limitations in the methodology of the GWAS approach itself and differences between various studies in data design and analysis [2]. The machine learning (ML) approaches were found to be quite promising in this field [3].

Genotyping by microarrays is efficient and cost-effective, but missing data appear. GWAS is based on a comparison of frequencies of genetic variants among patients and healthy people. It assumes that all genotypes are provided (usually, their percentage is defined by a genotype calling threshold). In this article, we demonstrate that missing data can affect not only statistical analysis but also the ML algorithms. The classifiers can fail because of missing values (uncalled genotypes) being distributed non-randomly. We assume that each set of DNA-microarray can possess a specific pattern of missing values marking both the dataset of patients and healthy people. Therefore, the missing data needs to be carefully estimated and processed without dropping too many SNPs that may contain crucial genetic information.

To overcome the problem of missing data, we aimed to apply a technique capable of discovering some groupings in a dataset by looking at the similarity across all individuals and their genotypes. The raw datasets can be converted into an integer matrix, where individuals are in rows, SNPs are in columns, and cells contain genotypes. For each SNP, the person can have either AA, AB, or BB genotype, where A and B are the alleles. Thus the genotypes can be coded as 0, 1, and 2, representing the counts of allele B.

The proposed method can simultaneously cluster rows and columns in a data matrix to find homogeneous submatrices [4], which can overlap. Each of these submatrices is called a bicluster [5], and the process of finding them is called biclustering [4–9].

Biclustering in genotype data allows identifying sets of individuals sharing SNPs with missing genotypes. A bicluster arises when there is a strong relationship between a specific set of objects and a specific set of attributes in a data table. A particular kind of bicluster is a formal concept in Formal Concept Analysis (FCA) [10]. A formal concept is a pair of the form (extent, intent), where extent consists of all objects that share the attributes in intent, and dually the intent consists of all attributes shared by the objects in extent. Formal concepts have a desirable property of being homogeneous and closed in the algebraic sense, which resulted in their extensive use in Gene Expression Analysis (GEA) [11–14].

A concept-based bicluster (or object-attribute bicluster) [15] is a scalable approximation of a formal concept with the following advantages:

1. Reduced number of patterns to analyze;
2. Reduced computational cost (polynomial vs. exponential);
3. Manual (interactive) tuning of bicluster density threshold;
4. Tolerance to missing (object, attribute) pairs.

In this paper, we propose an extended biclustering algorithm of [16] that can identify large biclusters with missing genotypes for categorical data (many-valued contexts with a selected value). This algorithm can generate a smaller amount of dense object-attribute biclusters than that of existing exact algorithms for formal concepts like concept miner In-Close4 [17], and is, therefore, better suited for large datasets. Moreover, during experimentation with the ischemic stroke dataset, we found that the number of large dense biclusters identified by our algorithm is significantly lower than the number of formal concepts extracted by In-Close4[1] and Concept Explorer (ConExp[2]) [18].

The paper is organized as follows. In Sect. 2, we recall basic notions from Formal Concept Analysis and Biclustering. In Sect. 3, we introduce a method of FCA-based biclustering and its variants along with bicluster post-processing schemes, consider discussing the complexity of the proposed algorithm. In Sect. 4, we describe a dataset that consists of a sample of patients and their SNPs collected from various (independent) groups of patients. Then we present the results obtained during experiments on this dataset in Sect. 5 and mention the used hardware and software configuration. Section 6 concludes the paper.

## 2  Basic Notions

### 2.1  Formal Concept Analysis

**Definition 1.** *A **formal context** in FCA [10] is a triple $\mathbb{K} = (G, M, I)$ consisting of two sets, $G$ and $M$, and a binary relation $I \subseteq G \times M$ between $G$ and $M$. The triple can be represented by a cross-table consisting of rows $G$, called* **objects**, *and columns $M$, called* **attributes**, *and crosses representing incidence relation $I$. Here, $gIm$ or $(g, m) \in I$ means that the object $g$* **has** *the attribute $m$.*

**Definition 2.** *For $A \subseteq G$ and $B \subseteq M$, let*

$$A' \stackrel{\text{def}}{=} \{m \in M \mid gIm \text{ for all } g \in A\}, \text{ and } B' \stackrel{\text{def}}{=} \{g \in G \mid gIm \text{ for all } m \in B\}.$$

*These two operators are the* **derivation operators** *for $\mathbb{K} = (G, M, I)$.*

**Proposition 1.** *Let $(G, M, I)$ be a formal context, for subsets $A, A_1, A_2 \subseteq G$ and $B \subseteq M$ we have*

1. $A_1 \subseteq A_2$ if $A_2' \subseteq A_1'$,
2. $A \subseteq A''$,
3. $A = A''$ (hence, $A'''' = A''$),
4. $(A_1 \cup A_2)' = A_1' \cap A_2'$,
5. $A \subseteq B' \Leftrightarrow B \subseteq A' \Leftrightarrow A \times B \subseteq I$.

*Similar properties hold for subsets of attributes.*

---

[1] https://sourceforge.net/projects/inclose/.
[2] http://conexp.sourceforge.net.

**Definition 3.** *A* **closure operator** *on set $S$ is a mapping $\varphi : 2^S \to 2^S$ with the following properties:*
*Let $X \subseteq S$, then*

1. $\varphi(\varphi(X)) = \varphi(X)$ *(**idempotency**),*
2. $X \subseteq \varphi(X)$ *(**extensity**),*
3. $X \subseteq Y \Rightarrow \varphi(X) \subseteq \varphi(Y)$ *(**monotonicity**).*

*For a closure operator $\varphi$ the set $\varphi(\varphi(X))$ is called* **closure** *of $X$, while a subset $X \subseteq S$ is called* **closed** *if $\varphi(\varphi(X)) = X$.*

It is evident from properties of derivation operators that for a formal context $(G, M, I)$, the operators

$$(\cdot)'' : 2^G \to 2^G \text{ and } (\cdot)'' : 2^M \to 2^M$$

are closure operators.

**Definition 4.** *$(A, B)$ is a* **formal concept** *of formal context $\mathbb{K} = (G, M, I)$ iff*

$$A \subseteq B, \ B \subseteq M, \ A' = B, \text{ and } A = B'.$$

*The sets $A$ and $B$ are called the* **extent** *and the* **intent** *of the formal concept $(A, B)$, respectively.*

This definition says that every formal concept has two parts, namely, its extent and intent. It follows an old tradition in philosophical concept logic, as expressed in the *Logic of Port Royal, 1662* [19].

**Definition 5.** *The set of all formal concepts $\mathfrak{B}(B, M, I)$ is partially ordered, given by relation $\leq_{\mathbb{K}}$:*

$$(A_1, B_1) \leq_{\mathbb{K}} (A_2, B_2) \iff A_1 \subseteq A_2 \text{ (dually } B_2 \subseteq B_1)$$

*$\mathfrak{B}(B, M, I)$ is called* **concept lattice** *of the formal context $\mathbb{K}$.*

In case an object has properties like colour or age the corresponding attributes should have values themselves.

**Definition 6.** *A* **many-valued context** *$(G, M, W, J)$ consists of sets $G$, $M$ and $W$ and a ternary relation $J \subseteq G \times M \times W$ for which it holds that*

$$(g, m, w) \in J \text{ and } (g, m, v) \in I \text{ imply } w = v.$$

*The elements of $M$ are called* **(many-valued) attributes** *and those of $W$* **attribute values**.

Since many-valued attributes can be considered as partial maps from $G$ in $W$, it is convenient to write $m(g) = w$.

## 2.2   Biclustering

In [6], *bicluster* is defined as a homogeneous submatrix of an input object-attribute matrix of real values in general. Consider a dataset as a matrix, $A = (X, Y) \in \mathbb{R}^{n \times m}$, with a set of rows/objects/individuals $X = \{x_1, \ldots, x_n\}$ and set of columns/attributes/SNPs $Y = \{y_1, \ldots, y_m\}$. A submatrix constructed from a subset of rows $I \subseteq X$ and that of columns $J \subseteq Y$ is denoted as $(I, J)$ is called a **bicluster** of $A$ [6]. The bicluster should satisfy some specific homogeneity properties, which varies from one method to another.

For instance, for the purpose of this research, we use the following FCA-based definition of a bicluster [15,16,20].

**Definition 7.** *For a formal context* $\mathbb{K} = (G, M, I)$ *any biset* $(A, B) \subseteq I$ *with* $A \neq \emptyset$ *and* $B \neq \emptyset$ *is called a* **bicluster**. *If* $(g, m) \in I$, *then the bicluster* $(A, B) = (m', g')$ *is called an object-attribute or* **OA-bicluster** *with density* $\rho(A, B) = \frac{|I \cap (A \times B)|}{|A| \cdot |B|}$.

The density $\rho(m', g')$ of a bicluster $(m', g')$ is the bicluster quality measure that shows how many non-empty pairs the bicluster contains divided by its size.

Several basic properties of OA-biclusters are below.

**Proposition 2**

1. *For any bicluster* $(A, B) \subseteq 2^G \times 2^M$ *it is true that* $0 \leq \rho(A, B) \leq 1$,
2. *OA-bicluster* $(m', g')$ *is a formal concept iff* $\rho = 1$,
3. *If* $(A, B)$ *is a OA-bicluster, there exists (at least one) its* **generating pair** $(g, m) \in A \times B$ *such that* $(m', g') = (A, B)$,
4. *If* $(m', g')$ *is a OA-bicluster, then* $(g'', g') \leq (m', m'')$.
5. *For every* $(g, m) \in I$, $(h, n) \in [g]_M \times [m]_G{}^3$, *it follows* $(m', g') = (n', h')$.

In Fig. 1, you can see the example of OA-bicluster, for a particular pair $(g, m) \in I$ of a certain context $(G, M, I)$. In general, only the regions $(g'', g')$ and $(m', m'')$ are full of non-empty pairs, i.e. have maximal density $\rho = 1$, since they are object and attribute formal concepts respectively. The black cells indicate non-empty pairs, which one may found in less dense white regions.

**Definition 8.** *Let* $(A, B) \in 2^G \times 2^M$ *be a OA-bicluster and* $\rho_{min} \in (0, 1]$, *then* $(A, B)$ *is called dense if it satisfies the constraint* $\rho(A, B) \geq \rho_{min}$.

The number of OA-biclusters of a context can be much less than the number of formal concepts (which may be $2^{\min(|G|,|M|)}$), as stated by the following propositions.

**Proposition 3.** *For a formal context* $\mathbb{K} = (G, M, I)$ *the largest number of OA-biclusters is equal to* $|I|$ *and all OA-biclusters can be generated in time* $\mathcal{O}(|I|)$.

**Proposition 4.** *For a formal context* $\mathbb{K} = (G, M, I)$ *and* $\rho_{min} > 0$ *the largest number of dense OA-biclusters is equal to* $|I|$, *all dense OA-biclusters can be generated in time* $\mathcal{O}(|I||G||M|)$.

---

[3] The equivalence classes are $[g]_M = \{h \mid h \in G, g' = h'\}$ and $[m]_G = \{n \mid n \in M, n' = m'\}$.

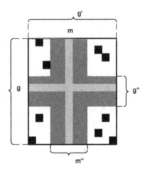

**Fig. 1.** OA-bicluster based on object and attribute primes.

## 3    Model and Algorithm Description

### 3.1    Parallel OA-Biclustering Algorithm

Algorithm 1 is a straightforward implementation, which takes an initial many-valued formal context and minimal density threshold as parameters and computes dense biclusters for each $(g, m)$ as pair in the relation $I$ that indicates which objects have SNP with missing values. However, since OA-biclusters for many-valued contexts were not formally introduced previously, we use a derived formal context with one-valued attributes denoting missing attribute-values of an original genotype matrix to correctly apply the definition of dense OA-bicluster.

**Definition 9.** *Let* $\mathbb{K} = (G, M, W, J)$ *is a many-valued context and* $v \in W$ *is a selected value (e.g., denoting the absence of an SNP value), then its* **derived context for the value** $v$ *is* $\mathbb{K}_v = (G, M, I)$ *where* $gIm$ *iff* $(g, m, v) \in J$.

For genotype matrices with missing SNP values as many-valued contexts, similar representation can be expressed in terms of co-domains of many-valued attributes (the absence of $m(g)$ means that of the corresponding SNP value) or by means of nominal scaling with a single attribute for the missing value $v$ [10].

If we compare the number of output pattern for formal concepts and dense OA-biclusters, in the worst case these values are $2^{min(|G|,|M|)}$ versus $|I|$. The time complexity of our algorithm is polynomial, $\mathcal{O}(|G||M||I|)$, versus exponential in the worse case for BiMax [21], $\mathcal{O}(|G||M||L| \log |L|)$, or $\mathcal{O}(|G|^2|M||L|)$ for CbO algorithms family [22], where $|L|$ is a number of generated concepts (also considered as biclusters) and is exponential in the worst case $|L| = 2^{min(|G|,|M|)}$.

For calculating biclusters that fulfil a minimum density constraint, we need to perform several steps (see Algorithm 1). Steps 5–8 consists of applying the Galois operator to all objects in $G$ and steps 9–12 then to all attributes in $M$ within the induced context. The outer for loops are parallel (the concrete implementation may differ), while the internal ones are ordinary for loops. Then all biclusters are enumerated in a parallel manner as well, and only those that fulfil the minimal density requirement are retained (Steps 13–16). Again, efficient implementation of set data-structure for storing biclusters and duplicate elimination on the fly in parallel execution mode are not addressed in the pseudo-code.

The novelties of this algorithm include the usage of parallelization to generate the OA-bicluster taking as input a medium-sized dataset (e.g. $10^3 \times 10^4$) and the possibility to work with selected values reducing many-valued context to contexts with one-valued attributes

---

**Algorithm 1:** OA-bicluster generation for a many-valued context.

---

**Data:** $\mathbb{K} = (G, M, W, J)$ is a many-valued formal context, $\rho_{min}$ is a threshold density value of bicluster density and $v \in W$ is a selected value

**Result:** $B = \{(A, B)|(A, B)$ is an OA-bicluster for value $v\}$

```
1  begin
2  │   Obj.Size := |G|
3  │   Attr.Size := |M|
4  │   B ⟵ ∅
5  │   parallel for g ∈ G do
6  │   │   for m ∈ M do
7  │   │   │   if m(g)=v then
8  │   │   │   │   Obj[g].Add(m)
9  │   parallel for m ∈ M do
10 │   │   for g ∈ G do
11 │   │   │   if m(g)=v then
12 │   │   │   │   Attr[m].Add(g)
13 │   parallel for (g, m, w) ∈ J do
14 │   │   if w=v then
15 │   │   │   if ρ(Attr[m], Obj[g]) ≥ ρ_min then
16 │   │   │   │   B := B ∪ {(Attr[m], Obj[g])}
```

---

## 3.2 One-Pass Version of the OA-Biclustering Algorithm

Let us describe the online problem of finding the set of prime OA-biclusters based on the online OAC-Prime Triclustering [23]. Let $\mathbb{K} = (G, M, I)$ be a context. The user has no a priori knowledge of the elements and even cardinalities of $G$, $M$, and $I$. At each iteration, we receive a set of pairs ("batch") from $I$: $J \subseteq I$. After that, we must process $J$ and get the current version of the set of all biclusters. It is important in this setting to consider every pair of biclusters different if they have different generating pairs even if their extents and intents are equal, because any other pair can change only one of them, thus making them different.

Also, the algorithm requires that the dictionaries containing the prime sets are implemented as hash-tables or similar efficient key-value structures. Because of this data structure, the algorithm can efficiently access prime sets.

The algorithm itself is also straightforward (Algorithm 2). It takes a set of pairs ($J$) and current versions of the biclusters set ($\mathcal{B}$) and the dictionaries containing prime sets ($PrimesO$ and $PrimesA$) as input and outputs the modified versions of the bicluster set and dictionaries. The algorithm processes each pair $(g, m)$ of $J$ sequentially (line 1). On each iteration the algorithm modifies the corresponding prime sets: it adds $m$ to $g'$ (line 2) and $g$ to $m'$ (line 3).

Finally, it adds a new bicluster to the bicluster set. Note that this bicluster contains pointers to the corresponding prime sets (in the corresponding dictionaries) instead of their copies (line 4).

In effect, this algorithm is very similar to the original OA-biclustering algorithm with some optimizations. First of all, instead of computing prime sets at the beginning, we modify them on spot, as adding a new pair to the relation modifies only two prime sets by one element. Secondly, we remove the main loop by using pointers for the bicluster' extents and intents, as we can generate biclusters at the same step as we modify the prime sets. And third, it uses only one pass through the pairs of the binary relation $I$, instead of enumeration of different pairwise combinations of objects and attributes.

---

**Algorithm 2:** Online generation of OA-biclusters

---

**Input:**  $J$ is a set of object-attribute pairs;
$\qquad \mathcal{B} = \{\mathbf{b} = (*X, *Y)\}$ is the current set of OA-biclusters;
$\qquad PrimesO, PrimesA$;
**Output:**  $\mathcal{B} = \{\mathbf{b} = (*X, *Y)\}$;
$\qquad PrimesO, PrimesA$;
1: **for all** $(g, m) \in J$ **do**
2: $\quad PrimesO[g] := PrimesO[g] \cup \{m\}$
3: $\quad PrimesA[m] := PrimesAC[m] \cup \{g\}$
4: $\quad \mathcal{B} := \mathcal{B} \cup \{(\&PrimesA[m], \&PrimesO[g])\}$
5: **end for**

---

Each step requires constant time: we need to modify two sets and add one bicluster to the set of biclusters. The total number of steps is equal to $|I|$; the time complexity is linear $\mathcal{O}(|I|)$. Beside that the algorithm is one-pass.

The memory complexity is the same: for each of $|I|$ steps the size of each dictionary containing prime sets is increased either by one element (if the required prime set is already present), or by one key-value pair (if not). Since each of these dictionaries requires $\mathcal{O}(|I|)$ memory, the memory complexity is also linear.

## 3.3   Post-processing Constraints

Another important step, in addition to this algorithm, is post-processing. Thus, we may want to remove additional biclusters with the same extent and intent from the output. Simple constraints like minimal support condition can be processed during this step without increasing the original complexity. It should be done only during the post-processing step, as the addition of a pair in the main algorithm can change the set of biclusters, and, respectively, the values used to check the conditions. Finally, if we need to fulfil more difficult constraints like minimal density condition, the time complexity of the post-processing will be higher than that of the original algorithm, but it can be efficiently implemented.

To remove the same biclusters we need to use an efficient hashing procedure that can be improved by implementing it in the main algorithm. For this, for all prime sets, we need to keep their hash-values with them in the memory. And finally, when using hash-functions other than LSH function (Locality-Sensitive Hashing) [24], we can calculate hash-values of prime sets as some function of their elements (for example, exclusive disjunction or sum). Then, when we modify prime sets, we just need to get the result of this function and the new element. In this case, the hash-value of the bicluster can be calculated as the same function of the hash-values of its extent and intent.

Then it would be enough to implement the bicluster set as a hash-set in order to efficiently remove the additional entries of the same bicluster.

Pseudo-code for the basic post-processing (Algorithm 3).

---

**Algorithm 3:** Post-processing for the online OA-biclustering algorithm.

---

**Input:** $\mathcal{B} = \{b = (*X, *Y)\}$ is a full set of biclusters;
**Output:** $\overline{\mathcal{B}} = \{b = (*X, *Y)\}$ is a processed hash-set of biclusters;
1: **for all** $b \in \mathcal{B}$ **do**
2:     Calculate $hash(b)$
3:     **if** $hash(b) \notin \overline{\mathcal{B}}$ **then**
4:         $\overline{\mathcal{B}} := \overline{\mathcal{B}} \cup \{b\}$
5:     **end if**
6: **end for**

---

If the names (codes) of the objects and attributes are small enough (the time complexity of computing their hash values is $\mathcal{O}(1)$), the time complexity of the post-processing is $\mathcal{O}(|I|)$ if we do not need to calculate densities, and $\mathcal{O}(|I||G||M|)$ otherwise. Also, the basic version of the post-processing does not require any additional memory; so, its memory complexity is $O(1)$.

Finally, the algorithm can be easily paralleled by splitting the subset of input pairs into several subsets, processing each of them independently, and merging the resulting sets afterwards, which may lead to distributed computing schemes for larger datasets (cf. [25]).

In case the output of the post-processing step is stored in a relational database along with the computed statistics and generating pairs, further usage of selection operators [26] is convenient to consider only a specific subset of biclusters.

We use the following operator resulting in a specific subset of biclusters

$$\sigma_{(\alpha_{min} \leq |A| \leq \alpha_{max}) \wedge (\beta_{min} \leq |B| \leq \beta_{max}) \wedge (\rho_{min} \leq \rho(A,B) \leq \rho_{max})}(\mathcal{B}),$$

where $|A|$ is the extent size, $|B|$ is the intent size, and $\rho(A, B)$ is the density of OA-bicluster $\mathbf{b} \in \mathcal{B}$, respectively. One more reason to use postprocessing is neither monotonic nor anti-monotonic character of the minimal density constraint in the sense of constraints pushing in pattern mining [11,16].

## 4   Data Collection

Collection of patients with ischemic stroke and their clinical characterisation were made at the Pirogov Russian National Research Medical University. The DNA extraction and genotyping of the samples were described previously [27].

The dataset contains samples corresponding to individuals with a genetic portrait for each and a group label. The former represents the genotypes determined at many SNPs all over the genome. The latter takes values 0 or 1 depending on whether a person did not have or had a stroke. Each SNP is a vector that components can take values from $\{0, 1, 2, -1\}$, where 0, 1, and 2 denote the genotypes, and $-1$ indicates a missing value.

We represent the dataset as a many-valued formal context. In the derived context $\mathbb{K} = (G, M, I)$, where objects from $G$ stand for samples and attributes from $M$ stand for SNPs, $gIm$ means that an individual $g$ has a missing SNP $m$. The context has the following parameters $|G| = 1,323$, $|M| = 85,142$, and $|I| = 45,075$ which represents the total number of attributes with missing values in the dataset and cover 0.491% of the whole data matrix. The number of attributes without missing values is 40,067.

The genotypic data were obtained with DNA-microarrays. The dataset was compiled from several experiments where different types of microarrays were applied. Not all genotypes are equally measured during the experiment. Thus, there is a certain instrumental error. The quality of DNA can also affect the output of the experiments. Figure 2 shows how many individuals have exactly $N$ missing genotypes per SNP in the dataset.

For instance, many individuals have about 85 missing genotypes per SNP.

## 5   Experiments

### 5.1   Hardware and Software Configuration

The experimental results with OA-biclustering generation and processing were obtained on an Intel(R) Core(TM) i5-8265U CPU @ 1.80 GHz with 8 GB of RAM and 64-bit Windows 10 Pro operating system. We used the following software releases to perform our experiments: Python 3.7.4 and Conda 4.8.2.

### 5.2   Identification of Biclusters with Missing SNPs

The following experiment was performed with ischemic stroke data collection: first of all, 383,733 OA-biclusters, with duplicates, were generated after applying the parallel biclustering algorithm to the dataset.

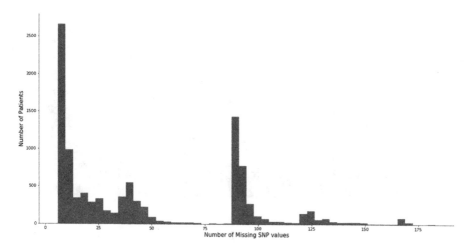

**Fig. 2.** The distribution of the number of missing SNP values by columns **before** elimination.

As we can see from the graph in Fig. 3, there is a reasonable amount of biclusters with a density value greater than 0.9. The distributions of biclusters by extent and intent show that the majority of biclusters have about 90 samples and 2,600 SNPs, respectively.

For the selection of large dense biclusters, we set the density constraint to be $\rho_{min} = 0.9$. Additional constraints were set as follows: $3 \leq |m'| \leq 1,500$ for the extent size and $3 \leq |g'| \leq 80,000$ for the intent size. In total, we selected 98,529 OA-biclusters with missing values. For this selection, the graph in Fig. 4 shows the selected peaks of large dense biclusters for different extent sizes.

**Example 1.** Biclusters in the form $(patients, SNPs)$.

For generating pair $(g, m) = (1102, rs6704827_A)$ we have that

$$(m', g') \in \sigma_{(3 \leq |A| \leq 1,500) \wedge (3 \leq |B| \leq 80,000) \wedge (0.9 \leq \rho(A,B) \leq 1)}(\mathcal{B}), \text{ where}$$

$(m'g') = (\{1101, 1102, \ldots, 1114\}, \{rs10915587_G, rs284267_A, \ldots, rs12171249_A\})$, $\rho(m', g') \approx 0.91$, $|m'| = 14$ individuals, $|g'| = 758$ SNPs, 9,657 pairs out of 10,612 correspond to missing SNP values.

We studied further large dense biclusters and chose the densest ones with possibly larger sizes of their extents and intents from each of the peaks identified in their distributions, respectively (Fig. 3).

Here are some examples of these subsets with their associated graphs.

**Example 2.** We can further narrow down the number of patterns in the previous selection by looking at the distribution of biclusters by their extent size and choosing proper boundaries. Thus, in Fig. 4, there is the third largest peak of the number of biclusters near the extent size 125.

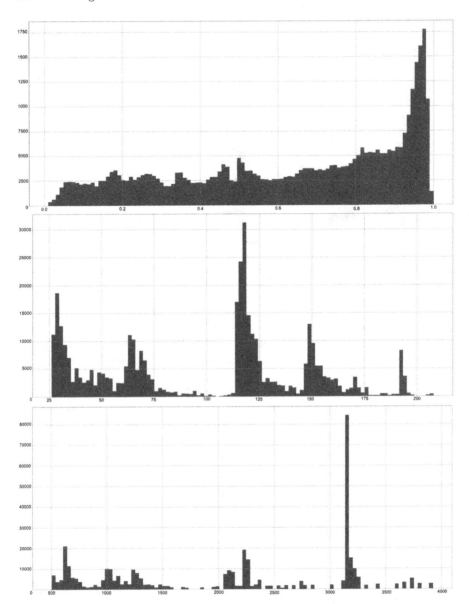

**Fig. 3.** The distribution of the number of biclusters by their density (top), extent (middle) and intent sizes (bottom).

For the constraints below

$$\rho_{min} = 94.08\% \wedge \rho_{max} = 100\% \wedge |g'| = 122 \wedge (3 \leq |m'| \leq 80,000)$$

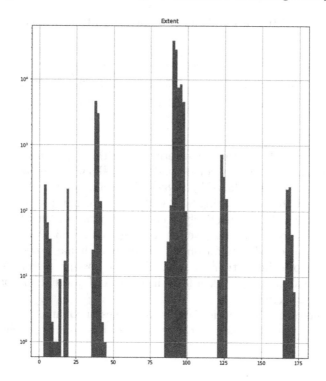

**Fig. 4.** The distribution of dense biclusters ($\rho_{min} = 0.9$) by their extent size.

the large dense bicluster with its intent size of 455 is identified and selected. Such bicluster has a large number of missing genotypes, which are subject to be eliminated later on.

**Example 3.** The selection around the rightmost peak (*see Fig.* 4) and further refining of the minimal value density

$$\rho_{min} = 95.4\% \wedge \rho_{max} = 100\% \wedge 160 \le |g'| \le 175 \wedge (3 \le |m'| \le 80,000)$$

resulted in the large bicluster with the extent size of 108 and the intent size of 166.

## 5.3   Elimination of Large Biclusters with Missing Genotypes

After applying the proposed biclustering algorithm to the collected dataset, all large biclusters with missing genotypes were identified and eliminated. That resulted in a new data matrix ready for further analysis[4]. We consolidate the evolution of the two datasets before and after removing missing values in Table 1.

---

[4] https://github.com/dimachine/OABicGWAS/.

**Table 1.** Basic statistics of the datasets **before** and **after** elimination of missing values.

|  | No. samples | No. SNPs | No. NaNs | NaNs fraction |
|---|---|---|---|---|
| Before elimination | 1,223 | 85,142 | 553,430 | 0.49% |
| After elimination | 1,472 | 82,690 | 388,052 | 0.31% |

As seen from Table 1, the biclustering algorithm application resulted in improvement in terms of entries corresponding to SNPs with missing genotypes, a fraction of such entries is reduced by 29.88%. The total number of biclusters generated before and after eliminating SNP with missing genotypes is 383,733 (with duplicates) and 259,440, respectively. The total amount of time for generating these biclusters before and after deleting missing data is 3433.2 and 2293.7 s (by Algorithm 1), respectively. As for online Algorithm 2, it has processed the original context (before elimination) in 1.5 s, while the post-processing Algorithm 3 for density computation has taken 907 s in sequential and 651 s in parallel (six cores) modes, respectively.

Figure 5 shows the distribution of missing values in columns in the new data set (after elimination of missing data), which now has less ragged character.

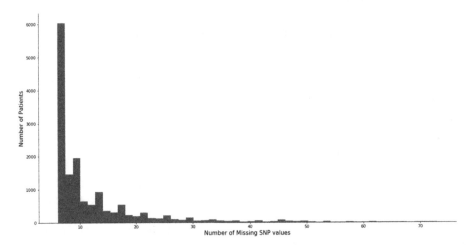

**Fig. 5.** The distribution of SNPs with missing genotypes by columns **after** elimination.

### 5.4 Large Dense Biclusters Elimination and Classification Quality

We have conducted a number of machine learning experiments on our datasets to check the impact of eliminating missing data. Our proposed algorithm handled on the quality measures of supervised learning algorithms.

We choose to use gradient boosting on decision trees (GBDT). For this purpose, we selected two libraries where it is already implemented, CATBOOST and LIGHTGBM. Both implementations can handle missing values.

A genome can essentially be interpreted as a sequence of SNPs, so we made a decision to also use LONG-SHORT TERM MEMORY NETWORK (LSTM) [28] as a strong approach to handle sequential data.

***First Dataset Experiments.*** Firstly, we applied GBDT algorithm from CAT-BOOST library to our initial dataset (before elimination of SNPs with missing genotypes). The following parameters were taken for the classifier:

- Maximum number of trees: 3;
- Tree depth limit: 3;
- Loss function: binary cross-entropy (log-loss/binary cross-entropy).

We also applied LSTM approach the following way: the initial sequence was resized to 100 elements by a fully-connected layer, then the layer output was passed to the LSTM module element-wise. The hidden state of LSTM after the last element was passed to a fully-connected classification layer.

The scores on this dataset were evaluated with 3-fold cross-validation with stratified splits. Basic classification metrics' scores are present in Table 2.

**Table 2.** Classification scores on the test set **before** elimination of missing SNP values

|  | Accuracy | F1-score | Precision | Recall |
|---|---|---|---|---|
| CatBoostClassifier | 0.966 | 0.9758 | 0.9558 | 0.9967 |
| FC+LSTM | 0.890 | 0.926 | 0.880 | 0.982 |

These unexpectedly high scores were unrealistic since the GBDT model complexity had one of the lowest possible configurations, and the LSTM model, which is handling the data in a different way, also achieved high accuracy. For a lot of samples, the model learned to "understand" on which chip it was analyzed by looking at the patterns of missing genotypes, so the data leak was present.

***Second Dataset Experiments.*** This dataset was obtained after the identification of large dense biclusters by application of our proposed algorithm with subsequent elimination. Table 3 recaps the experiments conducted on the dataset. For the first and second experiments, we used *CatBoost classifier* with train/test split in the proportion of 8:2 and 3-fold cross-validation, respectively, while maintaining the balance of classes for model validation. In the third experiment, we used *LGBMClassifier* classifier with 3-fold cross-validation while maintaining the balance of classes for model validation. In the fourth experiment, the described earlier *LSTM* classifier was used with the aforementioned cross-validation.

**Table 3.** Results of different machine learning classifiers applied to the dataset **after** elimination of SNP with missing genotypes.

|  | No. trees | Depth | Accuracy | F1-score | Precision | Recall |
|---|---|---|---|---|---|---|
| CatBoostClassifier | 2 | 2 | 0.715 | 0.834 | 0.715 | **1.000** |
|  | 5 | 2 | **0.773** | **0.862** | **0.761** | 0.995 |
|  | 5 | 3 | **0.773** | **0.862** | **0.761** | 0.995 |
| CatBoostClassifier | 4 | 3 | **0.768** | **0.859** | **0.990** | **0.759** |
|  | 5 | 3 | **0.768** | **0.859** | **0.990** | **0.759** |
| LGBMClassifier | 5 | 3 | 0.753 | 0.852 | **0.997** | 0.744 |
|  | 5 | 5 | 0.753 | 0.852 | 0.996 | 0.744 |
|  | 4 | 4 | 0.751 | 0.851 | **0.997** | 0.742 |
|  | 4 | 3 | 0.749 | 0.850 | **0.997** | 0.741 |
|  | 5 | 4 | **0.756** | **0.854** | 0.996 | **0.747** |
| FC+LSTM | – | – | 0.731 | 0.839 | 0.735 | 0.981 |

From Table 3, one can see that scores are more realistic in comparison to those of Table 2, thus showing us that data leak and subsequent overfitting effects are gone. We realize that our proposed biclustering algorithm successfully identified large submatrices with missing data, which we eliminated and successfully removed the impact of data leak and overfitting.

### 5.5 Detecting Concepts of Missing SNP Values Under Size Constraints

In-Close4 is an open-source software tool [29], which provides a highly optimised algorithm from CBO family [22,30] to construct the set of concepts satisfying given constraints on sizes of extents and intents. In-Close4 takes as input a context and outputs a reduced concept lattice: all concepts satisfying the constraints given by parameter values ($|A| \geq m$ and $|B| \geq n$, where $A$ and $B$ are extent and intent of an output formal concept, and $m, n \in \mathbb{N}$).

To deal with our large real-world dataset, we changed the maximum default values used in the executable of In-Close4 parameters as follows:

```
#define MAX_CONS 30000000 //max number of concepts
#define MAX_COLS 90000     //max number of attributes
#define MAX_ROWS 2000      //max number of objects
```

From Tables 4 and 5, one can see that the number of concepts generated by In-Close4 becomes several times larger than that of OA-biclusters, in our case study. When we set the extent size constraint to 5 with the input context before and the extent and the intent size constraint to 20 and 0, respectively, after the elimination of missing data, the software crashed.

As the author of InClose suggested in private communication, the tool was optimised for "tall" contexts with a large number of objects rather than attributes, while in bioinformatics the contexts are often "wide" like in our case when the number of SNPs is almost 57 times larger than that of individuals. So, the results on the transposed context along with properly set compilation parameters allowed to process the whole context for $m = 0$ and $n = 0^5$.

**Table 4.** The number of concepts and elapsed time generated by In-Close4 algorithm **before** eliminating SNPs with missing genotypes.

| Min intent size | Min extent size | Total time, s | No. of concepts |
|---|---|---|---|
| 0 | 45 | 21.2 | 18,617 |
| 0 | 40 | 23.6 | 34,400 |
| 0 | 30 | 35.8 | 68,477 |
| 0 | 20 | 46.1 | 165,864 |
| 0 | 10 | 64.3 | 214,007 |
| 0 | 5 | 188.3 | 1,220,576 |
| 0 | 0 | 143.43 | 1,979,439 |

**Table 5.** The number of concepts and elapsed time generated by In-Close4 algorithm **after** eliminating SNPs with missing genotypes.

| Min intent size | Min extent size | Total time, s | Number of concepts |
|---|---|---|---|
| 0 | 40 | 10.4 | 2,743 |
| 0 | 30 | 10.6 | 4,196 |
| 0 | 20 | 12.6 | 19,620 |
| 30 | 0 | 5.8 | 352,257 |
| 20 | 0 | 7.4 | 695,962 |
| 10 | 0 | 18.3 | 3,226,277 |
| 5 | 5 | 22.7 | 7,833,203 |
| 0 | 0 | 22.3 | 15,410,516 |

Finally, we have managed to perform extra experiments on a machine with core i7-8750H and 16 Gb RAM under Win10[6]. The total number of output concepts for the context after elimination of missing SNP values is about 15M, i.e. more than 59 times larger than that of OA-biclusters.

---

[5] The last line in Table 4 and the last five lines in Table 5 corresponds to the experiments conducted for the final version of the paper on the transposed contexts.
[6] the last two lines in Table 5.

202    D. I. Ignatov et al.

# 6   Conclusion

A new approach to process the missing values in datasets of SNP genotypes obtained with DNA-microarrays is proposed. It is based on OA-biclustering. We applied the approach to the real-world datasets representing the genotypes of patients with ischemic stroke and healthy people. It allowed us to estimate and eliminate the SNPs carefully with missing genotypes. Results of the OA-biclustering algorithm showed the possibility of detecting relatively large dense biclusters, which significantly helped in removing the effects of data leaks and overfitting while applying ML algorithms.

We compared our algorithm with In-Close4. The number of OA-biclusters generated by our algorithm is significantly lower than the number of concepts (or biclusters) generated by In-Close4. Besides, our algorithm has the advantage of using OA-bicluster without the need to experiment with finding the best minimum support, as in the case of using In-Close4 for generating formal concepts.

Since survey [31] mentioned frequent itemset mining (FIM) as a tool to identify strong associations between allelic combinations associated with diseases, the proposed algorithm needs further comparison with other approaches from FIM like DeBi [32] and anytime discovery approaches like Alpine [33] tested on GEA datasets as well; though their use may get complicated if we need to keep information about object names for decision-makers. It also requires further time complexity improvements to increase the scalability and quality of the extensive bicluster finding process for massive datasets.

Another venue for related studies delve in Boolean biclustering [34] and factorisation techniques [35].

Speaking about other possible applications of biclustering, we suggest the development of a new imputation technique. Since biclustering has been recently applied to impute the missing values in gene expression data [36] and both GED and SNP genotyping data are obtained with DNA-microarrays and represented as an integer matrix, it can be potentially applied to impute the genotypes that facilitates statistical analyses and empowers ML algorithms.

**Acknowledgements.** This study was implemented in the Basic Research Program's framework at the HSE University and the Laboratory of Models and Methods of Computational Pragmatics in 2020. The authors thank prof. Alexei Fedorov (University of Toledo College of Medicine, Ohio, USA) and prof. Svetlana Limborska (Institute of Molecular Genetics of National Research Centre "Kurchatov Institute", Moscow, Russia) for insightful discussions of the results obtained, and anonymous reviewers.

**Funding.** The study was funded by RFBR (Russian Foundation for Basic Research) according to the research project No 19-29-01151. The foundation had no role in study design, data collection and analysis, writing the manuscript, and decision to publish.

# References

1. Bumgarner, R.: Overview of DNA microarrays: types, applications, and their future. Curr. Protoc. Mol. Biol. **101**(1), 1–11 (2013). Chapter 22
2. Dehghan, A.: Genome-wide association studies. In: Evangelou, E. (ed.) Genetic Epidemiology. MMB, vol. 1793, pp. 37–49. Springer, New York (2018). https:// doi.org/10.1007/978-1-4939-7868-7_4
3. Nicholls, H.L., John, C.R., Watson, D.S., Munroe, P.B., Barnes, M.R., Cabrera, C.P.: Reaching the end-game for GWAS: machine learning approaches for the prioritization of complex disease loci. Front. Genet. **11**, 350 (2020)
4. Tanay, A., Sharan, R., Shamir, R.: Discovering statistically significant biclusters in gene expression data. Bioinform. **18**(suppl_1), S136–S144 (2002)
5. Mirkin, B.: Mathematical Classification and Clustering. Kluwer, Dordrecht (1996)
6. Madeira, S.C., Oliveira, A.L.: Biclustering algorithms for biological data analysis: a survey. IEEE/ACM Trans. Comput. Biol. Bioinform. **1**(1), 24–45 (2004)
7. Cheng, Y., Church, G.M.: Biclustering of expression data. In: Bourne, P.E. (ed.) Proceedings of the Eighth International Conference on Intelligent Systems for Molecular Biology, AAAI 2000, pp. 93–103 (2000)
8. Tanay, A., Sharan, R., Shamir, R.: Biclustering algorithms: a survey. In: Handbook of Computational Molecular Biology, vol. 9, no. 1–20, pp. 122–124 (2005)
9. Busygin, S., Prokopyev, O., Pardalos, P.M.: Biclustering in data mining. Comput. Oper. Res. **35**(9), 2964–2987 (2008)
10. Ganter, B., Wille, R.: Formal Concept Analysis: Mathematical Foundations, 1st edn. Springer, New York (1999). https://doi.org/10.1007/978-3-642-59830-2
11. Besson, J., Robardet, C., Boulicaut, J., Rome, S.: Constraint-based concept mining and its application to microarray data analysis. Int. Data Anal. **9**(1), 59–82 (2005)
12. Blachon, S., Pensa, R.G., Besson, J., Robardet, C., Boulicaut, J., Gandrillon, O.: Clustering formal concepts to discover biologically relevant knowledge from gene expression data. Silico Biol. **7**(4–5), 467–483 (2007)
13. Kaytoue, M., Kuznetsov, S.O., Napoli, A., Duplessis, S.: Mining gene expression data with pattern structures in formal concept analysis. Inf. Sci. **181**(10), 1989–2001 (2011)
14. Andrews, S., McLeod, K.: Gene co-expression in mouse embryo tissues. Int. J. Intell. Inf. Technol. **9**(4), 55–68 (2013)
15. Ignatov, D.I., Kaminskaya, A.Y., Kuznetsov, S., Magizov, R.A.: Method of biclusterzation based on object and attribute closures. In: Proceedings of the 8th International Conference on Intellectualization of Information Processing (IIP 2011), Cyprus, Paphos, 17–24 October 2010, pp. 140–143. MAKS Press (2010). (in Russian)
16. Ignatov, D.I., Kuznetsov, S.O., Poelmans, J.: Concept-based biclustering for internet advertisement. In: 2012 IEEE 12th International Conference on Data Mining Workshops, pp. 123–130. IEEE (2012)
17. Andrews, S.: In-close2, a high performance formal concept miner. In: Andrews, S., Polovina, S., Hill, R., Akhgar, B. (eds.) ICCS 2011. LNCS (LNAI), vol. 6828, pp. 50–62. Springer, Heidelberg (2011). https://doi.org/10.1007/978-3-642-22688-5_4
18. Yevtushenko, S.A.: System of data analysis "concept explorer". In: Proceedings of the 7th National Conference on Artificial Intelligence (KII 2000), pp. 127–134 (2000)
19. Arnauld, A., Nicole, P.: La logique ou l'art de penser (Logique de Port Royal). Archives de la linguistique française. Ch. Savreuf, Guignart (1662)

20. Ignatov, D.: Models, algorithms, and software tools of biclustering based on closed sets. Ph.D. thesis, HSE University, Moscow (2010)
21. Prelic, A., et al.: A systematic comparison and evaluation of biclustering methods for gene expression data. Bioinform. **22**(9), 1122–1129 (2006)
22. Kuznetsov, S.O.: Mathematical aspects of concept analysis. J. Math. Sci. **80**(2), 1654–1698 (1996). https://doi.org/10.1007/BF02362847
23. Gnatyshak, D., Ignatov, D.I., Kuznetsov, S.O., Nourine, L.: A one-pass triclustering approach: is there any room for big data? In: Bertet, K., Rudolph, S. (eds.) Proceedings of the 11th International Conference on Concept Lattices and Their Applications (CLA 2014). Volume 1252 of CEUR Workshop Proceedings, pp. 231–242. CEUR-WS.org (2014)
24. Leskovec, J., Rajaraman, A., Ullman, J.D.: Finding similar items. In: Mining of Massive Datasets, 3nd edn, pp. 73–134. Cambridge University Press (2020)
25. Ignatov, D.I., Tochilkin, D., Egurnov, D.: Multimodal clustering of Boolean tensors on mapreduce: experiments revisited. In: D.C., et al. (eds.) Suppl. Proceedings of ICFCA 2019 Conference and Workshops. Volume 2378 of CEUR Workshop Proceedings, pp. 137–151. CEUR-WS.org (2019). http://ceur-ws.org/Vol-2378/longBDE4.pdf
26. Codd, E.F.: A relational model of data for large shared data banks. Commun. ACM **13**(6), 377–387 (1970)
27. Shetova, I.M., et al.: The association between the DNA marker rs1842993 and risk for cardioembolic stroke in the Slavic population. Zh. Nevrol. Psikhiatr. Im. S S Korsakova **112**(3 Pt 2), 38–41 (2012)
28. Hochreiter, S., Schmidhuber, J.: Long short-term memory. Neural Comput. **9**, 1735–1780 (1997)
29. Andrews, S.: Making use of empty intersections to improve the performance of CbO-type algorithms. In: Bertet, K., Borchmann, D., Cellier, P., Ferré, S. (eds.) ICFCA 2017. LNCS (LNAI), vol. 10308, pp. 56–71. Springer, Cham (2017). https://doi.org/10.1007/978-3-319-59271-8_4
30. Janostik, R., Konecny, J., Krajca, P.: LCM is well implemented CbO: study of LCM from FCA point of view. In: Valverde-Albacete, F.J., Trnecka, M. (eds.) Proceedings of the Fifteenth International Conference on Concept Lattices and Their Applications. Volume 2668 of CEUR Workshop Proceedings, pp. 47–58. CEUR-WS.org (2020)
31. Naulaerts, S., et al.: A primer to frequent itemset mining for bioinformatics. Briefings Bioinform. **16**(2), 216–231 (2015)
32. Serin, A., Vingron, M.: DeBi: discovering differentially expressed biclusters using a frequent itemset approach. Algorithms Mol. Biol. **6**(1), 18 (2011)
33. Hu, Q., Imielinski, T.: ALPINE: progressive itemset mining with definite guarantees. In: Chawla, N.V., Wang, W. (eds.) Proceedings of the 2017 SIAM International Conference on Data Mining, pp. 63–71. SIAM (2017)
34. Michalak, M., Slezak, D.: On Boolean representation of continuous data biclustering. Fundam. Informaticae **167**(3), 193–217 (2019)
35. Belohlávek, R., Outrata, J., Trnecka, M.: Factorizing Boolean matrices using formal concepts and iterative usage of essential entries. Inf. Sci. **489**, 37–49 (2019)
36. Chowdhury, H.A., Ahmed, H.A., Bhattacharyya, D.K., Kalita, J.K.: NCBI: a novel correlation based imputing technique using biclustering. In: Das, A.K., Nayak, J., Naik, B., Pati, S.K., Pelusi, D. (eds.) Computational Intelligence in Pattern Recognition. AISC, vol. 999, pp. 509–519. Springer, Singapore (2020). https://doi.org/10.1007/978-981-13-9042-5_43

# Bitcoin Abnormal Transaction Detection Based on Machine Learning

Elena V. Feldman[1]($\boxtimes$), Alexey N. Ruchay[1,2], Veronica K. Matveeva[1],
and Valeria D. Samsonova[1]

[1] Department of Mathematics, Chelyabinsk State University, Chelyabinsk, Russia
[2] Department of Information Security, South Ural State University
(National Research University), Chelyabinsk, Russia
`mila008.is@gmail.com, ran@csu.ru, veronicasomniator@gmail.com,`
`samsonova147@gmail.com`

**Abstract.** This paper is devoted to the development of a reliable abnormal bitcoin transaction detection that may be involved in money laundering and illegal traffic of goods and services. The article proposed an algorithm of abnormal bitcoin transaction detection based on machine learning. For training and evaluation of the model, the Elliptic dataset is used, consisting of 46564 Bitcoin transactions: 4545 of "illegal" and 42019 of "legal". The proposed algorithm for detecting abnormal bitcoin transactions is based on various machine learning algorithms with the selection of hyperparameters. To evaluate the proposed algorithm, we used the metric of accuracy, precision, recall, F1 score and index of balanced accuracy. Using the resampling algorithm in conditions of class imbalance, it was possible to increase the reliability of the classification of abnormal bitcoin transactions in comparison to the best known result on the Elliptic dataset.

**Keywords:** Bitcoin transactions · Classification · Detection of abnormal transactions · Machine learning

## 1 Introduction

The launch of first bitcoin in 2008 drew technological and business interest to payments through cryptographic methods (digital signature and hash function) and distributed transaction retention (blockchain). The main advantage of BTC is to provide anonymous and cheap money transactions both in the native country and abroad. However, criminals started actively using anonymity of BTC for illegal trade.

Anti-money laundering, AML plays the key role not only in providing safety for financial systems, but also determining illegal trade. BTC launch aroused a paradox: anonymity makes it possible for criminals to hide, though open BTC transaction database allows to perform forensic analysis or AML analytics. The objective of AML analytics is determining abnormal transactions which might

© Springer Nature Switzerland AG 2021
W. M. P. van der Aalst et al. (Eds.): AIST 2020, CCIS 1357, pp. 205–215, 2021.
https://doi.org/10.1007/978-3-030-71214-3_17

take place via reliable classification of minor illegal transactions in massive constantly increasing datasets. Manual or semi-automatic transaction analysis produces a high error rate. Meanwhile, success in machine learning shows great perspectives to use it for AML analysis [1].

The current paper is devoted to the development of a reliable abnormal bitcoin transaction detection that may be involved in money laundering and illegal traffic of goods and services. The Elliptic dataset is used [2], consisting of over 200 kt bitcoin transactions (nodes), 234 kt directed payments (edges) and 166 nodal functions, incl. the ones based on secretive data.

Methods of machine learning for binary classification which forecasts illegal transactions via logistic regression, random forest, multilayer perceptrons and convoluted network charts are used in the article [1] to detect abnormal bitcoin transactions. Convoluted network charts have appeared as a potential tool for AML analysis and it is especially attractive as a new method to gather and analyze transactions [1]. The results of work [1] reflect the advantage of random forest, though F1 score equal to 0.796 for the classifier set up on random forest is not sufficient to speak about reliable classification for bitcoin abnormal transactions.

To detect abnormal bitcoin transactions the following algorithms of machine learning were used in this article: linear regression, quadratic regression, logistic regression, k-nearest neighborhood, decision trees, random forest, naive Bayesian classifier, Support Vector Machines method, classifier based on multilayer perceptrons, linear discriminant analysis, quadratic discriminant analysis, adaptive boosting. Besides, methods of hyperparameter optimization for machine learning algorithms are used in this work which allowed to increase classification reliability.

It was important to observe that the Elliptic dataset is imbalanced (4545 "illegal" and 42019 "legal"). Therefore experiments on classification reliability for abnormal bitcoin transactions were performed with resampling algorithms under conditions of imbalanced classes. Due to it, it was possible to increase classification reliability for bitcoin abnormal transactions as compared to the best result in the Elliptic dataset [1].

## 2    Review of Approaches to Bitcoin Transaction Analysis

A set of tools for forensic analysis in bitcoin transactions is described in the article [3]. The set of tools comprises modular scalable framework including : Bitcore Node—"completely nodular" Bitcoin client which keeps information about transactions; Bitcoin Addresses Scraper, a tool for searching Bitcoin addreses on web pages to deanonymize their owners; Mixing Services Detector, a detector of mixing transactions services; BlockChainVis, a tool for transaction data visualization to simplify criminalists job, and also Bitcoin Addresses Clusteriser which is necessary to search group of clusters belonging to the same user. Bitcoin Addresses Clusteriser uses in its work a number of heuristics and their combinations. The dataset under research is not submitted.

The authors [4] describe algorithm of treatment major non-structured data within the task of detecting an illegal cycle of money. However, the dataset under research is not submitted.

The method of detecting transactions used in the services of mixing transactions blockchain (Blockchain Mixer) of a special type is described in the work [5]. It is performed via building assigned acyclic transaction graph, where nodes are transactions, and edges are bitcoin streams between transactions. To address this problem, a mathematical model has been developed, which showed in the course of research that it allows detecting the fact of participation of a user in arranging a transaction which gives extra advantages in detecting transactions connected with money laundering schemes. The dataset under research is not submitted.

The authors [6] proposed a mathematical model to analyze bitcoin transactions based on a hidden Markov model, a method of determining certainty values of transitions between bitcoin addresses. Markov model is a stochastic random process in the form of a direct graph where certainty values of state transitions depend only on the previous condition. The hidden Markov model belongs to the model where conditions are either hidden or not obvious. Regarding bitcoin transactions the observations give time-series of transactions between nodes. The research did not use real data, there was generated an imitative blockchain where the required transactions always enter the block regardless the amount. The dataset under research is not submitted.

The authors [7] analyze methods of money laundering detection based on Big Data treatment technology, instantiated, for example, in complex systems of fulfilling tasks at counteracting against money legalization and laundering which was obtained illegally (SAS Anti-Money Laundering System, SAS AML). The data are submitted as a graph. The necessary program modules are instantiated. The dataset under research is not submitted.

The results of using machine learning to solve a problem of deanonymizing bitcoin users are described in the article [8]. Data by Chainalysis company are used in the research. The dataset went through preliminary treatment and included 56 categories concerning illegal activities, and 957 various categories in total. The task comes down to searching the most effective algorithm of classification for data treatment. As a result, the method of gradient boosting is considered the best. The dataset under research is not submitted.

## 3   The Elliptic Dataset

There are several possible ways to represent bitcoin blockchain as a graph. The simplest of them is a graph where nodes are transactions, and edges are a stream of bitcoins between one and another transaction. In this representation bitcoin blockchain is an assigned acyclic graph. Excluding bordering cases when several outputs of transaction are spent as inputs of the same transaction, the degree of node input makes the number of transaction outputs, the spent number of outputs. The only nodes without inbound edges are coinbase transactions, i.e. newly made bitcoins included into the blockchain for the first time.

Each transaction is linked with a time feature, i.e. approximate time when transaction is streamed into bitcoin net. It allows to include temporal information into graph visualization. Payoff for obtaining a new block is currently BTC 12.5 (as of April 2020). Nowadays bitcoin transaction graph consists of over 438 million nodes and 1.1 billion edges. It is a constantly increasing graph as almost every day over 350,000 new confirmed BTC transactions take place.

The Elliptic dataset is a subgraph of total bitcoin transactions consisting of 203,769 nodes and 234,355 edges. Besides graph information, the Elliptic dataset includes information about node class: "legal", "illegal" or "unknown". A node is considered either legal or illegal if the current transaction was arranged by a legal party (broker's board, wallets suppliers, miners, financial services providers, and etc.) or illegal (fraud, malicious software, terrorist groups, ransomware, pyramid investment schemes, drug dealers, and etc.) class correspondingly.

The task is to classify "legal" and "illegal" transactions taking into account a set of objects and a graph topology. As not all the nodes are marked, it is possible to approach the problem in a semi-control mode which includes data transmitted via non-marked nodes.

Just 2% (4545) of nodes belong to "illegal" class, 21% (42019) belong to "legal" class, 77% (157205) other transactions are "unknown".

Timestamp in the graph is coded with 1 to 49 pitch which measures approximate time feature of the transaction. Timestamps are evenly distributed with about a 2-week interval; each of them contains one transaction organization unit which appears in the blockchain with the interval less than three hours. Edges connecting various time pitches also can be missing in the graph. Each associated graph component consists of 1,000–8,000 nodes. Most nodes are referred as "unknown".

Each node is associated with 166 features. The first 94 features comprise local data on transaction including: time pitch, input and output qty, transaction fee, issue volume and aggregates, such as average bitcoin number received and spent in inputs and outputs, average number of incoming and outcoming transactions associated with inputs and outputs. Other 72 features named aggregate features are received with transaction information aggregation at one pitch backward/forward from the central node, giving max, min, standard deviation and correlation coefficients of neighboring transactions for the same data (number of inputs and outputs, transaction fee, and etc.).

The article [1] is presented by IBM research workers Weber and Domeniconi at the seminar on detection anomaly in financial flows named Knowledge Discovery and Data Mining Conference (KDD) on August 5, 2019, the Head of the Conference on digital intelligence. The article sets out results of early experiments with various algorithms at machine learning: from conventional classification models (logistic regression, random forest, multilayer perceptrons) to more complex, such as graph convoluted networks (GCN). The above-mentioned various models are thoroughly compared in the article, and the effect of graph data inclusion into conventional classification methods is studied and presented as a table. This work brings several conclusions:

1. Non-local information implication, in particular, information about central node's neighbors always improves models' productivity;
2. Conventional (non-graph) models of classification win from extra functions provided by GCN attachments;
3. Random forest is the best classification model for this task.

The fact that GCN model did not prove as the best one is a curious discovery. At the Elliptic dataset GCN operates much better than logistic regression, but worse than random forest. As it was proposed in the article [1], insignificant improvement of results can be reached at combining abilities learning graph algorithms (such as GCN) with decision trees or random forest. One more problem of the Elliptic dataset is temporal dynamics of appearance and disappearance of new matters in a blockchain. Dark Market closure can serve as an example, when after this event no model operate normally due to sudden changes in basic system behavior.

## 4 Bitcoin Abnormal Transaction Detection Based on Machine Learning

Bitcoin abnormal transaction detection can be built with supervised machine learning [9–11]. Supervised machine learning uses marked training data $(x, y)$ to determine objective function $f$, where $x$ is a vector of input features, and $y$ is an output mark. Algorithm of learning generalizes correlation between the vector of features and the mark in training data for new examples to determine their marks correctly.

Assume that $\{(x_1, y_1), \ldots, (x_n, y_n)\}$ is a marked training dataset, where $x_i = (x_{i,1}, \ldots, x_{i,d}) \in X$ for $1 \leq i \leq n$ which is a vector of features for $x_i$ and $y_i \in Y$ is its relevant mark. The goal in supervised learning is to detect an objective function for a set of marks using a training data set. Assume that $(x_i, y_i)$ are independent and equally distributed. The supposed objective function can be estimated with its forecast ability on test data. If either $Y \in R$ or $Y \in R^d$ marks are continuous actual levels or a vector above real levels, then the machine learning model is regressive. If marks come as discrete states or symbols, then the machine learning model is named classification.

The task of supervised machine learning is to search objective function $f : X \to Y$, where $X$ is an input dataset, and $Y$ is a set of output variables. The process of searching this objective function $f$ is called supervised learning or modeling. The objective function can be detected only if enough marked data are available. As it's difficult to detect precise objective function in practically it is close to approximate function. Learning actually includes function h search which at its best approximates the unknown objective function $f$. Model training is performed with a test set which features a representative dataset randomly taken from the general dataset. As a rule, 70% of data are used for training. Then model parameters are given with a validation set; usually 20% of training data are taken for the validation set. At last, actual forecast capacity of the model is checked with a test set, as a rule 30% of data are taken for a test set.

There is a dataset $M$ in bitcoin abnormal transaction detection which consists of $n = 46564$ transactions, and a mark is a binary variable ("illegal" or "legal"). We are not considering "unknown" class of transactions. We can apply a semi-supervised machine learning approach to bitcoin abnormal transaction detection using unknown transactions. We leave the execution of this idea as future investigation. Assume that mark 0 points at "legal" transaction, and mark 1 points at "illegal" transaction. A number of vector features consisting of 94 features of local information and 72 aggregate features are associated with each transaction $m$. The objective function $f : M \in L$ is to determine if a definite node $m$ is "illegal" 1 or "legal" 0. Function $f : M \in 0, 1$ can be detected with one of machine learning algorithms for a set of $n$ marked nodes $\{(m_1; l_1), (m_2, l_2), (m_n, l_n)\}$, where $m_i \in M$ and $l_i \in 0, 1$ for $1 \le i \le n$.

The following machine learning algorithms are used in the article to detect bitcoin abnormal transactions: linear regression; quadratic regression; logistic regression; $k$-nearest neighbourhood, decision trees, random forest, naive Bayesian classifier, Support Vector Machines method, classifier based on multi-layer perceptrons, linear discriminant analysis, quadratic discriminant analysis, adaptive boosting, gradient boosting.

Consider general evaluation metrics used in the machine learning models in case the classification is binary as positive ("illegal") and negative ("legal") class.

One important observation took place concerning the data in the Elliptic dataset, which proves imbalanced (4545 "illegal" and 42019 "legal"). Assume this class minority, when example share of a certain class in the dataset is too small, and the other class is majority which is broadly represented in the dataset.

Among the approaches to solve the imbalance problem is to apply various resampling strategies. There exist two ways to restore class balance. In the first case some examples of major class are removed (undersampling), in the other the examples of minor class are increased with synthetic data (oversampling).

TomekLinks [12] algorithm was selected to solve the imbalance problem, where all major records incoming in TomekLinks must be removed from the dataset. TomekLinks can be determined as follows. Assume that $E_i = (x_i, y_i)$ and $E_j = (x_j, y_j)$ are two examples from different classes, where $y_i \ne y_j$, and $d(E_i, E_j)$ is the distance between $E_i$ and $E_j$, then a pair $(E_i, E_j)$ is defined as TomekLinks if there is no such example $E_l$, as either $d(E_i, E_l) < d(E_i, E_j)$ or $d(E_j, E_l) < d(E_i, E_j)$. TomekLinks algorithm succeeds in noisy record removing.

Geometric mean (Gmean) is a parameter $\sqrt{TP + TN}$, which is used to maximize correct and positive, and correct and negative classification results at keeping the balance between them. Where TP is a number of correct and positive results when the example of a positive class is correctly forecast with the model, and belonging to a positive class; and TN is a number of correct and negative results when the example of a negative class is correctly forecast with the model as belonging to a negative class. It's worth mentioning that geometric mean leads to minimizing a negative impact of class distributions errors, though it is unable to explain the contribution of each class to overall index, giving the same result for various TP and TN combinations.

Dominance can be defined as $TP - TN$, which is used to estimate relationship between TP and TN.

The article [13] proposes Index of Balanced Accuracy (IBA) to estimate binary classifier in case with data imbalanced which can be obtained as

$$IBA = (1 + Dominance) \cdot Gmean^2.$$

At replacement of Dominance and Gmean the final equation gives useful information for better understanding how IBA supports a compromise between Dominance and Gmean, besides a weighted parameter $0 \leq \alpha \leq 1$ can be added for Dominance

$$IBA = (1 + \alpha(TP - TN)) \cdot TP \cdot TN.$$

However, if $\alpha = 0$, then IBA turns exactly into $Gmean^2$.

## 5   Experiments

To implement the algorithm of bitcoin transaction classification with machine learning Python 3 programming language was used as it is the best suitable one for fulfilling this task, and it has a perfect productivity at data treatment. Besides, Python contains many frameworks and libraries which simplify the process of writing a code and reduce the time for implementation.

Repository Pandas with high-level data structure was selected for data handling. It possesses in-built methods to group, combine and filter data. Pandas makes it possible to take data from different sources, such as SQL dataset, CSV, Excel, JSON files, and manipulate with the data to operate them.

Algorithms of machine learning from repository SciKit-Learn (SKlearn) [14] are used for abnormal transactions classification. The following algorithm types of machine learning were used to classify abnormal transactions: LogisticRegression, RandomForestClassifier, DecisionTreeClassifier, SVC, KNeighborsClassifier, MLPClassifier, AdaBoostClassifier.

Some part of machine learning algorithms gave poor results, therefore further on results of the following machine learning algorithms are not presented: linear regression, quadratic regression, linear discriminant analysis, quadratic discriminant analysis, naive Bayesian classifier.

We also used the following algorithms of machine learning for classification: CatBoostClassifier from the open repository implementing unique patented algorithm in constructing models of machine learning based on original schemes of gradient boosting, CatBoost [15]; XGBClassifier from repository of scaled gradient boosting XGBoost [16]; LGBMClassifier from the repository of gradient boosting LightGBM [17].

As test and training samples there were selected transactions marked as either "legal" or "illegal" in ratio 3:7 correspondingly. The bitcoin abnormal transactions detection was supervised with the help of selected algorithms, and to estimate the reliability of supervised models the metrics are calculated: accuracy, precision, recall, F1 Score and IBA. Table 1 provides the results of classification reliability based on a test set, where metrics precision, recall, F1 Score are given only for "illegal" transactions, as to estimate effectiveness of transaction detection is in most cases important to analyze errors for "illegal" transactions.

From results represented in Table 1 it is obvious that the best classification algorithm is CatBoostClassifier with F1 score equal to 0.791 and accuracy 0.9780. However, the obtained results are slightly lower than the best result from article [1], where RandomForest algorithm has F1 score equal to 0.796 and accuracy 0.9780. Whereas, extra features obtained via GCN were used for a supervised model based on RandomForest in the article [1].

F1 score is equal to 0.796 and accuracy is 0.9780 which is not an acceptable result in effective detection of abnormal transactions. Our experiments with models involves testing different combinations of related hyperparameters to find the optimal response within a given set of values. To automate the process of obtaining the best combination of hyperparameters, the GridSearchCV algorithm from sklearn library is used. For the best CatBoostClassifier model, under different values of depth, iterations, l2_leaf_reg, learning_rate, border_count are analyzed. In the end, the hyperparameters of depth = 9, iterations = 1000, l2_leaf_reg = 4, learning_rate = 0.1, border_count = 10 yield the optimal response for the best CatBoostClassifier model.

**Table 1.** Results of algorithm classification reliability based on a test set with accuracy, precision, recall, F1 score and IBA estimation. Metrics of precision, recall, F1 score are presented only for "illegal" transaction class.

| Algorithm | "Illegal" transactions | | | Accuracy | IBA |
|---|---|---|---|---|---|
| | Precision | Recall | F1 Score | | |
| RandomForestClassifier | 0.985 | 0.653 | 0.785 | 0.9775 | 0.6688 |
| AdaBoostClassifier | 0.873 | 0.594 | 0.707 | 0.9690 | 0.6111 |
| SVC | 0.862 | 0.588 | 0.699 | 0.9681 | 0.6052 |
| MLPClassifier | 0.814 | 0.612 | 0.699 | 0.9668 | 0.6041 |
| KNeighborsClassifier | 0.634 | 0.603 | 0.618 | 0.9531 | 0.6081 |
| DecisionTreeClassifier | 0.507 | 0.681 | 0.582 | 0.9383 | 0.6668 |
| LogisticRegression | 0.454 | 0.633 | 0.529 | 0.9290 | 0.6168 |
| LGBMClassifier | 0.928 | 0.589 | 0.721 | 0.9713 | 0.6056 |
| XGBClassifier | 0.984 | 0.636 | 0.773 | 0.9764 | 0.6601 |
| CatBoostClassifier | 0.985 | 0.661 | 0.791 | 0.9780 | 0.6801 |
| RandomForest [1] | 0.971 | 0.675 | 0.796 | 0.9780 | – |

To solve the imbalance problem the following resampling algorithms were used from the imbalanced-learn library [18]: TomekLinks, ClusterCentroids, RandomUnderSampler, SMOTE, RandomOverSampler, SMOTETomek. TomekLinks was selected as the best one.

Table 2 holds results of algorithm classification based on a test set, where precision, recall, F1 Score are given only for "illegal" transactions. Preliminary

TomekLinks algorithm was implemented to classify a dataset, and it removed 226 "legal" transactions. Then the dataset was randomly split into a test and a training part in ratio 3:7 correspondingly.

The results presented in Table 2 show that the best classification algorithm is XGBClassifier with F1 Score equal to 0.957, IBA 0.9599 and accuracy 0.9921. Obtained metrics significantly outperform random forest algorithm with F1 score equal to 0.796 and accuracy equal to 0.9780 given in the article [1].

F1 Score 0.957 and accuracy 0.9921 on a test set are acceptable for effective detection of abnormal transactions, so the conclusion comes that with Tomek-Links resampling algorithm in imbalanced-learn classes it was possible to increase reliability of bitcoin abnormal transaction classification.

**Table 2.** Results of algorithm classification reliability based on a test set at estimating accuracy, precision, recall, F1 Score and IBA. Precision, recall, F1 Score metrics are given only for "illegal" transactions.

| Algorithm | "Illegal" transactions | | | Accuracy | IBA |
|---|---|---|---|---|---|
| | Precision | Recall | F1 Score | | |
| RandomForestClassifier | 0.997 | 0.882 | 0.936 | 0.9885 | 0.9395 |
| AdaBoostClassifier | 0.916 | 0.858 | 0.886 | 0.9790 | 0.9225 |
| SVC | 0.908 | 0.777 | 0.837 | 0.9712 | 0.8778 |
| MLPClassifier | 0.901 | 0.768 | 0.825 | 0.9646 | 0.8712 |
| KNeighborsClassifier | 0.889 | 0.840 | 0.864 | 0.9747 | 0.9115 |
| DecisionTreeClassifier | 0.881 | 0.899 | 0.890 | 0.9787 | 0.9419 |
| LogisticRegression | 0.823 | 0.767 | 0.794 | 0.9290 | 0.8679 |
| LGBMClassifier | 0.994 | 0.918 | 0.954 | 0.9916 | 0.9579 |
| XGBClassifier | 0.995 | 0.922 | 0.957 | 0.9921 | 0.9599 |
| CatBoostClassifier | 0.992 | 0.920 | 0.954 | 0.9916 | 0.9586 |
| RandomForest [1] | 0.971 | 0.675 | 0.796 | 0.9780 | – |

Our experiments with models involves testing different combinations of related hyperparameters to find the optimal response within a given set of values. To automate the process of obtaining the best combination of hyperparameters, the GridSearchCV algorithm from sklearn library is used. For best XGBClassifier model, under different values of max_depth, min_child_weight, n_estimators, learning_rate are analyzed. In the end, the hyperparameters of max_depth=10, min_child_weight = 1, n_estimators = 200, learning_rate = 0.05 yield the optimal response for XGBClassifier model.

The accuracy and IBA scores in our experiments are also validated through 10-fold cross-validation. The results of 10-fold cross-validation for the best performing methods (RandomForestClassifier, LGBMClassifier, XGBClassifier, CatBoostClassifier) for various evaluation measures are shown in Table 3. For all 10 iterations, the values of evaluation measures remain almost the same, indicating the stability of the XGBClassifier method for bitcoin abnormal transaction detection. Thus, we can say that the XGBClassifier method performed better than all other models used in this study for detecting abnormal bitcoin transactions.

**Table 3.** The results of 10-fold cross-validation for the best performing methods (RandomForestClassifier, LGBMClassifier, XGBClassifier, CatBoostClassifier) for accuracy and IBA evaluation measures. Mean: mean value of various evaluation measures. SD ($\times 10^{-4}$): standard deviation of various evaluation measures.

| Algorithm | Accuracy | | IBA | |
|---|---|---|---|---|
| | Mean | SD | Mean | SD |
| RandomForestClassifier | 0.9874 | 6.7864 | 0.8835 | 6.1886 |
| LGBMClassifier | 0.9911 | 7.5099 | 0.9199 | 6.6320 |
| XGBClassifier | 0.9917 | 7.1377 | 0.9203 | 6.6133 |
| CatBoostClassifier | 0.9910 | 7.7603 | 0.9198 | 6.6755 |

## 6   Conclusion

The algorithm of bitcoin abnormal transaction detection based on machine learning was proposed in this paper. Bitcoin abnormal transactions are determined as transactions which can participate in money laundering and illegal traffic of goods and services. For training and evaluation of the proposed algorithm the Elliptic dataset is used comprising over 200,000 bitcoin transactions: 4545 are "illegal", 42019 "legal" and 157205 "unknown". The proposed model for bitcoin abnormal transaction detection used various algorithms of machine learning with the selection of hyperparameters, however the productivity of the proposed model equal to 0.9780 is not acceptable for effective detection of abnormal transactions. With TomekLinks resampling algorithm in imbalanced-learn conditions we managed to increase reliability of bitcoin abnormal transaction classification as compared to the best result equal to 0.9780 from the article [1]. Subsequently, the reliability of the proposed model for bitcoin abnormal transaction detection based on XGBClassifier algorithm equals to 0.9921.

## References

1. Weber, M., et al.: Anti-money laundering in bitcoin: experimenting with graph convolutional networks for financial forensics (2019)
2. Elliptic data set. https://www.kaggle.com/ellipticco/elliptic-data-set
3. Bistarelli, S., Mercanti, I., Santini, F.: A suite of tools for the forensic analysis of bitcoin transactions: preliminary report. In: Mencagli, G., et al. (eds.) Euro-Par 2018. LNCS, vol. 11339, pp. 329–341. Springer, Cham (2019). https://doi.org/10.1007/978-3-030-10549-5_26

4. Kedharewsari, K., Maria Anu, M., Rajalakshmi, V.: Integration of big data & cloud computing to detect black money rotation with range - aggregate queries. Int. J. Eng. Technol. **8**, 768–773 (2016)
5. Maksutov, A.A., Alexeev, M.S., Fedorova, N.O., Andreev, D.A.: Detection of blockchain transactions used in blockchain mixer of coin join type. In: 2019 IEEE Conference of Russian Young Researchers in Electrical and Electronic Engineering (EIConRus), pp. 274–277 (2019)
6. Oakley, J., Worley, C., Yu, L., Brooks, R., Skjellum, A.: Unmasking criminal enterprises: an analysis of bitcoin transactions. In: 2018 13th International Conference on Malicious and Unwanted Software (MALWARE), pp. 161–166 (2018)
7. Plaksiy, K., Nikiforov, A., Miloslavskaya, N.: Applying big data technologies to detect cases of money laundering and counter financing of terrorism. In: 2018 6th International Conference on Future Internet of Things and Cloud Workshops (FiCloudW), pp. 70–77 (2018)
8. Yin, H., Langenheldt, K., Harlev, M., Mukkamala, R.R., Vatrapu, R.: Regulating cryptocurrencies: a supervised machine learning approach to de-anonymizing the bitcoin blockchain. J. Manage. Inf. Syst. **36**, 37–73 (2019)
9. Thomas, T., P. Vijayaraghavan, A., Emmanuel, S.: Machine Learning Approaches in Cyber Security Analytics. Springer, Singapore (2020). https://doi.org/10.1007/978-981-15-1706-8
10. Bishop, C.M.: Pattern Recognition and Machine Learning. Springer, Berlin (2006)
11. MacKay, D.J.C.: Information Theory. Inference & Learning Algorithms. Cambridge University Press, USA (2003)
12. Tomek, I.: Two modifications of CNN. IEEE Trans. Syst. Man Cybern. **6**, 769–772 (1976)
13. García, V., Mollineda, R.A., Sánchez, J.S.: Index of balanced accuracy: a performance measure for skewed class distributions. In: Araujo, H., Mendonça, A.M., Pinho, A.J., Torres, M.I. (eds.) IbPRIA 2009. LNCS, vol. 5524, pp. 441–448. Springer, Heidelberg (2009). https://doi.org/10.1007/978-3-642-02172-5_57
14. Pedregosa, F., et al.: SciKit-learn: machine learning in Python. J. Mach. Learn. Res. **12**, 2825–2830 (2011)
15. Dorogush, A.V., Ershov, V., Gulin, A.: CatBoost: gradient boosting with categorical features support. CoRR abs/1810.11363 (2018)
16. Chen, T., Guestrin, C.: XGBoost: a scalable tree boosting system. In: Proceedings of the 22nd ACM SIGKDD International Conference on Knowledge Discovery and Data Mining. KDD 2016, New York, NY, USA, Association for Computing Machinery, pp. 785–794 (2016)
17. Ke, G., et al.: LightGBm: a highly efficient gradient boosting decision tree. In: Guyon, I., et al. (eds.) Advances in Neural Information Processing Systems 30, pp. 3146–3154. Curran Associates, Inc. (2017)
18. Lemaître, G., Nogueira, F., Aridas, C.K.: Imbalanced-learn: a python toolbox to tackle the curse of imbalanced datasets in machine learning. J. Mach. Learn. Res. **18**, 559–563 (2016)

# Commutes and Contagions: Simulating Disease Propagation on Urban Transportation Networks

Ho Lum Cheung$^{(\boxtimes)}$ and Dimas Muñoz-Montesinos

HSE University, Moscow 101000, Russia
cheung.ho.lum@gmail.com, dmunosmontesinos@edu.hse.ru

**Abstract.** Public transportation plays a vital role in bringing people together. International trade and tourism is reliant on commercial aviation, and the most successful cities have buses, trams, or subways connecting workers to their workplaces. However, the efficiency and connectivity of modern transportation networks leaves us vulnerable to the spread of diseases.

In this paper, we introduce a general agent-based framework for modeling disease propagation on networks and use it to build models of major transportation networks. Using our models, we take a look at how disease spreads throughout a city. We pay particular attention to predicting the regions of a city which are likely to become disease hot-spots. This is of particular interest to city planners who want to allocate resources appropriately to combat an outbreak.

Using information about subway routes, ridership, and city-wide countermeasures, we were able to predict areas with high caseloads during the onset of the 2020 COVID-19 pandemic in New York City. Our findings suggest a correlation between these factors and the spread of COVID-19, Influenza-Like Illnesses, and other contagious diseases.

**Keywords:** Communicable · Disease · Transportation · Network

## 1 Introduction

When studying and discussing epidemics, news is often focused on national and international levels. However, there is significant empirical evidence that different regions and cities react differently to epidemics. And within individual cities, the spread of a disease such as the recent COVID-19 is uneven. Certain neighborhoods are much more likely to be affected than others.

While factors such as income and residential density are often studied for their correlation with this spread, in this paper we look at mobility data with a focus on subway systems. First, we built an agent-based framework to simulate infections in locations tied together by a transportation network. Localities (neighborhoods, postal codes, boroughs, etc.) are represented by an individual agent following a SEIR model, but modified to include an outside infection

© Springer Nature Switzerland AG 2021
W. M. P. van der Aalst et al. (Eds.): AIST 2020, CCIS 1357, pp. 216–228, 2021.
https://doi.org/10.1007/978-3-030-71214-3_18

chance. We base this outside chance primarily on the commute time from the neighborhood and the primary subway routes serving it. We then use this agent-based framework to build models for COVID-19 in cities such as New York and London.

To build the transportation networks for our model, we pulled from publicly available data stores from various city government websites. We were also able to find satisfactorily granular geographical and demographic data from publicly available national census data.

We found that with minimal tuning based on the overall case rate, we can forecast the case rate of COVID-19 across all localities with a median MAPE less than 0.25. With better tuning, or location-specific hyper-parameters, we can further reduce this general result.

We also uncovered many interesting paths for future study such as creating a fuller mobility picture for individual cities, and leave some thoughts in the discussion. But the main conclusion is that commute times and subway routes are correlated with the propagation of disease within a city, and should be taken into consideration when planning for future outbreaks.

The paper is organized as follows. Section 2 contains a background on mathematical modeling of epidemics including SEIR model. Section 3 describes the used data sources. The proposed methodology is discribed in Sect. 4. The model fitting and obtained results are summarized in Sect. 5, while the discussion and future prospects are given in Sect. 6. Finally, Sect. 7 concludes the paper.

## 2 Background and Prior Research

### 2.1 Epidemics and COVID-19

Coronavirus disease 2019 (COVID-19) is a disease caused by the SARS-CoV-2 coronavirus. Since being identified in December 2019, it has been labelled by the WHO as a pandemic [9], and spread around the world. Epidemics such as the coronavirus have been a subject of research for centuries, and is of special interest to those working in public health. Recent waves of new research came in 2002 (SARS), 2009 (H1N1), and 2014 (Ebola). However, in these prior epidemics, researchers did not have access to the data and tools we have now.

### 2.2 Mathematical Modeling of Epidemics

**SEIR Model.** The SEIR compartmental model [17] is a mathematical modeling of infectious diseases where a closed population of people move successively from compartment to compartment (from susceptible to exposed, to infected, and to removed). While statistical modeling has typically given more accurate forecasts for a well-known situation, mechanistic models such as SEIR do a better job with exploring general phenomena and explaining the impact of policy decisions. Below, we provide an explanation of each SEIR compartment and give the corresponding system of equations.

- **Susceptible** ($S$) - These people are susceptible to infection from infected people.
- **Exposed** ($E$) - These people are no longer susceptible to the disease, and do not infect others. After a latent period, they become infectious.
- **Infected** ($I$) - These people will spread the disease to susceptible people. After a period of time they are removed by isolation, recovery, hospitalization, or death.
- **Removed** ($R$) - Sometimes known as resistant or recovered. We will do our modeling with the term 'removed'. These people no longer spread the disease.
- **Contact Rate** ($\beta$) - Rate at which infected people infect susceptible people.
- **Latent Rate** ($\alpha$) - Rate at which exposed people become infected.
- **Removal Rate** ($\gamma$) - Rate at which infected people become removed.

$$S(t) + E(t) + I(t) + R(t) = N$$

$$s(t) = \frac{S(t)}{N}, e(t) = \frac{E(t)}{N}, i(t) = \frac{I(t)}{N}, r(t) = \frac{R(t)}{N}$$

$$\frac{ds}{dt} = -\beta \cdot s(t) \cdot i(t)$$

$$\frac{de}{dt} = \beta \cdot s(t) \cdot i(t) - \alpha \cdot e(t)$$

$$\frac{di}{dt} = \alpha \cdot e(t) - \gamma \cdot i(t)$$

$$\frac{dr}{dt} = \gamma \cdot i(t)$$

The system of equations described above can be numerically solved given $\beta, \alpha, \gamma$ and initial values $S(0)$, $E(0)$, $I(0)$, and $R(0)$. And if we have values for $S, E, I$, and $R$ at certain times, we can fit $\beta, \alpha, \gamma$ to better define the disease's epidemiological characteristics and predict its future course. Lastly, we note that $R_0$, an important characteristic known as the basic reproductive rate, can be calculated for the SEIR model as $R_0 = \beta/\gamma$.

**Other Compartmental Models.** While we have done research into simpler and more advanced models and are interested in cases such as super-spreaders, we believe the SEIR model to be sufficient for our needs. Basic SIR is insufficient because public health officials often make policy decisions based on positive case numbers. For example, an official may decide to impose strict isolation only after 100 positive cases. But by the time there are 100 cases of 'infected' people, there may be 1000 exposed people who will meaningfully impact epidemic statistics.

### 2.3   Epidemics on Transportation Networks

Epidemics on transportation networks have been modeled in many different ways depending on the needs of the researcher. The most important differences are usually the type of transportation network, the duration of interest, the granularity of population data, and passenger flow characteristics. A popular sub-area of research is exploring countermeasure policies. We refer the reader to the following papers [3,10,11], and discuss two of the most relevant ones below.

**Flu on the London Underground.** In [14], the researchers modeled the "contact rate" of riders of the London Underground (subway system) by breaking down the stages of subway travel (entering, waiting, riding, exiting), and concluded that riders of some boroughs were at higher risk than riders from other boroughs. Their analysis was consistent with PHE data for influenza-like illnesses (ILI). The researchers also made additional observations such as that rush hour contributes to infection due to higher passenger density and longer waiting times.

**A Simulation of New York City.** In [12], the researchers make a very thorough model of New York City. While the title suggests they focused on subways, they modeled hospitals, schools, and major hubs with the subways delivering people to their places of work and education. They fit their work to historical flu data as well as infection numbers from prior work and also investigated the effect of countermeasures (interventions). They estimated that the subway was directly responsible for 12.5% of all infections in NYC.

### 2.4    Agent-Based Models

An agent-based model (ABM) is a computational model used to simulate the overall effects of individual agents on a system. Some famous prior uses include Conway's Game of Life and Schelling's Segregation Model [22]. ABMs consist of one or several types of agents interacting with an environment. For example, in Schelling's Segregation Model, the agents have a race and a tolerance level (of other races). If the agent finds the surrounding environment intolerable, they will independently move away. Given certain hyper-parameters, segregated communities eventually form.

ABMs offer a number of benefits over traditional mathematical models. Complex systems which cannot be easily solved mathematically can be simulated. These simulations help policy makers with decisions when mathematical results are not available and real world experiments are impractical [19].

### 2.5    Subway Nomenclature

Below we define some common terms used in subway systems.

- Station - Passengers enter subway stations in order to ride the subway to an exit station.
- Turnstiles - Barriers at the entrance and exit of stations which count people entering and exiting.
- Line - The train tracks on which services and routes run.
- Service/Route - Trains follow specific routes between stations based on a timetable.

# 3 Data Sources

## 3.1 London

**Demographics.** The finest granularity for which we found London COVID-19 data was by borough. Since March 1, the city government of London has been combining data from Public Health England and the Office of National Statistics (United Kingdom) to publish case, death, and recovery data for COVID-19 by boroug [1].

For visualization, we used a shapefile for London boroughs also available from the city government of London [5].

**Subway System.** For the subway system, we used a parsed list of London Underground stations from Wikimedia [6], and verified the results against an official map [18].

## 3.2 New York City

**Demographics.** The finest granularity for which we found New York City COVID-19 data was by MODZCTA (Modified Zip Code Tabulation Area). Since March 26, the NYC Health Department has been releasing and updating this data on Github [21]. Some data is incomplete or unavailable due to technical or privacy issues. Detailed case, death, and recovery numbers by MODZCTA only became available on May 18, 2020. However, a rudimentary record of positive tests for COVID-19 by MODZCTA has been available since April 1, 2020. The dataset also contains the estimated population of each MODZCTA.

For visualization, we used a shapefile for New York City from NYC Open Data [16].

**Subway System.** The New York City MTA (Metropolitan Transportation Authority), in charge of the New York City subway system, has a dataset containing basic subway station information [8]. It has stations, longitude, latitude, line, route, borough, and other information. While it does not have MODZCTA, we found it by using longitude and latitude.

In this dataset, station 167 is listed twice, so we took care to combine the route data. We also split service 'S' into 3 different services as it represents 3 different shuttle services. We also split services on the 'A' line into 3 different services depending on the destination, and did the same for the branching '5' line.

# 4 Methodology

## 4.1 MESA

To implement our ideas, we chose to use MESA [7], an ABM framework written in Python. MESA provides a simple framework with basic Agent, Model, and Schedule classes from which we can build more complex behavior.

## 4.2    NetworkX

NetworkX is a standard Python library used to model and visualize networks [15]. Its documentation can be found online.

## 4.3    Modeling Framework

First, we built a general framework for modeling various transportation problems. This allowed us to see how subways (and subway commuters) differ from other types of transportation networks. It also allows us to extend our initial research into other networks and environments. In Fig. 1, the UML diagram of our framework is provided.

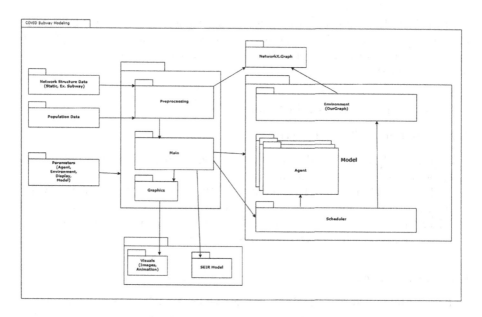

**Fig. 1.** The UML diagram of our framework

## 4.4    Special Considerations for Subways

We next create the main components of the ABM using the following ideas:

– Infection on a subway network spreads along routes.
– Infection on a subway network depends on the average commute time of commuters using the station.

## 4.5  SubwayModel

This class inherits from the base TransportationModel. The main additional functionality is based on the model using SubwayAgents to model agents and a SubwayGraph to model the environment. It also has functionality related to deploying countermeasures given a certain threshold of infection.

**SubwayAgent.** This class inherits from the base SEIRAgent and represents the population living around a specific subway station. The base agent simulates a disease spreading through a closed population. The main additional functionality is adjustment of infection (and removal) rates based on outside exposure to the disease.

**SubwayGraph.** This class represents the environment of our agents. Each node represents a subway station. Each edge represents a connecting subway line or passage between stations. It serves as a wrapper around a NetworkX Graph with additional functionality.

## 4.6  Hyper-parameters and Values

### Basic Epidemic Characteristics

- DEFAULT_BETA - The default $\beta$ of the modeled disease in the SEIR model. It can be affected by countermeasures.
- DEFAULT_GAMMA - The default $\gamma$ of the modeled disease in the SEIR model. It can be affected by countermeasures.
- DEFAULT_ALPHA - The default $\alpha$ of the modeled disease in the SEIR model. It can be affected by countermeasures.

### Countermeasures

- ISOLATION_COUNTERMEASURE_START - Models civic and government countermeasures to combat a disease. The countermeasures start after this number of people are infected.
- INITIAL_REDUCTION_TIME - Models the amount of time it takes to implement an initial set of countermeasures.
- INITIAL_REDUCTION_TGT - Models the effectiveness of the initial countermeasures. The $R_0$ of the virus will decrease to this number linearly over the INITIAL_REDUCTION_TIME.
- FULL_REDUCTION_TIME - Models the amount of time it takes to implement a full set of countermeasures.
- FULL_REDUCTION_TGT - Models the effectiveness of full countermeasures. The $R_0$ of the virus will decrease to this number linearly over the FULL_REDUCTION_TIME.
- STARTING_PERCENTAGE - Models the initial percentage of the population infected by the virus. Initial infected are evenly distributed.

- STARTING_RATIO - Models the initial percentage of the population exposed to the virus. Initial exposed are evenly distributed.
- CONTACT_MODIFIER - A parameter used to linearly modify the exposure from contacts by subway commuters. Note that comparing this modifier between different cities is meaningless unless all other exposure factors are normalized.

### 4.7 Algorithmic Description

Below is an algorithmic description of our agent-based model.

---

**Algorithm 1.** Simulation of Disease Spread on Subways

---
1: **for** $i = 1; i < TIMESPAN; i + +$ **do**
2:     Check conditions ($i$, number of infected) to see if we should deploy COUNTER-MEASURES
3:     **for** Station in SubwayModel.Environment.Nodes **do**
4:         Add 'Exposure' from infected at this station to routes stopping at this station.
5:     **end for**
6:     **for** Agent in SubwayNetwork.Agents **do**
7:         Get 'Exposure' from routes at this station.
8:         Get City-wide COUNTERMEASURES
9:         Get Percentage of commuters
10:        Update actual $\beta$ and $\gamma$ based on conditions
11:        Update (self) SEIR numbers
12:    **end for**
13: **end for**

---

Alternatively, the model can be described as changing SEIR to work as follows:

$$\Delta E = \frac{\beta SI}{N} + S\rho\mu, \tag{1}$$

where $\rho$ is the contact 'density' of the specific population and is directly linearly correlated to the factors listed in the algorithm (exposure, percentage of commuters), and $\mu$ is a global normalization hyperparameter. Note, of course, that $\Delta S$ is respectively altered.

## 5 Fitting and Results

Overall, in Table 1, we chose the following values for the hyper-parameters.

**Table 1.** The choice of hyper-parameters

|  | London | NYC |
|---|---|---|
| DEFAULT_BETA | 1.00 | 1.00 |
| DEFAULT_ALPHA | 0.24 | 0.24 |
| DEFAULT_GAMMA | 0.16 | 0.16 |
| ISOLATION_COUNTERMEASURE_START | 200 | 500 |
| INITIAL_REDUCTION_TIME | 7 | 7 |
| INITIAL_REDUCTION_TGT | 1.00 | 1.50 |
| FULL_REDUCTION_TIME | 60 | 60 |
| FULL_REDUCTION_TGT | 0.60 | 0.60 |
| STARTING_PERCENTAGE | 1e−6 | 1e−6 |
| STARTING_RATIO | 2.5 | 2.5 |
| CONTACT_MODIFIER | 0.10 | 0.22 |
| COMMUTER_RATIO | 0.50 | 0.50 |

### 5.1 Choosing Epidemiological Characteristics

Default values for $\alpha$, $\beta$, and $\gamma$ were chosen to approximate the epidemiological characteristics of COVID-19. Other sources [2,13,20,23] have investigated the best numbers more thoroughly, but for our modeling it is only necessary that the values are reasonable.

### 5.2 Fitting to SEIR

Next, we fit overall SEIR numbers to case, death, and recovery numbers. We approximated the start of the epidemic in both cities to March 1 and picked STARTING_PERCENTAGE and ISOLATION_COUNTERMEASURE_START to match this. We chose INITIAL_REDUCTION_TIME and FULL_RED UCTION_TIME also based on these numbers. We chose a STARTING_RATIO of 2.5 based on default $R_0$ numbers. We chose a COMMUTER_RATIO of 0.50 based on ridership numbers compared to population [4].

We next sought to fit the total cases to $I(t) + R(t)$ and adjusted the remaining hyper-parameters (CONTACT_MODIFIER, INITIAL_REDUCTION_TGT, FULL_REDUCTION_TGT) to do so. The results are provided in Fig. 2.

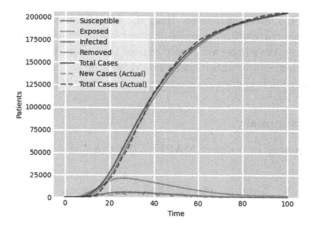

**Fig. 2.** Fitting the model to NYC Case Total: MAPE $(t \geq 30) = 0.0145$

## 5.3   Comparison by Localities

Next we compare the case numbers of our agents against their respective localities. In New York City, we compare the case numbers for a MODZCTA against the infected and recovered numbers for agents representing the MODZCTA. Due to a lack of data, we are only able to compare starting on April 1st $(t = 32)$. In Fig. 3 we show a visual comparison of data on April 14.

**Fig. 3.** Predicted and actual case rates - NYC April 14: Left – the model prediction of cases with subway overlay, Middle - the model prediction of cases without subway overlay, Right - the actual case rates.

The value of MAPE $(t \geq 32, Stations \geq 1)$ is 0.233 and the scale is from 0.00 to 0.20 cumulative cases/person.

A MAPE of 0.233 is much worse than the overall error, but it still suggests that there is some correlation. It is also a great improvement over simply fitting the default value everywhere. We would expect that even with additional tuning and more granular subway data, there would be a large amount of error due to infection spread which cannot be modeled with currently available data.

## 5.4   Results Summary

The results summary in terms of MAPE is given in Table 2.

**Table 2.** The modeling results summary

|  | London (by Borough) | New York City (by MODZCTA) |
|---|---|---|
| Overall MAPE | 0.056 | 0.071 |
| MAPE by region (no model) | 0.571 | 0.623 |
| Median MAPE by region | 0.310 | 0.233 |
| Mean MAPE by region | 0.457 | 0.310 |
| Highest MAPE by region | 1.40 | 0.946 |

# 6   Discussion and Further Research

## 6.1   Additional Parameters

While creating our model, we proposed and experimented with numerous other hyper-parameters which would increase the accuracy of our predictions. However, we excluded them to reduce the specificity and complexity of the model or omitted them due to a lack of reliable data. We list some of the most useful below:

- Awareness Increase Rate - An additional countermeasure modeling increasing public awareness of an outbreak.
- Defiance - The local population will defy government orders if very few people are infected in their area.
- Locality-specific Commuting Modifier - This modifier approximates the percent of the population which commutes in each locality.
- Global Exposure Rate - This adds an additional infection chance based on globally infected. The value of 0.7 indicates that we think approximately 30% of viral propagation is subway-based and the other 70% is not subway based.

## 6.2   Other Modeling Improvements

While there is good correlation between empirical infection rates and our modeling, there is also likely to be good correlation between these rates and demographic data like population density and income. We have not tried to control for this.

Next, our research into New York City suggests that there is a significant amount of commuting from outside the 5 boroughs [4], and we haven't even included Staten Island as there is no subway there. This phenomenon is undoubtedly true for London and all other major cities as well. We have not tried to model these commuters.

Lastly, although we have done similar modeling for Madrid's commuter rails and the world airline network, we have not formally tried to fit our model to other cities.

# 7   Conclusion

In this paper, we outlined how to create an agent-based framework to model the spread of diseases through transportation networks. We then built specific models for the spread of diseases through subway networks. Taking into account various factors such as commute time, subway routes, and city-wide countermeasures, we found that our model could fairly accurately predict New York City and London case data.

Our results suggest that city regions with longer commute times and commute routes linked to highly infected areas have a higher prevalence of disease than more central areas. This should be of interest to public health officials looking to allocate resources appropriately during the early days of a new disease outbreak.

# References

1. Coronavirus (Covid-19) cases. https://data.london.gov.uk/dataset/coronavirus-covid-19-cases
2. Covid-19: Data. https://www1.nyc.gov/site/doh/covid/covid-19-data.page
3. Infection scenarios for Russian cities. https://story.tutu.ru/scenarii-zarazhenija-gorodov-rossii/
4. The ins and outs of NYC commuting. https://www1.nyc.gov/assets/planning/download/pdf/planning-level/housing-economy/nyc-ins-and-out-of-commuting.pdf
5. London shapefile by ward. https://data.london.gov.uk/dataset/statistical-gis-boundary-files-london
6. London underground geographic maps. https://commons.wikimedia.org/wiki/London_Underground_geographic_maps/CSV
7. Mesa docuementation. https://mesa.readthedocs.io/en/master/overview.html
8. MTA station data. http://web.mta.info/developers/data/nyct/subway/Stations.csv
9. Who coronavirus information page. https://www.who.int/emergencies/diseases/novel-coronavirus-2019

10. World airport traffic rankings. https://aci.aero/news/2019/03/13/preliminary-world-airport-traffic-rankings-released/
11. Mo, B., et al.: Modeling epidemic spreading through public transit using time-varying encounter network. Preprint, ResearchGate. https://www.researchgate.net/publication/340541290_Modeling_Epidemic_Spreading_through_Public_Transit_using_Time-Varying_Encounter_Network
12. Cooley, P., et al.: The role of subway travel in an influenza epidemic: a New York city simulation. J. Urban Health **88**(5), 982–995 (2011). https://doi.org/10.1007/s11524-011-9603-4
13. Gonzalez-Reiche, A.S., et al.: Introductions and early spread of SARS-CoV-2 in the New York City area. Science **369**, 297–301 (2020). https://doi.org/10.1101/2020.04.08.20056929
14. Goscé, L., Johansson, A.: Analysing the link between public transport use and airborne transmission: mobility and contagion in the London underground. Environ. Health **17**(1), 1–11 (2018). https://doi.org/10.1186/s12940-018-0427-5
15. Hagberg, A.A., Schult, D.A., Swart, P.J.: Exploring network structure, dynamics, and function using NetworkX. In: Proceedings of the 7th Python in Science Conference (SciPy2008), pp. 11–15. Pasadena, CA USA, August 2008
16. Department of Health (DOHMH): Modified zip code tabulation areas (MOD-ZCTA): NYC open data, May 2020. https://data.cityofnewyork.us/Health/Modified-Zip-Code-Tabulation-Areas-MODZCTA-/pri4-ifjk
17. Hethcote, H.W.: The mathematics of infectious diseases. SIAM Rev. **42**(4), 599–653 (2000). https://doi.org/10.1137/s0036144500371907
18. for London — Every Journey Matters, T.: Tube map. https://tfl.gov.uk/maps/track/tube
19. Medicine, I.O., Geller, A.B., Ogawa, V.A., Wallace, R.B.: Assessing the Use of Agent-based Models for Tobacco Regulation. National Academies Press, appendix A: Considerations and Best Practices in Agent-Based Modeling to Inform Policy (2015)
20. Ndaïrou, F., Area, I., Nieto, J.J., Torres, D.F.M.: Mathematical modeling of COVID-19 transmission dynamics with a case study of Wuhan, April 2020. https://www.ncbi.nlm.nih.gov/pmc/articles/PMC7184012/
21. NYCHealth: NYC coronavirus-data repository. https://github.com/nychealth/coronavirus-data
22. Schelling, T.C.: Dynamic models of segregation†. J. Math. Sociol. **1**(2), 143–186 (1971). https://doi.org/10.1080/0022250x.1971.9989794
23. Xu, F., Mccluskey, C.C., Cressman, R.: Spatial spread of an epidemic through public transportation systems with a hub. Math. Biosci. **246**(1), 164–175 (2013). https://doi.org/10.1016/j.mbs.2013.08.014

# Interpretation of 3D CNNs for Brain MRI Data Classification

Maxim Kan, Ruslan Aliev, Anna Rudenko, Nikita Drobyshev,
Nikita Petrashen, Ekaterina Kondrateva$^{(\boxtimes)}$, Maxim Sharaev,
Alexander Bernstein, and Evgeny Burnaev

Skolkovo Institute of Science and Technology, Moscow, Russia
ekaterina.kondrateva@skoltech.ru

**Abstract.** Deep learning shows high potential for many medical image analysis tasks. Neural networks can work with full-size data without extensive preprocessing and feature generation and, thus, information loss. Recent work has shown that the morphological difference in specific brain regions can be found on MRI with the means of Convolution Neural Networks (CNN). However, interpretation of the existing models is based on a region of interest and can not be extended to voxel-wise image interpretation on a whole image. In the current work, we consider the classification task on a large-scale open-source dataset of young healthy subjects—an exploration of brain differences between men and women. In this paper, we extend the previous findings in gender differences from diffusion-tensor imaging on T1 brain MRI scans. We provide the voxel-wise 3D CNN interpretation comparing the results of three interpretation methods: Meaningful Perturbations, Grad CAM and Guided Backpropagation, and contribute with the open-source library.

**Keywords:** MRI · Deep learning · 3D CNN · CNN interpretation · Meaningful perturbation · Grad CAM · Guided Back-propagation

## 1 Introduction

Deep learning and specifically Convolutional Neural Networks (CNNs) has recently found many applications in the area of medical diagnostics and image processing [15,20]. For example, processing Magnetic Resonance Images (MRI) with CNN allow reducing the dose of gadolinium used for contrast by an order of magnitude [8]. Another example is the detection of cerebral microbleeds using a 3D-CNN [4]. In the following work [26] authors apply convolutional networks to multi-modal (T1, T2 and FA) MRI images in order to segment infant brain tissue images into Gray Matter (GM), White Matter (WM), and Cerebrospinal Fluid (CSF). CNNs are used in early-stage Alzheimer's disease detection in MRI and PET images [22]. Finally, these type of networks are applied to a variety of predictive regression tasks in brain imaging, see [16].

Conventionally, the brain data is firstly processed to get the lower dimensional meaningful features [10] or goes with whole-brain statistical analysis or Voxel-Based Morphometry (VBM). For the diffusion tensor imaging (DTI) extracted

© Springer Nature Switzerland AG 2021
W. M. P. van der Aalst et al. (Eds.): AIST 2020, CCIS 1357, pp. 229–241, 2021.
https://doi.org/10.1007/978-3-030-71214-3_19

features conventionally include fraction anisotropy (FA), mean, axial and radial diffusivity values [23]. For functional T2* MRI images functional connectivity features are commonly extracted, and for T1 structural imaging—morphometry features. Data analysis and machine learning usually follow this feature extraction step [19].

From the other side, deep learning approaches, especially those for processing 3D data, are shown to be more accurate in many applications [23] as they use full-sized data without information loss during aforementioned extensive pre-processing and feature extraction.

However, the stable and reliable interpretation of 3D CNNs is still a big topic of discussion. For example, recent studies of age and gender brain differences point out that despite its high prediction quality "the localised predictions of age and gender do not yield easily interpretable insights into the workings of the neural network" [14].

Deep learning models interpretation in MRI implies training on large databases of healthy subjects. One of the most common and highly explored databases available in open-access is a Human Connection Project (HCP)[1]. A conventional task being extensively explored within this database is a task of gender patterns recognition between men and women [1,3].

However, previous studies on morphological difference between specific brain regions show interpretable results only on the feature or region-of-interest level [23]. On the contrary, the state-of-the-art deep-learning interpretation methods allow visualization of the decision rule in a pixel-wise fashion. Or in the case of 3D convolution models—voxel-wise [14]. The contributions of the proposed paper are as follows:

- we reproduce and extend the state-of-the-art 3D CNN model [23] to investigate the difference between men and women brain on T1 images and confirm previous findings on DTI;
- we apply several network interpretation methods to the 3D CNN model: Meaningful Perturbations, Grad CAM and Guided Backpropagation to find gender-specific patterns and compare their performances;
- we compare the obtained attention maps to the conventional machine learning classification models on morphometry data and discuss the differences as well as with previous findings;
- lastly we provide the code for MRI 3D CNN interpretation as an open-source library.

The source code is open and available at https://github.com/maxs-kan/InterpretableNeuroDL.

---

[1] https://db.humanconnectome.org.

# 2   Data

The database Human Connectome Project (HCP) contains MRI data from 1113 subjects, including 507 men and 606 women of ages 22–36. We explored T1 images, preprocessed with HCP-pipelines[2].

For the morphometry data analysis we used Freesufrer[3] preprocessed features from section **Expanded FreeSurfer Data** for the same 1113 subjects. The morphometry characteristics as number of vertices, volumes, surface areas, and others were computed for 34 cortical regions according to Desikan-Killiany Atlas and for 45 subcortical areas according to Automated Subcortical Segmentation Atlas [6] summing up in 935 vectorized features.

# 3   Methods

## 3.1   Morphometry Data Analysis and Interpretation

We used the morphometry data for gender classification with machine learning models in comparison as one of the most popular methods for data analysis in neuroimaging. The best performing model is chosen among different classifiers: XGBoost, k-Nearest Neighbors (KNN) and Logistic Regression (LR) with a grid-search. All considered models were validated with 10-fold cross-validation technique to give an understanding how the model will generalize to an independent dataset.

## 3.2   Full-Size Data Analysis: 3D CNN

In this work we reproduced 3D CNN model architecture [23] for the images sized [58,70,58]. The proposed network consists of three hidden layers, thus it is light, fast in training, and easy to interpret. To ensure the model stability, we performed 10-fold cross-validation with stratification to estimate model performance.

As the baseline for the 3D CNN network results we chose the support vector machine classifier with `rbf` kernel (SVM) on full sized images. This classifier is conventionally used as the network reference for image classification, for example on `MNIST`. The SVM classifier from `sklearn` was trained on the full-size data reshaped to the 1-dimensional vector, as proposed in [23]. We performed all 3D CNN experiments with `pytorch` framework, on Google Colab work station (Tesla K80 GPU).

## 3.3   3D CNN Interpretation

**Meaningful Perturbations for 3D CNN Results Interpretation.** The goal of the method [7] is to perturb the smallest possible region of the MRI such

---

[2] https://github.com/Washington-University/HCPpipelines.
[3] https://surfer.nmr.mgh.harvard.edu/.

that the model significantly changes its output probability for MR image class, which means that this region is the most important for model decision and it is the most informative part of the image. In this work we perturbed original image $x_0$ by replacing the corresponding region with Gaussian blurring of the image. Let $m : \Lambda \to [0,1]$ be a mask which map each voxel $u \in \Lambda$ of MRI to $m(u)$. Then the operation of perturbation of the MRI has the form:

$$P(x_0; m) = x_0 \odot m + (g_{\sigma_0} * x_0) \odot (1 - m), \qquad (1)$$

where $g_{\sigma_0}$ is a 3D Gaussian kernel with standard deviation $\sigma_0$. The closer $m(u)$ is to 1, the less perturbed the input image. The goal of the algorithm is to find the mask $m$ that leads to $f_c(P(x_0; m)) \ll f_c(x_0)$, where $f_c(\cdot)$ is the probability of $x_0$ belonging to a class $c$, while perturbing the smallest part of the input image, i.e. with a bigger amount of values $m(u)$ close to 1. To avoid the artifacts [9], we paded the $x_0$ with $j$ zeros and the mask $m$ applied to $x_0^K$:

$$x_0^K = x_0[K : H + j + K, K : W + j + K, K : D + j + K] \qquad (2)$$

with integer $K$ drawn from the discrete uniform distribution on $[0, j)$, where $H, W, D$—size of the image. Also, we regularized $m$ in total-variation (TV) norm in low-resolution version, to force it had a more simple structure, upsampled it by factor $s$ to image size and applied a Gaussian filter on the full size mask. Let $M = g_{\sigma_m} * (Up(m, s))$, where $g_{\sigma_m}$ is a 3D Gaussian kernel with standard deviation $\sigma_m$. and $Up(\cdot, s)$ is a trilinear upsampling algorithm by factor of $s$. Finding $m_c$ for class $c$ can be formulated as the following optimization problem:

$$m_c = \arg\min_m \mathbb{E}_{K \sim U[0,j)} \left[ f_c(P(x_0^K, M)) \right] + \lambda_2 \sum_u \|\nabla M(u)\|^\beta + \lambda_1 \|1 - m\|_1. \quad (3)$$

We perform ablation study and find following hyper-parameters which provide stable convergence: $\sigma_0 = \sigma_m = 10$, $\lambda_1 = 3$, $\lambda_2 = 1$, $\beta = 7$, $s = 4$, $j = 5$, and we drawn $K$ from the discrete uniform distribution 10 times. The score is optimized by Adam optimizer with learning rate $\alpha = 0.3$, with exponential decay rates equals $\beta_1 = 0.9$, $\beta_2 = 0.99$.

**Guided Backpropagation for 3D CNN Results Interpretation.** In order to obtain saliency map for each voxel importance we used Guided Backpropagation method [21]. This approach computes the gradient of the score $y^c$ for class $c$ with respect to the input MRI image $x$ based on the network:

$$\hat{m}_c = \frac{dy^c}{dx}. \qquad (4)$$

The gradient is computed with specific backpropagation through the ReLU nonlinearity. In Guided Backpropagation, we backpropagate the positive values of the gradient and set the negative ones to zero. Let $G^l$ be a gradient backpropagated through layer $l$ and $f_i^{l+1} = ReLU(f_i^l)$, then the used backpropagation process is given by the equation:

$$G_i^l = (f_i^l > 0) \cdot (G_i^{l+1} > 0) \cdot G_i^{l+1}. \qquad (5)$$

After obtaining the saliency map in accordance with this specific backpropagation, we obtain an attention map $m_c = ReLU(\hat{m}_c)$ showing the voxels, which has only positive impact to score $y^c$.

**Grad CAM for 3D CNN Results Interpretation.** Grad CAM [18] interprets the model, assuming that the deep CNN layers capture higher-level visual constructs [2]. In the neural network terminal layers we expected features maps to capture higher-level patterns which could be responsible for class discrimination. Grad CAM computes the gradient for score $y^c$ of class $c$ before terminal layer with respect to filter activations of the last convolutional layer $F^k$. Then it computes the importance weights for each filter, i.e.

$$\alpha_k^c = \frac{1}{H \cdot W \cdot D} \sum_{i=1}^{H} \sum_{j=1}^{W} \sum_{k=1}^{D} \frac{\partial y^c}{\partial F_{i,j,k}^k}, \tag{6}$$

where $H, W, D$—size of a filter activation tensor. $\alpha_k^c$ captures the "importance" of filter $k$ for a target class $c$. To obtain the class-discriminative localization mask $m_c$, we computed a weighted combination of filter activations, and follow it by a ReLU:

$$m_c = ReLU \left[ \sum_{k=1} \alpha_k^c \cdot F^k \right]. \tag{7}$$

Also we upsampled $m_c$ to the input image resolution using trilinear interpolation.

## 4 Results

### 4.1 Morphometry Data

The results of accuracy metrics for 10-folds cross-validation and 1113 subjects are in Table 1. Feature importances ($\beta$ scores) for Logistic Regression model chosen via a grid search are provided in Fig. 1. Parameters of model are follow: C = 1.3, penalty = 'l2', max iter = 10000. Importances were selected by coefficients of features in the decision function. Here on $x$ axis we provide the features scores and on $y$ axis—names of features.

The classification models were chosen to explore the classification baseline accuracy across several different machine learning methods.

**Table 1.** Results for baseline morphometry data classification models: 10-fold cross-validation

|               | XGB  | KNN  | LR   |
|---------------|------|------|------|
| Mean accuracy | 0.89 | 0.85 | 0.92 |
| STD           | 0.02 | 0.04 | 0.03 |

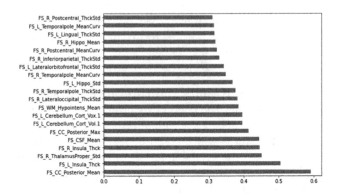

**Fig. 1.** Feature importances for Logistic Regression model.

As can be seen from Fig. 1, the most important features for the Logistic Regression model belong to the following brain regions volumes and intensities: corpus callosum, left and right insula and thalamic regions, as well as whole brian metrics for white matter hyperintensities.

### 4.2 3D CNN Model Results

The 3D CNN model yielded the mean accuracy of $0.92 \pm 0.03$ on 10-fold cross-validation. SVM achieved $0.90 \pm 0.02$ accuracy on 10-fold cross-validation. So 3D-CNN model slightly outperformed standard SVM.

### 4.3 Meaningful Perturbations for 3D CNN

For the two target classes in the gender differences classification, we got two different feature maps correspondingly with the meaningful perturbations algorithm. Two feature masks for men and women appeared to highlight different regions of interest and were then explored separately.

We completed 10 fold cross-validation to check the 3D CNN performance with the images restricted on masks. Then we multiplied every validation sample by averaging men mask, by averaging women mask or by the sum of these masks voxel-wise. The accuracy for the mask belonging to the men group is $0.59 \pm 0.18$, for the women mask—$0.59 \pm 0.07$ and for the conjoined mask—$0.82 \pm 0.12$ respectively. Thus, we can conclude that all necessary information for classification task is both in men and women masks (in their conjunction). The difference in masks for man and women may be explained by the specifics of the algorithm: we need to find the smallest region in the input image, deletion of which will decrease the probability of being a specific class. In Fig. 2a we show the final mask which contains regions for men and women.

Next, to find the correspondence of the attention maps with specific brain regions we performed a segmentation of the MR images on 246 grey matter regions with the Human Brainnettome Atlas [5], as well as 50 white matter

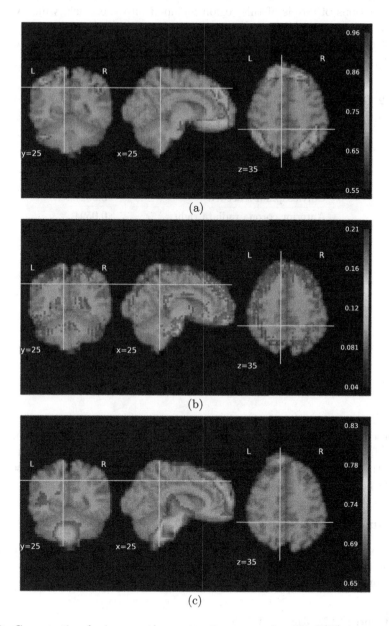

**Fig. 2.** Cross-sectional view on three attention maps for 3D CNN interpretation obtained with: a. Meaningful Perturbation (conjoined men and women attention mask), b. Guided Backpropagation, c. Grad CAM. The greater the voxel's value of each mask, the more important this voxel for classification.

regions with ICBM-81 Atlas [13]. For each region of each brain atlas, we estimated fractions of voxels of this region included into the mask, which we have
obtained via Meaningful Perturbations. We normalized these fractions, so the
values for all regions would sum up to 1. The top-5 scored regions of each atlas
with the largest values are presented in Table 2.

**Table 2.** The most discriminative regions of each atlas obtained with Meaningful
Perturbations method.

ICBM White-Matter Labels Atlas

| Regions | Score |
| --- | --- |
| Corticospinal tract right | 0.1273 |
| Corticospinal tract left | 0.0927 |
| Anterior corona radiata right | 0.0594 |
| Pontine crossing tract | 0.0580 |
| Cerebral peduncle left | 0.0488 |

Human Brainnettome Atlas

| Regions | Score |
| --- | --- |
| A13 Orbital gyrus Right | 0.0131 |
| A11L_R, Orbital gyrus Right | 0.0124 |
| A39RD_R, Inferior parietal lobule Right | 0.0120 |
| A9_46D_R, Middle frontal gyrus Right | 0.0118 |
| TI_L, Parahippocampal gyrus Left | 0.0117 |

These findings partially overlap with the morphometry results, showing common white matter regions in the corpus callosum (Anterior corona radiata), as
well as the cerebellum (Cerebral peduncle). The grey matter region in common
overlaps on frontal gyri (Middle frontal gyrus Right).

### 4.4   Guided Back-Propagation for 3D CNN

We computed a saliency map for every person in the dataset and then took mean
over the dataset. As we have two classes in our dataset, the final map contains
the regions of interest for each class, see Fig. 2b.

### 4.5   Grad CAM for 3D CNN

We computed corresponding localization masks, containing information about
both men and woman discriminative regions of interest. The cross-sectional view
of the result is shown in Fig. 2c. To compare the resulting masks, we calculated
the DICE [12] score between them. For this, we chose a threshold value $t$ to obtain
binary masks in the way to nullify all values outside the brain. The threshold

of $t = 0.65$ was chosen for Grad CAM and Meaningful Perturbations masks and $t = 0.04$ for Guided Backpropagation. For Guided Backpropagation, such small threshold is used, due to the fact that the resulting mask is the gradient of the input image, and the exact values in the mask are much lower compared to other methods. The cross-sectional DICE score is shown in Table 3.

**Table 3.** DICE score between masks, MP - Meaningful Perturbations mask, GC - Grad CAM mask, GB - Guided Backpropagation mask

|        | MP and GC | MP and GB | GC and GB |
|--------|-----------|-----------|-----------|
| DICE   | 0.94      | 0.94      | 0.93      |

## 5   Discussion

In the current work we aimed at studying gender-related differences in human brain by creating a predictive model to solve a classification task and to interpret its decisions.

In order to localize the most informative brain areas for classification task, we created attention maps for 3D CNN output in three different ways. Using these maps we were able to denote which brain regions play the most important role in gender classification. According to Meaningful Perturbations method the brain regions with the highest classification accuracy Table 2 was Orbital Gyrus Right (A13_R and A11l_R) what corresponds with the results obtained from [23]. Moreover we got that Inferior parietal lobule (A39RD_R) is the region with high classification accuracy. That goes in line with the previous studies [17], where it was shown that parietal lobe activity is biased to the right hemisphere in men. We got that the parahippocampal gyrus (TI_L) plays part in gender differentiation. This result corresponds to the previous findings, where it was shown that parahippocampal gyrus activates more intensely in the female brain while executing large-scale spatial tasks [24].

Right, and left corticospinal tracts showed high classification accuracy. This can be explained by the fact that the relative volume of the corticospinal tract on the left is larger in the female brain [11]. Anterior corona radiata right and cerebral peduncle are also responsible for gender differentiation what goes in line with [23]. That confirms the most discriminative brain regions in morphometry model results, showing the importance of Corpus colosseum posterior part voxels mean (FS_CC_Posterior_Mean) as the most discriminative in the Logistic Regression model.

It is also worth noting, that attention maps in Fig. 2a show the spatial pattern of frontoparietal resting-state brain network, which was initially discovered from resting-state fMRI activity and is thought to be involved in a wide variety of tasks by initiating and modulating cognitive control abilities [25]. It might be interesting in future research to look specifically at this network and explore it in terms of gender-related brain differences.

We show that the classification result on T1 images is lower than the result on FA images comparing the cross-validation mean scores (0.933), thus suggesting that DWI imaging might be more predictive for these type of structural differences. However, it should be mentioned that in the original approach as well, as in current paper, all T1 images were compressed to smaller tensors from [260, 311, 260] to [58, 70, 58] addressing memory issues. This compression may result in information loss and can explain similar classification score in neural networks comparing to the baseline methods.

# 6  Conclusion

This work is an extension of the study on FA images from [23]. Here we create a 3D CNN model on T1 images of the same subjects and reveal similar gender-related differences in brain structure. The model exhibits the mean accuracy of $0.92 \pm 0.03$, which corresponds to the morphometry data classification and higher than SVM classification ($0.90 \pm 0.02$).

We apply several network interpretation methods to the 3D CNN model: Meaningful Perturbations, Grad CAM and Guided Backpropagation to find gender-specific patterns, and to compare their performances. We compare the results of interpretation in terms of the Dice coefficient. High scores on paired mask comparisons confirm that all three methods reveal similar patterns and thus belong to stable and trustworthy predictions. We found that Grad CAM and Guided Backpropagation are very sensitive to the slightest changes in the model weights, therefore, when using such layers as Dropout, it is necessary to fix the random seed. Then the Meaningful Perturbations method is less sensitive to small changes in model weights. We found that the Grad CAM method is the fastest one (executed within one minute) and ready plug-and-play method, Guided Backpropagation is slower (up to several minutes), as it propagates the gradient across the entire network, and has a more sparse attention map. The Meaningful Perturbations method is the slowest one (executes for several hours) yet showing the most anatomic-like attention maps. We recommend using the Grad CAM method for network interpretation at first in similar tasks.

Our deep learning model interpretation results are in line with the results from machine learning classification based on morphometry data. These findings are in line with previous studies as well, confirming the assumption that men's and women's differences more probably exist in the whole-brain range.

We also publish the code to the open-source library for public use. The proposed interpretation tool could be successfully used in various MRI pathology detection applications like epilepsy detection, Alzheimer's disorder diagnosis, Autism Spectrum disorder classification, and others.

**Acknowledgements.** The reported study was funded by RFBR according to the research project №20-37-90149. Also we acknowledge participation of Ruslan Rakhimov in development of meaningfull perturbation method on MRI data.

# A    The First Hidden Layer of 3D CNN Attention Analysis

We analyzed features obtained in First Hidden Layer of 3D CNN as in [23]. Even though we used T1 modality MRI images in contorary to DWI modality and fractional anisotropy (FA) images in previous studies.

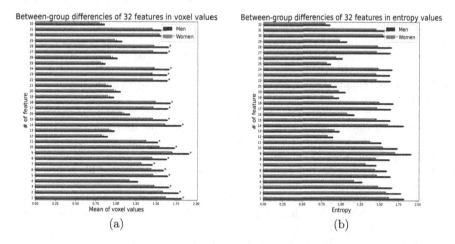

Fig. 3. (a) Mean voxel values for each feature in men/women groups. Features that are significantly large for men are marked with *, features that are significantly large for women are marked with +. (b) Mean entropy values for each feature in men/women groups.

Similar to results shown on FA images, we found that mean voxel values for 31 features have a significant difference in men-women groups, with 10 features larger for women, and 21 features larger for men (see Fig. 3) accounting the multiple-comparisons correction. That reproduces the previously stated result, assuming that "men's brains likely have more complex features as reflected by significantly higher entropy." As well as that important gender-related patterns are likely to be spread in the whole-brain grey and white matter. That highlights the importance of the results discussed in the main paper, as the attention maps compared from different approaches are extracted from the whole brain imagery, without any region-of-interest removal, as in [23].

# References

1. Cahill, L.: Why sex matters for neuroscience. Nature Rev. Neurosci. **7**(6), 477–484 (2006)
2. Chen, X., et al.: Microsoft COCO captions: data collection and evaluation server. arXiv preprint arXiv:1504.00325 (2015)
3. Cosgrove, K.P., Mazure, C.M., Staley, J.K.: Evolving knowledge of sex differences in brain structure, function, and chemistry. Biol. Psychiatr. **62**(8), 847–855 (2007)

4. Dou, Q., et al.: Automatic detection of cerebral microbleeds from MR images via 3D convolutional neural networks. IEEE Trans. Med. Imaging **35**(5), 1182–1195 (2016)
5. Fan, L., et al.: The human Brainnetome Atlas: a new brain atlas based on connectional architecture. Cereb. Cortex **26**(8), 3508–3526 (2016)
6. Fischl, B.: Freesurfer. Neuroimage **62**(2), 774–781 (2012)
7. Fong, R.C., Vedaldi, A.: Interpretable explanations of black boxes by meaningful perturbation. In: Proceedings of the IEEE International Conference on Computer Vision, pp. 3429–3437 (2017)
8. Gong, E., Pauly, J.M., Wintermark, M., Zaharchuk, G.: Deep learning enables reduced gadolinium dose for contrast-enhanced brain MRI. J. Magn. Reson. Imaging **48**(2), 330–340 (2018)
9. Kurakin, A., Goodfellow, I., Bengio, S.: Adversarial examples in the physical world. arXiv preprint arXiv:1607.02533 (2016)
10. Lasič, S., Szczepankiewicz, F., Eriksson, S., Nilsson, M., Topgaard, D.: Microanisotropy imaging: quantification of microscopic diffusion anisotropy and orientational order parameter by diffusion MRI with magic-angle spinning of the q-vector. Front. Phys. **2**, 11 (2014)
11. Liu, Y., et al.: Gender differences in language and motor-related fibers in a population of healthy preterm neonates at term-equivalent age: a diffusion tensor and probabilistic tractography study. Am. J. Neuroradiol. **32**(11) (2011)
12. Milletari, F., Navab, N., Ahmadi, S.A.: V-Net: fully convolutional neural networks for volumetric medical image segmentation. In: 2016 fourth international conference on 3D vision (3DV), pp. 565–571. IEEE (2016)
13. Mori, S., Wakana, S., Nagae-Poetscher, L., Van Zijl, P.: MRI atlas of human white matter. Am. J. Neuroradiol. **27**(6), 1384 (2006)
14. Pawlowski, N., Glocker, B.: Is texture predictive for age and sex in brain MRI? arXiv preprint arXiv:1907.10961 (2019)
15. Pominova, M., Artemov, A., Sharaev, M., Kondrateva, E., Bernstein, A., Burnaev, E.: Voxelwise 3D convolutional and recurrent neural networks for epilepsy and depression diagnostics from structural and functional MRI data. In: 2018 IEEE International Conference on Data Mining Workshops (ICDMW), pp. 299–307. IEEE (2018)
16. Pominova, M., et al.: Ensemble of 3D CNN regressors with data fusion for fluid intelligence prediction. In: Pohl, K.M., Thompson, W.K., Adeli, E., Linguraru, M.G. (eds.) ABCD-NP 2019. LNCS, vol. 11791, pp. 158–166. Springer, Cham (2019). https://doi.org/10.1007/978-3-030-31901-4_19
17. Rescher, B., Rappelsberger, P.: Gender dependent EEG-changes during a mental rotation task. Int. J. Psychophysiol. **33**(3), 209–222 (1999)
18. Selvaraju, R.R., Cogswell, M., Das, A., Vedantam, R., Parikh, D., Batra, D.: Grad-CAM: visual explanations from deep networks via gradient-based localization. Int. J. Comput. Vis. **128**(2), 336–359 (2019). https://doi.org/10.1007/s11263-019-01228-7.
19. Sharaev, M., et al.: Pattern recognition pipeline for neuroimaging data. In: 8th IAPR TC3 Workshop on Artificial Neural Networks in Pattern Recognition, pp. 306–319 (2018)
20. Sharaev, M., et al.: MRI-based diagnostics of depression concomitant with epilepsy: in search of the potential biomarkers. In: 2018 IEEE 5th International Conference on Data Science and Advanced Analytics (DSAA), pp. 555–564. IEEE (2018)
21. Springenberg, J.T., Dosovitskiy, A., Brox, T., Riedmiller, M.: Striving for simplicity: the all convolutional net. arXiv preprint arXiv:1412.6806 (2014)

22. Suk, H.I., Lee, S.W., Shen, D., Initiative, A.D.N., et al.: Hierarchical feature representation and multimodal fusion with deep learning for AD/MCI diagnosis. NeuroImage **101**, 569–582 (2014)
23. Xin, J., Zhang, X.Y., Tang, Y., Yang, Y.: Brain differences between men and women: evidence from deep learning. Front. Neurosci. **13**, 185 (2019)
24. Yuan, L., Kong, F., Luo, Y., Zeng, S., Lan, J., You, X.: Gender differences in large-scale and small-scale spatial ability: a systematic review based on behavioral and neuroimaging research. Front. Behav. Neurosci. **13**, 128 (2019)
25. Zanto Theodore, P., Gazzaley, A.: Fronto-parietal network: flexible hub of cognitive control. Trends Cogn. Sci. **17**, 602–603 (2013)
26. Zhang, W., et al.: Deep convolutional neural networks for multi-modality isointense infant brain image segmentation. NeuroImage **108**, 214–224 (2015)

# A Comparison of Neural Networks Architectures for Diacritics Restoration

Eduard Klyshinsky[1]([envelope]) [iD], Olesya Karpik[2] [iD], and Alexander Bondarenko[3] [iD]

[1] National Research University Higher School of Economics,
S. Basmannaya Street, 21, Moscow 105066, Russia
`klyshinsky@mail.ru`
[2] Keldysh Institute of Applied Mathematics,
Miusskaya sq., 4, Moscow 125047, Russia
`parlak@mail.ru`
[3] State Research Institute of Aviation Systems,
Viktorenko Street, 7, Moscow 125319, Russia
`cod@fgosniias.ru`

**Abstract.** Neural networks are widely used for the task of diacritics restoration last years. Authors use different architectures of neural network for selected languages. In this paper, we demonstrated that an architecture should be selected according to a language in hand. It also depends on a task one states: low and full resourced languages could use different architectures. We demonstrated that common used accuracy metric should be changed in this task to precision and recall due to the heavy unbalanced nature of the input data. The paper contains results for seven languages: Croatian, Slovak, Romanian, French, German, Latvian, and Turkish.

**Keywords:** Diacritics restoration · Neural Networks ·
Out-of-vocabulary words

## 1 Introduction

Let us consider an extended alphabet that contains characters with diacritics, e.g. the extended Latin or Cyrillic alphabet. Every character with a diacritical mark could be matched with a character without such a mark; e.g. Ä→A. In this paper, we do not consider the substitution of a single character with a diacritical mark for several characters, e.g. Ä→AE. Let us substitute all characters with diacritics for their corresponding characters without diacritics. After such replacements we could state the task of diacritics restoration: consider the resulting text and restore the diacritics only in those positions where they were omitted.

Originally, the task of diacritics restoration was set for a natural language text. Finding a way to automate the input of diacritics is necessary not only for old valuable texts stored in an electronic format, but also for modern electronic texts, since they continue to be created in non-diacritical form. For some

W. M. P. van der Aalst et al. (Eds.): AIST 2020, CCIS 1357, pp. 242–253, 2021.
https://doi.org/10.1007/978-3-030-71214-3_20

reasons, such as the lack of keyboards with diacritics, OCR mistakes, ergonomic factors, etc., the diacritics are omitted in a text, and they must be restored. Such situation is typical not only for languages that use the Latin script, but also for Cyrillic and Arabic writing.

Recently, this task has been solved by machine learning methods. Preparing a data set could be done in a following easy way. On the first step, one constructs a set of replacements for a given language. For example, the letter A with different diacritics (ÄÀÂÁĂ) will be replaced with the same letter – A. To create training and test sets, one could replace all characters with diacritics in a text. In order to create a test set, one should tag positions of proper characters without diacritics by corresponding classes of diacritical marks. The set of tagged words could be shuffled and separated into the training and test sets.

In such case, the most frequent words have a chance of occurrence in both the training and test sets. Therefore, a model is both trained and tested on the highly intersected sets of words, and the probability of overlap is obviously high. As a result, the algorithm could 'study answers' but not 'generalize data'. Such approach to the construction of training and test sets is similar to diacritics restoration using a dictionary.

The motivation of this paper is the following. According to the ICAO standard specification, all proper names in formal documents should be presented without diacritics. There is no complete dictionary of proper names for all languages yet; therefore this dictionary could not be used to restore diacritics in proper names. There is the same problem with low resourced languages. So, the more correct way to model such situation is to divide a vocabulary of a text and generate non-intersecting training and test sets. In this paper, we investigate the difference between these two approaches and compare several different models based on neural networks.

The rest of the paper is organized as following. In Sect. 2, we make a brief overview of current approaches to restoration of diacritics. In Sect. 3, we describe our data preparation method and machine learning methods used. Section 4 presents our experimental results and their brief discussion. Finally, Sect. 5 concludes this paper with a summary of outcomes.

## 2 Overview

One of the first papers in the area of diacritics restoration are Yarowsky [1] and Tufiş [2]. The main method used in these papers is searching a word in a dictionary of a given language. The word from a text was compared with the dictionary regardless of the accentuation. However, such approach could not be applied to a language that has majority of accentuated words. For example, in German the diacritical mark can be an indicator of the plural form (der Vater – die Väter). In Vietnamese and Igbo, there are words that differ in diacritics only, but have extremely different meanings. So, for example, the paper [3] reports the following chain:

*ákwà (cloth), àkwà (bed/bridge), ákwá (cry), àkwá (egg).*

The same problem is presented in the Arabic script, where diacritical marks are mostly omitted despite of having a semantic meaning.

Due to lack of accuracy, the described methods were replaced by machine learning methods. For example, Pauw [8] and Schlippe [4] propose to use character n-grams. The former article used the CRF method to determine the correct n-gram for the current position. The reported quality of this method is about 85%–95%.

The development of neural networks shifted the focus of investigations. The state-of-the-art implementations are using two architectures of neural networks - recurrent (RNN) and convolutional (CNN) neural networks. The results of both architectures are comparable. The article [5] reports accuracy metrics for twelve languages as high as 97–99%. The authors used a network constructed using BiLSTM units with input embedding layer. The authors of [6] used Acausal Temporal Convolutional Neural Network with two hidden layers; precision of diacritics restoration reported as 96.2% for the Yoruba language, 97% for Arabic, and 97.5% for the Vietnamese language. The difference between results of CNN and RNN reported less than 0.3%. The article [7] uses CNN with an attention layer. Note that all authors use a single neural network to restore all diacritical marks in a given language. All neural networks accept as an input the vector of the character n-gram in various forms.

As it was mentioned in the previous section, low resourced languages have a problem with training a neural network on a small data set. The article [6] demonstrates that though the precision of a neural network learned and tested on the same dictionary is about 96–98%, the precision of the same network for out-of-vocabulary words dramatically decreased to 70–85% and even less than 50%. Most of the authors (e.g. [3,5,7]) do not consider the class of out-of-vocabulary words at all. It should be noted that the accuracy metrics could not be used for such a project, since there is a disbalance between amount of words with and without diacritical marks. As it shown in [5], the amount of words with diacritics in different languages varies from 8% in German to 88% in Vietnamese. Our experience demonstrates that the relation between amount of occurrences of the same character, with and without diacritical marks, could be correlated as 1:100. In such conditions, the accuracy metric is useless, and one should prefer the precision metric.

According to the written above, we can state that most of the current methods of diacritics restoration have problems with processing of out-of-vocabulary words. To use machine learning methods, it is more correct to create non-intersected training and test sets. The precision and recall metrics should be prioritized for evaluation of the achieved results. Convolutional and recurrent neural networks demonstrate almost the same results in the same language, but these results differ among languages. Therefore, the aim of this article is to investigate the dependence of the neural network architecture on the selected language and the type of test and training sets. Moreover, we want to test a new approach to constructing a neural network solver, where instead of training single neural network that recognizes any character, we train a set of binary solvers where one network is responsible for only one character with a given diacritic.

# 3    Experimental Setup

This section discusses the used dataset, the data preparation method, and used architectures of neural networks.

## 3.1    Used Data and Data Preparation

In this paper, we used several news collections. French, German, and Turkish collections were downloaded from the news wire sites. Collections for Croatian, Latvian, Romanian, and Slovak languages were downloaded from the corpus presented in [5] (http://hdl.handle.net/11234/1-2607). The size of the former ranges from 63 to 217 mln of characters, the latter were cut to the first 200 mln characters.

We constructed a list of character replacements with diacritics to their ASCII counterparts for each language. We divided the collection into words not shorter than 4 characters; each word was aligned to the left and right side with three alignment symbols. Then we extracted all 7-grams with the character from the replacements list in the central position[1]. The 7-gram was tagged depending on whether the symbol in the central position had a diacritical mark or not. Finally, we replaced all the characters with diacritics with their counterparts.

For example, the Turkish word '*öncesi*' is aligned with three spaces in the starting and final positions: ' ‿‿‿ öncesi ‿‿‿'. Than the algorithm extracts all 7-grams containing an accentuated symbol or its counterpart in the central position; in our case the resulting list is [' ‿‿‿ önce', ' ‿ öncesi', 'ncesi ‿‿', 'cesi ‿‿‿ ']. Finally, symbols with diacritics are replaced by their counterparts, 7-grams are tagged by 0 (no diacritics) and 1 (with diacritics) according to the symbol in the central position: [(' ‿‿‿ once', 1), (' ‿ oncesi', 0), ('ncesi ‿‿', 0), ('cesi ‿‿‿', 0)].

We used two types of character embedding. In the first case, we replace the symbol with its position in the alphabet of the given language (dense representation); in the second case, we used one-hot encoding for this alphabet (one-hot representation). The alphabet was shaped automatically by the first million characters in the collection after word tokens were extracted.

## 3.2    Used Architectures

Unlike other authors, we constructed one binary classifier for each accentuated character in a given language. Such approach can be directly applied to languages like German, which has only three accentuated characters. Therefore, one could just extract all 7-grams with counterpart symbols and pass them to the proper classifier, which takes decision whether the diacritic mark should be placed here. For languages such as French, Romanian, and Latvian, which have several variant of diacritics for the same symbols, our approach should be

---

[1] We extracted 5-grams in our preliminary experiments, but achieved remarkably worse results not published here.

extended to the second step. A 7-gram with a candidate character is passed to several classifiers connected to the given character. After classification, we should find the maximum among classifiers' answers and make a conclusion about the necessary diacritical mark or its elimination. In this paper, we also compared two approaches: training separate classifiers without taking decision about different diacritics for the same base character vs training one classifier for every character with diacritics.

According to the chosen approach, any neural network contains an output dense layer consisting of two neurons with SoftMax activation function.

We used three different neural network architectures: dense, convolutional, and recurrent[2]; we also used the Random Forest method as a baseline. As mentioned above, we used two types of input vectors in our experiments: a vector of seven character codes and one-hot encoded vector of 7*(size of an alphabet) elements. The used architectures of neural networks are also listed in Table 1.

The dense network consists of three layers of 128 neurons with the ReLU activation function and an output layer of 2 neurons with the SoftMax function; the batch size was 2048. We experimented with the number of neurons and layers, but it decreased the results. We studied this type of networks for 7 character codes vectors only. The network converges in 30 epochs.

The convolutional network consists of two layers with 32 convolution matrices size of $3 \times 1$, a dense layer with 128 neurons with the ReLU activation function, and an output layer of 2 neurons with the SoftMax function. The batch size was 512; the network converges in 10 epochs. We experimented with the number of neurons, convolution matrices, and their size but it decreased the results.

For the recurrent networks we used three configurations. The first one consists of a layer with 64 biLSTM units and an output layer of 2 neurons with the SoftMax function. The first layer of the second configuration consists of 128 biLSTM units. The third configuration had a dense input layer with 32 neurons and the ReLU activation function, 128 biLSTM units, and an output layer of 2 neurons with SoftMax function. The batch size was 512; all networks converge in 10 epochs.

**Table 1.** Used architectures of neural networks.

| Dense network | CNN | RNN |
|---|---|---|
| Output vector: $< (0,1)^2 >$ | | |
| SoftMax layer | SoftMax layer | SoftMax layer |
| Dense 128 ReLU | Dense 128 ReLU | 32/64/128 biLSTM |
| Dense 128 ReLU | 32 convolution $3 \times 1$ | |
| Dense 128 ReLU | 32 convolution $3 \times 1$ | Dense 128 ReLU (optional) |
| Input vector: $< (charcode)^7 >$ or $< (0,1)^{7*|alphabet|} >$ | | |

---

[2] Links to used data and source codes are placed at GitHub: https://github.com/klyshinsky/diacritics_restoration.

In order to investigate the influence of the intersection of training and test vocabularies, we conducted two series of experiments. For the first series, we separated the input set of 7-grams into two non-intersecting sets; the sum of the test set's frequencies was approximately 20% of the total frequency sum. Thus, in this series we trained and tested a classifier on different vocabularies and simulated the situation of out-of-vocabulary words. The second series of experiments tested only convolutional and recurrent networks, since they shown better productivity. In this series, we simply shuffled the set of 7-grams and split the input set into training and test in the ratio 80%–20%. Our experiments demonstrated that 80% of the words of the test set could be found in the training set.

## 4   Results of Experiments

The results of our experiments are shown in the Tables 2, 3, 4, 5, 6, 8[3]. We did not use Word Error Rate (WER) and Diacritics Error rate (DER) since they are introduced as accuracy calculated for words and single diacritics, respectively. Instead of these metrics, we report precision and recall. As shown below, they evaluate results of the methods more correctly.

The tables are organized as following. Each table consists of results for a single language. We measured the results separately for every character with a diacritic mark. The first line for every character contains precision; the second line reports a recall. Figures marked with the bold font demonstrate maximal precision for the character among all results for separated training and test vocabularies. Figures marked with the underlined font demonstrate maximal precision for the character among all results for mixed training and test vocabularies. The last two rows in each table demonstrate the average precision and recall for the given classifier. Note that the best precision does not mean the best F1-value, but we did not calculate the F1-value due to the big size of resulting tables. Mostly, it is the case of low recall; however, such classifiers are not practically applicable. In the case of a high recall and a small difference between the values, both classifiers can be applied and should be compared by other features. In this article, we tried to draw the overall picture, but not to compare different neural networks in detail.

In the case of zero precision and recall, our system either failed to randomly select any 7-grams with a diacritic mark because of lack of words with diacritics, either evaluate all 7-grams as ones without diacritics due to the huge disbalance between classes. In case of Romanian Ș, it was true for every experimental run. For the French Ê, classes was unbalanced about 1:50. Its precision and recall were equal to zero but the accuracy was equal to 0.98. Thus, the accuracy metrics do not demonstrate the real situation here. That is why we used both precision and recall in this project.

---

[3] We want to thank Google Colab for the provided GPU time. It was not enough for all our calculations, but it was very useful.

**Table 2.** Results of experiments for the Croatian language (precision, recall). Bold script indicates the best solution for separated train and test, underline - for mixed ones.

| Characters | Separated train and test sets | | | | | | | Mixed train and test sets | | |
| | 7 character input vector | | | | | | One-hot encoded | 7 character input vector | | One-hot encoded |
| | Rand Forest | Dense 3 × 128 | Conv 2 × 32 × 3 | LSTM x128 | LSTM x64 | Dense × 32, LSTM x128 | Conv 2 × 32 × 3 | LSTM x128 | Conv 2 × 32 × 3 | Conv 2 × 32 × 3 |
| | 0 | 1 | 2 | 3 | 4 | 5 | 6 | 7 | 8 | 9 |
| Đ | 0.163/ 0.195 | 0.740/ 0.583 | 0.901/ 0.711 | 0.656/ 0.784 | 0.743/ 0.762 | 0.709/ 0.581 | **0.902/ 0.860** | 0.973/ 0.924 | 0.963/ 0.936 | 0.998/ 0.970 |
| Ć | 0.522/ 0.376 | 0.721/ 0.584 | 0.809/ 0.684 | 0.826/ 0.782 | 0.841/ 0.758 | **0.855/ 0.715** | 0.798/ 0.886 | 0.972/ 0.954 | 0.971/ 0.937 | 0.989/ 0.984 |
| Č | 0.694/ 0.462 | **0.917/ 0.649** | 0.850/ 0.658 | 0.892/ 0.576 | 0.860/ 0.772 | 0.831/ 0.638 | 0.881/ 0.815 | 0.969/ 0.979 | 0.972/ 0.955 | 0.968/ 0.977 |
| Š | 0.365/ 0.297 | 0.573/ 0.520 | 0.450/ 0.614 | 0.504/ 0.562 | **0.696/ 0.637** | 0.613/ 0.535 | 0.659/ 0.560 | 0.960/ 0.934 | 0.931/ 0.936 | 0.944/ 0.876 |
| Ž | 0.638/ 0.589 | 0.686/ 0.602 | 0.851/ 0.660 | 0.770/ 0.477 | **0.911/ 0.702** | 0.626/ 0.471 | 0.840/ 0.851 | 0.978/ 0.964 | 0.936/ 0.982 | 0.967/ 0.990 |
| avg | 0.476/ 0.383 | 0.727/ 0.587 | 0.772/ 0.665 | 0.729/ 0.636 | 0.810/ 0.726 | 0.726/ 0.588 | 0.816/ 0.794 | 0.970/ 0.951 | 0.954/ 0.949 | 0.973/ 0.959 |

**Table 3.** Results of experiments for the German language (precision, recall). Column names are presented in Table 2. Bold script indicates the best solution for separated train and test, underline - for mixed ones.

| | 0 | 1 | 2 | 3 | 4 | 5 | 6 | 7 | 8 | 9 |
| --- | --- | --- | --- | --- | --- | --- | --- | --- | --- | --- |
| Ä | 0.065/ 0.142 | 0.219/ 0.141 | **0.338/ 0.288** | 0.278/ 0.311 | 0.314/ 0.185 | 0.270/ 0.408 | 0.139/ 0.075 | 0.916/ 0.761 | 0.899/ 0.762 | 0.655/ 0.556 |
| Ö | 0.039/ 0.072 | 0.290/ 0.340 | 0.218/ 0.238 | 0.241/ 0.125 | 0.168/ 0.091 | 0.169/ 0.147 | **0.957/ 0.634** | 0.917/ 0.817 | 0.889/ 0.826 | 0.860/ 0.690 |
| Ü | 0.704/ 0.509 | 0.472/ 0.481 | 0.577/ 0.581 | 0.446/ 0.325 | 0.322/ 0.195 | 0.278/ 0.485 | **0.761/ 0.582** | 0.949/ 0.900 | 0.938/ 0.920 | 0.991/ 0.955 |
| avg | 0.269/ 0.241 | 0.327/ 0.320 | 0.377/ 0.369 | 0.321/ 0.253 | 0.268/ 0.157 | 0.239/ 0.346 | 0.619/ 0.430 | 0.927/ 0.826 | 0.908/ 0.836 | 0.835/ 0.733 |

As we can see, there is no 'silver bullet' among the neural network architectures even for a single language. For German and Turkish, the preferred architecture is convolutional network over one-hot encoded input vector, if we consider the number of winning characters (2 of 3 and 4 of 5 respectively). Other languages do not demonstrate any preferences for this feature. In the case of consideration the average precision for the ML model, the same network wins for Croatian, German, and Turkish; but for Romanian, Slovak, and Latvian the leading model is LSTM with 64 units; the convolutional network with dense vectors wins for French. Note that training CNN with a one-hot-encoded input vector takes about 10–50 times more calculation time. For mixed training and test vocabularies, the biLSTM is a default best choice, it wins in 39 out of 56 cases.

**Table 4.** Results of experiments for the French language (precision, recall). Column names are presented in Table 2. Bold script indicates the best solution for separated train and test, underline - for mixed ones.

| | 0 | 1 | 2 | 3 | 4 | 5 | 6 | 7 | 8 | 9 |
|---|---|---|---|---|---|---|---|---|---|---|
| À | 0.541/ | 0.0/ | **0.865/** | 0.222/ | 0.0/ | 0.0/ | 0.0/ | 0.992/ | 0.986/ | 0.010/ |
| | 0.748 | 0.0 | **0.115** | 0.000 | 0.0 | 0.0 | 0.0 | <u>0.751</u> | 0.821 | 0.061 |
| Â | 0.050/ | 0.135/ | 0.052/ | **1.0/** | 0.0/ | 0.347/ | 0.0/ | 0.996/ | 0.950/ | 0.0/ |
| | 0.028 | 0.001 | 0.247 | **0.011** | 0.0 | 0.065 | 0.0 | <u>0.289</u> | 0.950 | 0.0 |
| Ç | 0.595/ | 0.356/ | **0.837/** | 0.0/ | 0.638/ | 0.816/ | 0.536/ | 0.0/ | 0.994/ | 0.771/ |
| | 0.446 | 0.110 | **0.564** | 0.0 | 0.699 | 0.254 | 0.868 | 0.0 | <u>0.997</u> | 0.985 |
| È | 0.846/ | **0.968/** | 0.831/ | 0.910/ | 0.948/ | 0.916/ | 0.0/ | 0.995/ | 0.989/ | 0.138/ |
| | 0.298 | **0.379** | 0.869 | 0.604 | 0.423 | 0.385 | 0.0 | <u>0.958</u> | 0.986 | 0.603 |
| É | 0.158/ | 0.208/ | 0.329/ | 0.287/ | 0.283/ | 0.208/ | **0.677/** | 0.957/ | 0.921/ | 0.516/ |
| | 0.232 | 0.317 | 0.530 | 0.399 | 0.441 | 0.315 | **0.572** | <u>0.852</u> | 0.870 | 0.591 |
| Ê | 0.136/ | 0.0/ | 0.371/ | <u>0.694/</u> | 0.680/ | 0.708/ | 0.0/ | 0.997/ | 0.996/ | 0.873/ |
| | 0.046 | 0.0 | 0.379 | <u>0.356</u> | 0.152 | 0.348 | 0.0 | <u>0.978</u> | 0.986 | 0.003 |
| Î | 0.108/ | 0.0/ | 0.487/ | 0.151/ | 0.047/ | **0.839/** | 0.0/ | 0.801/ | 0.912/ | 0.0/ |
| | 0.033 | 0.0 | 0.282 | 0.019 | 0.000 | **0.353** | 0.0 | 0.941 | <u>0.857</u> | 0.0 |
| Ô | 0.034/ | **0.982/** | 0.569/ | 0.940/ | 0.643/ | 0.101/ | 0.559/ | 0.977/ | 0.979/ | 0.0/ |
| | 0.068 | **0.193** | 0.234 | 0.243 | 0.159 | 0.072 | 0.950 | 0.716 | <u>0.982</u> | 0.0 |
| Ù | 0.0/ | 0.0/ | 0.0/ | 0.0/ | 0.0/ | 0.0/ | 0.0/ | 0.0/ | 0.946/ | 0.0/ |
| | 0.0 | 0.0 | 0.0 | 0.0 | 0.0 | 0.0 | 0.0 | 0.0 | <u>0.963</u> | 0.0 |
| Û | 0.033/ | 0.0/ | 0.668/ | 0.773/ | 0.008/ | **0.989/** | 0.240/ | 0.991/ | 0.983/ | 0.0/ |
| | 0.016 | 0.0 | 0.973 | 0.102 | 0.016 | **0.600** | 0.782 | <u>0.919</u> | 0.971 | 0.0 |
| avg | 0.250/ | 0.264/ | 0.500/ | 0.497/ | 0.324/ | 0.492/ | 0.201/ | 0.770/ | 0.965/ | 0.230/ |
| | 0.191 | 0.1 | 0.419 | 0.173 | 0.189 | 0.239 | 0.317 | 0.640 | 0.938 | 0.224 |

**Table 5.** Results of experiments for the Romanian language (precision, recall). Column names are presented in Table 2. Bold script indicates the best solution for separated train and test, underline - for mixed ones.

| | 0 | 1 | 2 | 3 | 4 | 5 | 6 | 7 | 8 | 9 |
|---|---|---|---|---|---|---|---|---|---|---|
| Â | 0.355/ | 0.698/ | 0.714/ | 0.800/ | **0.865/** | 0.588/ | 0.386/ | 0.975/ | 0.966/ | 0.446/ |
| | 0.302 | 0.419 | 0.172 | 0.316 | **0.422** | 0.225 | 0.723 | <u>0.950</u> | 0.936 | 0.467 |
| Î | 0.095/ | 0.145/ | 0.311/ | **0.324/** | 0.241/ | 0.207/ | 0.136/ | 0.893/ | 0.892/ | 0.124/ |
| | 0.056 | 0.108 | 0.235 | **0.348** | 0.134 | 0.104 | 0.073 | <u>0.924</u> | 0.913 | 0.246 |
| Ă | 0.323/ | 0.388/ | 0.420/ | 0.315/ | **0.495/** | 0.411/ | 0.458/ | 0.870/ | 0.847/ | 0.337/ |
| | 0.175 | 0.282 | 0.227 | 0.203 | **0.230** | 0.236 | 0.809 | <u>0.934</u> | 0.904 | 0.993 |
| Ș | 0.0/ | 0.0/ | 0.0/ | 0.0/ | 0.0/ | 0.0/ | 0.0/ | 0.0/ | 0.0/ | 0.0/ |
| | 0.0 | 0.0 | 0.0 | 0.0 | 0.0 | 0.0 | 0.0 | 0.0 | 0.0 | 0.0 |
| Ț | 0.154/ | 0.521/ | **0.563/** | 0.393/ | 0.438/ | 0.473/ | 0.317/ | 0.942/ | 0.918/ | 0.0/ |
| | 0.184 | 0.526 | **0.480** | 0.392 | 0.412 | 0.405 | 0.484 | <u>0.915</u> | 0.909 | 0.0 |
| avg | 0.185/ | 0.350/ | 0.401/ | 0.366/ | 0.407/ | 0.335/ | 0.259/ | 0.736/ | 0.724/ | 0.181/ |
| | 0.143 | 0.267 | 0.222 | 0.251 | 0.239 | 0.194 | 0.417 | 0.744 | 0.732 | 0.341 |

We compared our approach, training of separate neural networks for each character, to the common one, training one neural network for all the characters of a given language, using convolutional networks over separated training and

**Table 6.** Results of experiments for the Slovak language (precision, recall). Column names are presented in Table 2. Bold script indicates the best solution for separated train and test, underline - for mixed ones.

| | 0 | 1 | 2 | 3 | 4 | 5 | 6 | 7 | 8 | 9 |
|---|---|---|---|---|---|---|---|---|---|---|
| Á | 0.329/ 0.290 | 0.492/ 0.401 | **0.618/ 0.544** | 0.532/ 0.460 | 0.604/ 0.479 | 0.525/ 0.483 | 0.408/ 0.742 | 0.930/ 0.918 | 0.910/ 0.722 | 0.645/ 0.843 |
| Ä | 0.719/ 0.158 | 0.362/ 0.000 | 0.904/ 0.335 | 0.714/ 0.108 | 0.843/ 0.703 | **0.919/ 0.052** | 0.0/ 0.0 | 0.988/ 0.454 | 0.945/ 0.962 | 0.388/ 0.258 |
| É | 0.191/ 0.138 | 0.481/ 0.380 | 0.535/ 0.515 | 0.514/ 0.205 | 0.499/ 0.524 | **0.591/ 0.312** | 0.500/ 0.700 | 0.873/ 0.896 | 0.876/ 0.854 | 0.684/ 0.682 |
| Í | 0.395/ 0.328 | 0.559/ 0.522 | **0.670/ 0.540** | 0.591/ 0.390 | 0.538/ 0.455 | 0.445/ 0.394 | 0.389/ 0.599 | 0.929/ 0.910 | 0.893/ 0.836 | 0.427/ 0.953 |
| Ó | 0.188/ 0.280 | **0.572/ 0.239** | 0.125/ 0.262 | 0.402/ 0.403 | 0.347/ 0.203 | 0.543/ 0.343 | 0.513/ 0.406 | 0.941/ 0.913 | 0.905/ 0.884 | 0.497/ 0.834 |
| Ô | 0.038/ 0.038 | 0.290/ 0.030 | 0.344/ 0.112 | 0.662/ 0.291 | **0.963/ 0.647** | 0.618/ 0.144 | 0.309/ 0.902 | 0.985/ 0.817 | 0.981/ 0.948 | 0.0/ 0.0 |
| Ú | 0.313/ 0.216 | 0.375/ 0.410 | 0.674/ 0.599 | 0.557/ 0.396 | 0.519/ 0.454 | 0.569/ 0.564 | **0.751/ 0.446** | 0.948/ 0.912 | 0.927/ 0.884 | 0.819/ 0.795 |
| Ý | 0.328/ 0.281 | 0.705/ 0.657 | 0.692/ 0.778 | 0.530/ 0.645 | 0.753/ 0.517 | **0.792/ 0.476** | 0.667/ 0.668 | 0.917/ 0.942 | 0.886/ 0.921 | 0.894/ 0.766 |
| Č | 0.640/ 0.373 | 0.849/ 0.619 | 0.718/ 0.641 | 0.828/ 0.692 | 0.849/ 0.649 | 0.786/ 0.740 | **0.877/ 0.860** | 0.979/ 0.979 | 0.888/ 0.962 | 0.957/ 0.980 |
| Ď | 0.055/ 0.151 | 0.0/ 0.0 | 0.794/ 0.204 | 0.538/ 0.023 | 0.525/ 0.068 | 0.387/ 0.009 | 0.327/ 0.797 | 0.965/ 0.829 | 0.959/ 0.964 | 0.940/ 0.993 |
| Ĺ | 0.425/ 0.556 | 0.687/ 0.471 | 0.351/ 0.758 | 0.876/ 0.654 | 0.783/ 0.510 | **0.967/ 0.689** | 0.0/ 0.0 | 0.981/ 0.755 | 0.981/ 0.950 | 0.998/ 0.974 |
| L' | 0.188/ 0.132 | 0.777/ 0.339 | 0.567/ 0.290 | 0.632/ 0.241 | **0.794/ 0.250** | 0.671/ 0.470 | 0.715/ 0.913 | 0.964/ 0.944 | 0.903/ 0.958 | 0.801/ 0.935 |
| Ň | 0.218/ 0.289 | 0.311/ 0.055 | 0.360/ 0.424 | 0.507/ 0.374 | 0.322/ 0.434 | **0.609/ 0.517** | 0.358/ 0.125 | 0.970/ 0.855 | 0.892/ 0.905 | 0.850/ 0.776 |
| Ŕ | 0.144/ 0.025 | 0.126/ 0.018 | **0.935/ 0.665** | 0.288/ 0.267 | 0.633/ 0.281 | 0.449/ 0.262 | 0.0/ 0.0 | 0.996/ 0.263 | 0.942/ 0.906 | 0.0/ 0.0 |
| Š | 0.408/ 0.553 | 0.537/ 0.393 | 0.559/ 0.584 | 0.477/ 0.476 | **0.726/ 0.558** | 0.430/ 0.468 | 0.620/ 0.524 | 0.962/ 0.940 | 0.941/ 0.901 | 0.868/ 0.838 |
| Ť | 0.242/ 0.105 | 0.752/ 0.298 | 0.716/ 0.484 | 0.655/ 0.247 | 0.671/ 0.216 | 0.619/ 0.394 | **0.840/ 0.951** | 0.952/ 0.951 | 0.954/ 0.955 | 0.892/ 0.978 |
| Ž | 0.794/ 0.464 | 0.794/ 0.497 | 0.844/ 0.772 | 0.803/ 0.608 | 0.837/ 0.733 | 0.821/ 0.662 | **0.925/ 0.968** | 0.988/ 0.970 | 0.973/ 0.941 | 0.999/ 0.980 |
| avg | 0.330/ 0.257 | 0.509/ 0.313 | 0.612/ 0.500 | 0.594/ 0.381 | 0.659/ 0.451 | 0.631/ 0.410 | 0.482/ 0.564 | 0.956/ 0.838 | 0.926/ 0.908 | 0.685/ 0.740 |

test sets. In the former case, the output layer consists of a number of neurons with SoftMax activation function equal to the length of the replacement list containing symbols both with and without a diacritical mark. For the Turkish language the average precision over accentuated symbols was as low as 0.645 (the average accuracy for one-hot CNN for our approach was 0.93), for the French language was 0.029 (0.5 for our approach), for the German language - 0.377 (0.619 in our case), 0.491 for Croatian (0.816 in our case), etc. Thus, it is easy to see that our method demonstrates better results. However, it takes more computations, the training of one deep neural network takes up to six hours on

**Table 7.** Results of experiments for the Latvian language (precision, recall). Column names are presented in Table 2. Bold script indicates the best solution for separated train and test, underline - for mixed ones.

| | 0 | 1 | 2 | 3 | 4 | 5 | 6 | 7 | 8 | 9 |
|---|---|---|---|---|---|---|---|---|---|---|
| A | 0.278/ 0.313 | 0.367/ 0.403 | 0.425/ 0.490 | **0.461/ 0.490** | 0.385/ 0.447 | 0.380/ 0.519 | 0.425/ 0.361 | 0.913/ <u>0.829</u> | 0.900/ 0.780 | 0.456/ 0.418 |
| Č | 0.154/ 0.213 | 0.633/ 0.422 | 0.271/ 0.197 | 0.209/ 0.205 | **0.948/ 0.789** | 0.508/ 0.408 | 0.809/ 0.443 | 0.964/ <u>0.877</u> | 0.962/ 0.899 | 0.821/ 0.996 |
| Ē | 0.524/ 0.434 | 0.683/ 0.689 | 0.710/ 0.706 | 0.737/ 0.709 | 0.650/ 0.637 | 0.784/ 0.777 | **0.791/ 0.808** | 0.966/ <u>0.936</u> | 0.951/ 0.941 | 0.892/ 0.902 |
| Ģ | 0.892/ 0.530 | 0.801/ 0.625 | 0.930/ 0.734 | 0.935/ 0.487 | 0.896/ 0.405 | **0.963/ 0.651** | 0.911/ 0.946 | 0.987/ 0.897 | 0.989/ 0.981 | <u>0.996/ 1.0</u> |
| Ī | 0.467/ 0.439 | 0.705/ 0.785 | 0.895/ 0.828 | **0.905/ 0.836** | 0.874/ 0.837 | 0.827/ 0.818 | 0.472/ 0.713 | 0.985/ <u>0.973</u> | 0.967/ 0.972 | 0.597/ 0.823 |
| Ķ | 0.529/ 0.302 | 0.923/ 0.526 | 0.724/ 0.594 | 0.867/ 0.736 | 0.809/ 0.777 | 0.773/ 0.703 | **0.943/ 0.873** | 0.972/ <u>0.949</u> | 0.963/ 0.947 | 0.969/ 0.924 |
| Ļ | 0.510/ 0.309 | 0.734/ 0.534 | 0.751/ 0.482 | 0.733/ 0.600 | **0.813/ 0.534** | 0.689/ 0.464 | 0.510/ 0.794 | 0.973/ <u>0.930</u> | 0.945/ 0.940 | 0.904/ 0.590 |
| Ņ | 0.367/ 0.264 | 0.737/ 0.635 | **0.797/ 0.598** | 0.691/ 0.629 | 0.793/ 0.668 | 0.740/ 0.538 | 0.726/ 0.593 | 0.967/ 0.927 | 0.950/ 0.919 | <u>0.979/ 0.962</u> |
| Š | 0.287/ 0.594 | **0.828/ 0.625** | 0.732/ 0.697 | 0.669/ 0.699 | 0.780/ 0.772 | 0.601/ 0.656 | 0.648/ 0.676 | 0.967/ <u>0.935</u> | 0.956/ 0.945 | 0.917/ 0.786 |
| Ū | 0.347/ 0.319 | 0.596/ 0.617 | 0.478/ 0.340 | **0.660/ 0.566** | 0.638/ 0.408 | 0.496/ 0.414 | 0.648/ 0.821 | 0.979/ <u>0.956</u> | 0.969/ 0.963 | 0.720/ 0.797 |
| Ž | 0.556/ 0.482 | 0.749/ 0.707 | 0.724/ 0.631 | 0.737/ 0.725 | 0.757/ 0.700 | 0.861/ 0.744 | **0.889/ 0.886** | 0.978/ <u>0.938</u> | 0.968/ 0.947 | 0.818/ 0.711 |
| avg | 0.446/ 0.381 | 0.705/ 0.597 | 0.676/ 0.572 | 0.691/ 0.607 | 0.758/ 0.634 | 0.692/ 0.608 | 0.706/ 0.719 | 0.968/ 0.922 | 0.956/ 0.930 | 0.824/ 0.809 |

Google Colab's T4. As result, the training of all the models for languages like Slovak takes a whole day and should be done several times.

Considering the difference between results with and without intersection of training and test vocabularies, we can state that results for mixed vocabularies are almost always better than for vocabularies without intersection. It is always true for a dense 7-character input vector, except the French Â with obviously erroneous results for LSTM on separate dictionaries, the French Ô, and the German Ö. Probably, the last two cases can be explained by the lack of statistics, since we had trained neural networks 3–6 times on separate data sets, while for mixed data sets - only once. The average difference here for the RNN network is 0.28, and for CNN is 0.33. Mixed vocabularies demonstrated the best result with the average difference of 0.12. Thus, we can state that methods of investigation of low resource languages should be changed.

Note, that we considered words not shorter than 4 characters. Such languages as Vietnamese, Chinese, and Irish have shorter names and prefixes. Our experiments demonstrated that the introduced method lacks of productivity; we suppose that it is mostly improbable to restore the correct diacritics having just

**Table 8.** Results of experiments for the Turkish language (precision, recall). Column names are presented in Table 2. Bold script indicates the best solution for separated train and test, underline - for mixed ones.

|   | 0 | 1 | 2 | 3 | 4 | 5 | 6 | 7 | 8 | 9 |
|---|---|---|---|---|---|---|---|---|---|---|
| Ç | 0.852/ 0.694 | 0.776/ 0.762 | **0.946/ 0.708** | 0.808/ 0.652 | 0.457/ 0.692 | 0.892/ 0.857 | 0.910/ 0.876 | 0.978/ 0.970 | 0.984/ <u>0.963</u> | 0.968/ 0.946 |
| Ö | 0.880/ 0.727 | 0.922/ 0.611 | 0.923/ 0.629 | 0.904/ 0.929 | 0.0/ 0.0 | 0.926/ 0.959 | **0.976/ 0.981** | 0.991/ <u>0.979</u> | 0.989/ 0.992 | 0.948/ 0.981 |
| Ü | 0.873/ 0.812 | 0.809/ 0.833 | 0.923/ 0.826 | 0.932/ 0.769 | 0.0/ 0.0 | 0.919/ 0.907 | **0.955/ 0.853** | 0.991/ <u>0.978</u> | 0.990/ 0.984 | 0.965/ 0.853 |
| Ğ | 0.974/ 0.854 | 0.950/ 0.992 | 0.973/ 0.995 | 0.990/ 0.986 | 0.727/ 0.689 | 0.955/ 0.994 | <u>0.953/ 0.998</u> | 0.995/ 0.998 | <u>0.997/ 0.999</u> | 0.993/ 0.996 |
| Ş | 0.699/ 0.549 | 0.637/ 0.730 | 0.809/ 0.814 | 0.622/ 0.713 | 0.451/ 0.510 | 0.575/ 0.486 | **0.859/ 0.752** | 0.961/ <u>0.938</u> | 0.958/ 0.921 | 0.790/ 0.760 |
| avg | 0.855/ 0.727 | 0.818/ 0.785 | 0.914/ 0.794 | 0.851/ 0.809 | 0.327/ 0.378 | 0.853/ 0.840 | 0.930/ 0.892 | 0.983/ 0.972 | 0.983/ 0.971 | 0.932/ 0.907 |

one symbol. Moreover, we can not be completely sure that a neural network trained on news wire will correctly restore diacritics in proper names due to specificity of testing data.

Some characters, such as French Î, German Ä, Romanian Ş, demonstrate very poor results. The reasons of such behavior need to be investigated in an independent research, but we can presuppose that these characters could be found in the same context as their counterparts. For example, this is true for French Î which is used to differentiate similarly sounding words. But this is not always the case of German Ä, since there are many words which do not have counterparts.

## 5   Conclusion

In this paper we demonstrated that there is no best neural network architecture for each considered language in the area of diacritics restoration. There are differences in the neural network design and their results in case of low and full resourced languages. For a full resourced language, the task is usually stated as searching for a word from the vocabulary and taking a decision about its diacritics. For a low resourced language, we could have in hand an out-of-vocabulary word. These tasks demand different approaches to their solution. In the former case, the slightly better results are achieved with the biLSTM networks; in the later case, the choice depends on the considered language.

We have investigated languages such as Croatian, Slovak, Romanian, French, German, Latvian, and Turkish. We found that there is a large difference between the results for the case of completely separated train and test vocabularies and the random distribution of tokens between these vocabularies. The average difference between precision for these two cases reaches 0.12–0.33.

We have presented a new approach based on several neural networks each trained for its own symbol. This approach wins over the previous one, where only one neural network is trained for all the symbols.

Almost always, there is a good option for any character with diacritics to be restored. However, the results for some characters are quite low. These are French Î, German Ä, and Romanian Ş. For the last character, any machine training algorithm has decided that it is easier to attribute all the characters as S since there are not so many examples of words with Ş; the same is true for many other cases. This means that we have to use other methods to form train and test sets and construct batches. The case of the German Ä needs a linguistic investigation of this phenomenon.

# References

1. Yarowsky, D.: Comparison of corpus-based techniques for restoring accents in Spanish and French text. Natural Language Processing using Very Large Corpora 1999, pp. 99–120. Springer, Dordrecht (1999)
2. Tufiş, D., Chiţu, A.: Automatic diacritics insertion in Romanian texts. In: Proceedings of COMPLEX 1999 International Conference on Computational Lexicography, Pecs, Hungary, pp. 185–194 (1999)
3. Ezeani, I., Hepple, M., Onyenwe, I.: Automatic restoration of diacritics for Igbo language. In: Sojka, P., Horák, A., Kopeček, I., Pala, K. (eds.) TSD 2016. LNCS (LNAI), vol. 9924, pp. 198–205. Springer, Cham (2016). https://doi.org/10.1007/978-3-319-45510-5_23
4. Schlippe, T., Nguyen, T.L., Vogel, S. E.: Diacritization as a machine translation problem and as a sequence labeling problem. In: Proceedings of the Eighth Conference of the Association for Machine Translation in the Americas, pp. 270–278 (2008)
5. Náplava, J., Straka, M., Straňák, P., Hajič, J.: Diacritics restoration using neural networks. In: Proceedings of the Eleventh International Conference on Language Resources and Evaluation (LREC 2018), pp. 1566–1573 (2018)
6. Alqahtani, S., Mishra, A., Diab, M.: Efficient convolutional neural networks for diacritic restoration. In: Proceedings of the 9th International Joint Conference on Natural Language Processing, pp. 1442–1448 (2019)
7. Orife, I.: attentive sequence-to-sequence learning for diacritic restoration of Yorùbá language text. In: Proceedings of the INTERSPEECH 2018, pp. 2848–2852 (2018). https://doi.org/10.21437/Interspeech.2018-42
8. De Pauw, G., Wagacha, P.W., de Schryver, G.-M.: Automatic diacritic restoration for resource-scarce languages. In: Proceedings of the Text, Speech and Dialogue, 10th International Conference, pp. 170–179 (2007). https://doi.org/10.1007/978-3-540-74628-7_24

# Theoretical Machine Learning
# and Optimization

# On Asymptotically Optimal Solvability of Euclidean Max $m$-$k$-Cycles Cover Problem

Edward Gimadi[1,2] and Ivan Rykov[1,2]([envelope])

[1] Sobolev Institute of Mathematics, Koptuga 4, Novosibirsk, Russia
[2] Department of Mechanics and Mathematics, Novosibirsk State University,
1 Pirogova Street, 630090 Novosibirsk, Russia
gimadi@math.nsc.ru, rykovweb@gmail.com
http://math.nsc.ru

**Abstract.** We consider the problem of finding $m$ edge-disjoint $k$-cycles covers formulated in $d$-dimensional Euclidean space. We construct a polynomial-time approximation algorithm for this problem and derive conditions of its asymptotical optimality.

**Keywords:** Cycles covering · m-TSP · Asymptotically optimal

## 1 Introduction

We consider the following problem: given a complete undirected weighted graph $G = (V, E)$, where the set $V$ consists of $n$ vertices represented by points in $d$-dimensional Euclidean space $R^d$ and the weight of each edge equals to the Euclidean distance between corresponding points.

A cycle cover of a graph is a spanning subgraph whose connected components are simple cycles.

Find a union $\mathcal{C} = \{C_1 \cup C_2 \cup \cdots \cup C_m\}$ of $m$ edge-disjoint cycle covers, such that each set $C_i$ is a spanning 2-factor in $G$, consisting of exactly $k$ cycles, so that the sum of weights of all the edges in the union $\mathcal{C}$ is maximized.

The paper is organized as follows. Section 2 summarizes related work on the studied problem. Section 3 contains necessary preliminary definitions and theorems. Section 4 introduces the proposed algorithm. Section 5 presents the theoretical analysis of the algorithm. Section 6 concludes the paper.

## 2 Related Works

We are considering the combinatorial optimization problem, which is closely related to the well-known Traveling Salesman Problem (TSP) and $k$-Cycles

The work was supported by the program of fundamental scientific researches of the SB RAS, project No 0314-2019-0014.

W. M. P. van der Aalst et al. (Eds.): AIST 2020, CCIS 1357, pp. 257–266, 2021.
https://doi.org/10.1007/978-3-030-71214-3_21

Cover Problem ($k$-CsCP). For a fixed natural number $k$ and a given complete weighted graph $G = (V, E)$, it is required to find an extremal (minimum, or maximum)-weight cover of the set $V$ by $k$ vertex-disjoint cycles.

TSP is a particular case of $k$-CsCP for $k = 1$.

It is known [14] that the TSP is NP-hard even in the Euclidean case, the optimal solution can not be found in polynomial time, unless P = NP. Although the TSP is hardly approximable [15] in the general case, for some special cases polynomial-time approximation algorithms are developed. For instance, the Metric TSP [3] can be approximated in polynomial time with a ratio 3/2, and, for Euclidean TSP, a polynomial-time approximations scheme [1] and an asymptotically correct algorithm are developed [16] and [5].

Other well-known generalizations of the TSP is $m$-PSP. Instances of this problem are given by undirected complete graphs with positive edge weights.

In $m$-PSP, the goal is to nd $m$ edge-disjoint Hamiltonian cycles $H_1, \ldots, H_m$ of minimun or maximum total edge weight. The problem remains intractable in metric and Euclidean special cases. For Euclidean Max $m$-PSP [12], there exist the asymptotically optimal algorithm [2].

In [10] and [11] it is shown that the Min $k$-CsCP is NP-hard in the strong sense, both in general and in particular cases, Metric and Euclidean. For the case $k = 2$ efficient 2-approximation algorithm is proposed, and for the Euclidean problem on the plane a PTAS is suggested.

Asymptotically optimal algorithms with running time $\mathcal{O}(n^3)$ were suggested:

1) for the Euclidean Max $k$-CsCP with given lengths cycles and $k = o(n)$ [6];
2) for the Random Min $k$-CsCP (on instances UNI(0,1)) with $k \leq n^{1/3} \ln n$ [7].

In [18] a problem of finding a maximum-cost cycle cover which satisfies an upper bound on the number of cycles and a lower bound on the number of edges in each cycle is considered. There is suggested a polynomial-time algorithm in the geometric case when the vertices of the graph are points in a multidimensional real space and the distances between them are induced by a positively homogeneous function whose unit ball is an arbitrary convex polytope with a fixed number of facets.

## 3 Preliminary Facts

The integer numbers $n, m, k$, and $d > 1$ are the input parameters of the problem Euclidean Max $m$-$k$-CsCP, together with points $v_1 = (v_{11}, \ldots v_{1d}), \ldots v_n = (v_{n1}, \ldots, v_{nd})$ defining the graph.

This problem is a generalization of two problems, considered earlier, both of them being generalizations of Euclidean Max TSP:

- problem of finding $m$ edge-disjoint Hamiltonian cycles in complete Euclidean graph (Euclidean Max $m$-PSP): Max TSP is a particular case with $m = 1$;
- problem of finding a maximum-weight single covering of complete Euclidean graph with $k$ cycles (the Max $k$-CsCP): Max TSP is a particular case with $k = 1$.

First of these problems was considered in [2], where an asymptotically optimal solvability for the problem Euclidean Max $m$-PSP was established.

**Theorem 1.** *[2] Euclidean Max m-PSP problem is solved with relative error*

$$\varepsilon_{\mathcal{A}}(n) \leq \frac{1}{n} + \alpha_d \left(\frac{m}{n}\right)^{\frac{2}{d+1}},$$

$\alpha_d$ *being a constant depending only on the dimension d.*

In [6] an algorithm for the latter problem (of finding $k$-cycles coverage of maximal length) was introduced, with the following estimation for the relative error:

**Theorem 2.** *[6] The Euclidean Max k-CsCP is solved with relative error*

$$\varepsilon_{\mathcal{A}}(n) \leq \frac{2k}{n} + \beta_d \left(\frac{1}{n}\right)^{\frac{2}{d+1}},$$

$\beta_d$ *being a constant depending only on the dimension d.*

Our goal is to explore a more general problem with both $m$ and $k$ being given as a part of input. Note that, using properties of the remote angle between two vectors in the normed space, introduced in [17], an asymptotically optimal approach was realized for the Normed Max $m$-PSP [8], and then for the Normed Max $m$-$k$-CsCP [9].

**Theorem 3.** *[9] The Normed Max m-k-CsCP is solved with relative error*

$$\varepsilon_{\mathcal{A}}(n) \leq \frac{2k+1}{n} + 2(d+1)\left(\frac{m}{n}\right)^{1/(d+1)}.$$

In the current work we present $\mathcal{O}(n^3)$-time Algorithm $\mathcal{A}$ for the Euclidean Max $m$-$k$-CsCP. It turns out that in Euclidean space the problem is solved more accurately than in a normed space.

Let $w(u, v)$ be the weight of edge $e = (u, v) \in E$ and $W(G') = \sum_{e \in E'} w(e)$ be the total weight of a subgraph $G' = (V; E')$ of the initial graph $G = (V; E)$ with the set of edges $E' \subseteq E$. The goal of the algorithm $\mathcal{A}$ for the maximum $m$-$k$-CsCP is to find a subset of edges $\widetilde{C} \subset E$, consisting of $m$ edge-disjoint $k$-Cycles Coverings $C_1, \ldots, C_m$. At the beginning of the algorithm, $\widetilde{C}$ is empty.

Let $\mathcal{M}^* = \{I_1, \ldots, I_\mu\}$ be the set of edges (intervals in $R^d$) of a maximum weight matching in $G$; $\mu = \lfloor n/2 \rfloor$.

**Definition 1.** *Two edges $e_1, e_2 \in E$ are linked (with respect to set $\widetilde{C}$), if there exists an edge $e \in E$ ($e \in \widetilde{C}$), that connects the end vertices of $e_1$ and $e_2$.*

**Definition 2.** *An I-chain is a sequence of edges, where each two neighboring edges are linked.*

**Definition 3.** *Two I-chains are linked (with respect to set $\widetilde{C}$) if their end edges are linked.*

We declare one of the end edges of an $I$-chain to be the master edge and the other one to be the inferior edge.

**Definition 4.** *An $\alpha$-chain is an $I$-chain, where the remote angle between any two neighboring edges of the chain is at most $\alpha$.*

For a description of the Algorithm it is convenient to use the spatial figure, which we will call a $\mu$-pseudo-prism P (Fig. 4), $\mu = \lfloor n/2 \rfloor$.

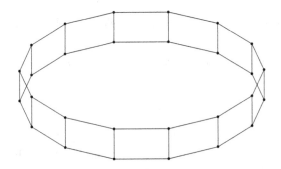

**Fig. 1.** The example of the $\mu$-pseudo-prism, $\mu = 14$.

**Definition 5.** *A subgraph of a graph $G$ forms the $\mu$-pseudo-prism if it consists of two edge-disjoint $\mu$-vertex circuits $(u_1, \ldots, u_\mu, u_1)$ and $(v_1, \ldots, v_\mu, v_1)$, connected between itself by $\mu$ edge jumpers $(u_j.v_j)$, $j = 1, \ldots, \mu$, where $\mu = \lfloor \frac{n}{2} \rfloor$. Thus, the prism consists of $2\mu$ vertices and $3\mu$ edges.*

In such pseudo-prism, unlike a regular spatial prism, the opposite sides of the quadrangles can be non-parallel, and the vertices of the cycles do not have to be in the same plane.

We will describe an asymptotically optimal algorithm $\mathcal{A}$, which

- builds $m$ maximum-total weight $\mu$-pseudo-prisms $P_1, \ldots, P_m$ in the Euclidean space, where the maximum matching edges of the original graph $G$ are used as the jumpers of the prisms, bat all of other edges are non adjacent.
- constructs the maximum $m$-$k$-CsCP.

## 4  Description of the Algorithm $\mathcal{A}$

**Preliminary Stage**
In the given graph $G$ find a matching $\mathcal{M}^* = \{I_1, \ldots, I_\mu\}$ of maximum weight, where $\mu = \lfloor n/2 \rfloor$ is the number of its edges (intervals).

Set $\widetilde{C} = \emptyset$ and fix a parameter $t \leq \mu/2$. Sort the edges of $\mathcal{M}^*$ in the non-increasing order. We will refer to the first $(\mu - t)$ edges of $\mathcal{M}^*$ as heavy edges, and the last $t$ edges as light.

**Stage 1: Designing pseudo-prisms $P_i$, $i = 1, \ldots, m$.**
Define angle $\alpha_i$ according to the relation :

$$\sin^2 \frac{\alpha_i}{2} = \gamma_d t_i^{-2/(d-1)}. \tag{1}$$

where $\gamma_d$ being a constant depending only on the dimension $d$;
$t_i$ is the number of admitible edges when building a prism $P_i$:

$$t_i = \begin{cases} t, & \text{if } i = 1; \\ t/(2i-2) & \text{for } 1 < i \leq m. \end{cases} \tag{2}$$

Set $i = 1$.

**Step 1.** Constructing a sequence $\mathcal{S} = \{S_1, \ldots, S_t\}$ of $t$ $\alpha$-chains.
Build a set $\mathcal{I}$ of $\alpha$-chains, which consist only of the heavy edges. (Note that an edge is a one-element $\alpha_i$-chain.)
We start with $\mathcal{I}_t$ consisting of the first $t$ heaviest edges of $\widetilde{\mathcal{M}^*}$: $\mathcal{I}_t = \{I_1, I_2, \ldots, I_t\}$.
Set $j = t$.
In the current $t$-chain $\mathcal{I}_j$, find a pair of non-linked (with respect to the set $\widetilde{C}$) $I$-chains such that the angle between their master edges is at most $\alpha_i$.
Join these chains into one $\alpha_i$-chain by setting their master edges to be neighbors and assign one of the end edges of the joined chain (one of the former inferior edges) to be the new master edge. Set $j := j + 1$. If $j < \mu - t$, then append one more heavy edge $I_j$ to the current set $\mathcal{I}$ and repeat Stage 1.
Otherwise, we have obtained a sequence $\mathcal{S} = \{S_1, \ldots, S_t\}$ of $t$ $\alpha$-chains such that each of them consists of a sequence of heavy edges with the angle between any consecutive (neighboring) pair of edges at most $\alpha_i = \alpha_d(t_i)$.

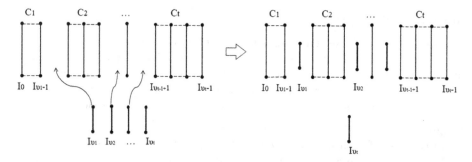

**Fig. 2.** The bold lines correspond to the edges of $\mathcal{M}^*$, while the dashed lines indicates the $\alpha$-chains. The $t$ light edges of $\mathcal{M}^*$ were placed to the positions between the $\alpha$-chains. The last light edge $I_{\nu_t}$ is placed between the first and the $t$-th $\alpha$-chains.

**Step 2.** Constructing a pseudo-prism $P_i$.

Let's regard the sequence $S$ as a cycle, i.e. the $\alpha_i$-chain $S_t$ is followed by the $\alpha_i$-chain $S_1$. Let the edges of the $\alpha_i$-chains $S_1, \ldots, S_t$ be enumerated so that $S_r = \{I_{\nu_{r-1}+1}, \ldots, I_{\nu_r-1}\}$, $1 \le r \le t$, where $\nu_1 < \nu_2 < \ldots < \nu_t$ are the numbers reserved for the remaining light edges of $\mathcal{M}^*$ ($\nu_0 = 0$, $\nu_t = \mu$).

Place $t$ light edges of $\mathcal{M}^*$ to the positions $\nu_1, \nu_2, \ldots, \nu_t = \mu$, such that no light edge is linked to the neighboring end edges of the $\alpha$-chains.

Construct a pseudo-prism $P_i$ such that $P_i \setminus \mathcal{M}^*$ is edge-disjoint with $P_1 \setminus \mathcal{M}^*, P_2 \setminus \mathcal{M}^*, \ldots, P_{i-1} \setminus \mathcal{M}^*$ in the following way. We assume that the sequence of edges $\{I_1, I_2, \ldots, I_\mu\}$ is given according to their order in the sequence $S$, $I_j = (x_j, y_j)$, $j = 1, \ldots, \mu$. Now we are going to construct a pseudo-prism $P_i$.

For $j = \nu_1, \ldots, \nu_t$ execute the following operator

| if $w(u_{j-1}, x_j, u_{j+1}) + w(v_{j-1}, y_j, v_{j+1}) \ge w(u_{j-1}, y_j, u_{j+1}) + w(v_{j-1}, x_j, v_{j+1})$, then set $u_j = x_j$; $v_j = y_j$; otherwise, set $u_j = y_j$ and $v_j = x_j$. |
|---|

As a result of the Step 2 we have obtained a pseudo-prism $P_i$, consisting of the two non-intersecting circuits $(u_1, u_2, \ldots, u_\mu.u_1)$ and $(v_1, v_2, \ldots, v_\mu.v_1)$, and of the maximum matching $\mathcal{M}^* = \{I_1, I_2, \ldots, I_\mu\}$ where $I_j = (u_j, v_j)$, $j = 1, \ldots, \mu$.

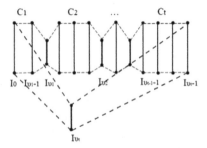

**Fig. 3.** The $\mu$-pseudo-prism: the dashed lines form two edge-disjoint $\mu$-vertex circuits; $\mu$ bold lines are edge jumpers of $\mathcal{M}^*$.

**Stage 3.** Constructing $m$-$k$-CsC.

**Common Step: Constructing $k$-Cycles Covering $C_i$, $i = 1, \ldots, m$.**

In order to obtain coverage with $k$ cycles, it (arbitrary) chooses $k > 1$ pairs of adjacent edges in the corresponding ordering of matching $\mathcal{M}^*$ and performs "reverse operation", i.e. returns edges of matching into the solution $C_i$ and removes pair of edges that connected endpoints of this pair of adjacent edges.

**Stage 4.** Cases of evenness and oddness of $n$.

In the case of even $n$ we have $k$-CsC $C_i$ of graph $G$ and go to Stage 5.

If $n$ is odd, there exists a vertex $x_0$ that is not in $\mathcal{M}^*$. In this case replace one of the matching edge $(u, v)$ of the constructed covering by the pair of edges $(u, x0)$ and $(v, x0)$ so that none of these edges intersects the set $\widetilde{C}$. The triangle inequality guarantees that the weight of the cycle will not decrease.

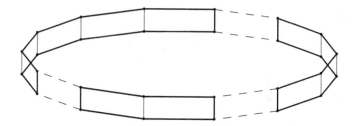

**Fig. 4.** The example of a solution on the 14- pseudo-prism $P_i$. Thick lines highlightes one of 3-Cycles Coverings.

**Stage 5.** Append the edges of the obtained $k$-covering $C_i$ to the set $\widetilde{C}$. Further these $2ik$ edges of matching $\mathcal{M}^*$ added to $C_1, \ldots C_i$ are marked as forbidden.

The description of the algorithm $\mathcal{A}$ is complete.

## 5   Algorithm Analysis

**Lemma 1.** *The admissible number of cycles in the Max $m$-$k$-CsCP satisfies the inequality*

$$k \leq \frac{n}{8m}.$$

*Proof.* We search for another $k$ pairs of adjacent edges. In the worst case $2ik$ edges are forbidden and $2ik - 1$ edges are located between them, one between each two forbidden and thus can't be used. So, in order to be able to finish this operation on m-th coverage, we need that $2k + 4(m-1)k - 1 \leq n/2$. From this it follows restriction on parameters of the problem: $k \leq \frac{n}{8m}$.

**Lemma 2.** *[2] For $d \geq 2$ the following inequality holds:*

$$\sum_{i=1}^{m} t_i^{-2/(d-1)} \leq \left(\frac{m}{t}\right)^{-2/(d-1)}. \tag{3}$$

**Theorem 4.** *The Max $m$-$k$-CsCP is approximately solved by the algorithm $\mathcal{A}$ in running time $\mathcal{O}(n^3)$ with relative error estimated by the inequality*

$$\varepsilon_{\mathcal{A}}(n) \leq \frac{2(t+k)+1}{n} + \gamma_d \left(\frac{m}{t}\right)^{-2/(d-1)}. \tag{4}$$

*Proof.* The running-time of the algorithm is determined by the time one needs to construct a maximum weight matching, which is $\mathcal{O}(n^3)$ [4].

The algorithm $\mathcal{A}$ produces $m$ $k$-coverings $C_1, C_2, \ldots, C_m$, and since each time we arrange the edges in $\mathcal{S}$ so that they are not linked with respect to the edges of $\widetilde{E}$, the obtained coverings are edge-disjoint.

In [16] it is shown, that

$$W(C_i) \geq 2\,W(\widetilde{\mathcal{M}^*})\left(1 - \frac{1}{n}\right)\cos\frac{\alpha_i}{2},$$

where for the total weight $W(\widetilde{\mathcal{M}^*})$ of the first $(\mu - t)$ heaviest edges of $\mathcal{M}^*$ the following inequality satisfies:

$$W(\widetilde{\mathcal{M}^*}) \geq W(\mathcal{M}^*)\left(1 - \frac{t}{\mu}\right). \tag{5}$$

It follows from this

$$W(C_i) \geq 2W(\mathcal{M}^*)\left(1 - \frac{1}{n}\right)\left(1 - \frac{t}{\mu}\right)\cos\frac{\alpha_i}{2} \geq 2W(\mathcal{M}^*)\left(1 - \frac{2t+1}{n} - \frac{\sin^2\alpha_i/2}{2}\right). \tag{6}$$

where the angle $\alpha_i$ is defined by (1).

On the other hand, the upper bound of optimum for the coverage problem is estimated as

$$2\,W(\mathcal{M}^*) \geq \left(1 - \frac{k}{n}\right)W(C_i^*), \tag{7}$$

(the worst case of all odd cycles), used in [6].

where $C^*$ is the solution of the Max $k$-CsCP in the given graph.

Using (4), (6) and (7), for the relative error $\varepsilon_{\mathcal{A}}(n)$ of algorithm $\mathcal{A}$ we have

$$\varepsilon_{\mathcal{A}}(n) = 1 - \frac{W_{\mathcal{A}}}{OPT} \leq 1 - \frac{W_{\mathcal{A}}}{mW(C^*)} = 1 - \frac{W(C_1) + \ldots + W(C_m)}{mW(C^*)}$$

$$\leq \frac{2(t+k)+1}{n} + \frac{\gamma_d}{2}\sum_{i=1}^{m}t_i^{-2/(d-1)} \leq \frac{2(t+k)+1}{n} + \frac{\gamma_d}{2}\left(\frac{m}{t}\right)^{-2/(d-1)}.$$

Theorem 4 is proved.

Now we can formulate the main result of the article.

**Theorem 5.** *A polynomial-time approximation algorithm $\mathcal{A}$ with parameters $t^* = m\left(\frac{n}{m}\right)^{\frac{d+1}{d-1}}$ and $k \leq \min\left(\frac{n}{8m}, o(n)\right)$ gives asymptotically optimal solutions for the Max $m$-$k$-CsCP in the Euclidean space with a fixed dimension $d$ and $m = o(n)$.*

*Proof.* Setting the given value of the parameter $t = t^*$ in the condition of Theorem, we obtain

$$\varepsilon_{\mathcal{A}}(n) \leq \frac{2k+1}{n} + \left(2 + \frac{\gamma_d}{2}\right)\left(\frac{m}{n}\right)^{2/(d+1)} \to 0$$

as $n \to \infty$.

**Remark.** *The specificity of the Euclidean space made it possible to obtain a solution of the Max $m$-$k$-CsCP with better accuracy in comparison with the solution in a Normed space in* [9].

Indeed, it follows directly from the estimations of the relative error $\varepsilon_{\mathcal{A}}(n)$ in Theorems 3 and 5:

$$\left(\frac{m}{n}\right)^{2/(d+1)} \leq \left(\frac{m}{n}\right)^{1/(d+1)}.$$

## 6  Conclusion

Using angle estimations obtained in [16], we construct an algorithm for the Euclidean Max $m$-$k$-CsCP, which gives asymptotically optimal solution for the problem in an Euclidean space of fixed dimension, given that $m = o(n)$ and $k \leq \min\left(\frac{n}{8m}, o(n)\right)$.

As a topic for further research it is interesting to extend this approach to different modifications of the considered problem. For example, it is of natural interest to construct an asymptotically optimal algorithm for the $m$-$k$-CsCP that would essentially rely on the specifics of problem statement and would have a better relative error or less tight condition for the numbers $m$ and $k$ of coverings.

## References

1. Arora, S.: Polynomial-programming: methods and applications. In: Reeves, C.R. (ed.) NATO Advanced Study Institute Series, Series C: Mathematics and Physics Science, vol. 19, pp. 1730–178. Reidel, Dordrecht (1975)
2. Baburin, A.E., Gimadi, E.Kh.: On the asymptotic optimality of an algorithm for solving the maximum m-PSP in a multidimensional Euclidean space. Proc. Steklov Inst. Math. **272**(Suppl. 1), 1–13 (2011)
3. Christodes, N.: Worst-case analysis of a new heuristic for the traveling salesman problem. In: Symposium on New Directions and Recent Results in Algorithms and Complexity, p. 441. Academic Press, New York (1976)
4. Gabow, H.N.: An efficient reduction technique for degree-restricted subgraph and bidirected network flow problems. In: Proceedings of the 15th Annual ACM Symposium on Theory of Computing 1983, Boston, USA, pp. 448–456. ACM, New York (1983)
5. Gimadi, E.Kh.: A new version of the asymptotically optimal algorithm for solving the Euclidean maximum traveling salesman problem. In: Proceeding of the 12th Baykal International Conference 2001, Irkutsk, vol. 1, pp. 117–123 (2009). (in Russian)
6. Gimadi, E.Kh., Rykov, I.A.: Asymptotically optimal approach to the approximate solution of several problems of covering a graph by nonadjacent cycles. Proc. Steklov Inst. Math. **295**, 57–67 (2016). https://doi.org/10.1134/S0081543816090078
7. Gimadi, E.Kh., Rykov, I.A.: On asymptotical optimality of solving maximum-weight Euclidean m-cycles covering problem. Doklady Math. **93**(1/2), 117–120 (2016)

8. Gimadi, E.Kh., Tsidulko, O.Yu.: Asymptotically optimal algorithm for the maximum $m$-peripatetic salesman problem in a normed space. In: Battiti, R., Brunato, M., Kotsireas, I., Pardalos, P.M. (eds.) LION 12 2018. LNCS, vol. 11353, pp. 402–410. Springer, Cham (2019). https://doi.org/10.1007/978-3-030-05348-2_33

9. Gimadi, E.Kh., Rykov, I.A.: On asymptotically optimal solvability of max $m$-$k$-cycles cover problem in a normed space. In: Kononov, A., Khachay, M., Kalyagin, V.A., Pardalos, P. (eds.) MOTOR 2020. LNCS, vol. 12095, pp. 85–97. Springer, Cham (2020). https://doi.org/10.1007/978-3-030-49988-4_6

10. Khachay, M., Neznakhina, E.: Approximation of Euclidean k-size cycle cover problem. Croatian Oper. Res. Rev. (CRORR) **5**, 177–188 (2014)

11. Khachay, M., Neznakhina, E.: Polynomial-time approximations scheme for a Euclidean problem on a cycle covering of agraph. Trudy instituta matematiki i mehaniki UrORAN **20**(4), 297–311 (2014). (in Russian)

12. Krarup, J.: The peripatetic salesman and some related unsolved problems. In: Roy, B. (eds.) Combinatorial Programming: Methods and Applications. NATO Advanced Study Institutes Series (Series C – Mathematical and Physical Sciences), vol. 19, pp. 173–178. Springer, Dordrecht (1975). https://doi.org/10.1007/978-94-011-7557-9_8

13. Manthey, B.: Minimum-weight cycle covers and their approximability. In: Brandstädt, A., Kratsch, D., Müller, H. (eds.) WG 2007. LNCS, vol. 4769, pp. 178–189. Springer, Heidelberg (2007). https://doi.org/10.1007/978-3-540-74839-7_18

14. Papadimitriou, C.: Euclidean TSP is NP-complete. Theoret. Comput. Sci. **4**, 237–244 (1977)

15. Sahni, S., Gonzales, T.: P-complete approximation problems. J. ACM **23**(3), 555–565 (1976)

16. Serdyukov, A.I.: An asymptotically optimal algorithm for the maximum traveling salesman problem in Euclidean space. Upravlyaemye sistemy, Novosibirsk **27**, 79–87 (1987). (in Russian)

17. Shenmaier, V.V.: Asymptotically optimal algorithms for geometric Max TSP and Max m-PSP. Discret. Appl. Math. **163**(2), 214–219 (2014)

18. Shenmaier, V.V.: An algorithm for the polyhedral cyclic covering problem with restrictions on the number and length of cycles. Tr. IMM UrORAN. **24**(3), 272–280 (2018). https://doi.org/10.21538/0134-4889-2018-24-3-272-280

# An Effective Algorithm for the Three-Stage Facility Location Problem on a Tree-Like Network

Edward Kh. Gimadi[1,2] and Aleksandr S. Shevyakov[2(✉)]

[1] Sobolev Institute of Mathematics SB RAS, Novosibirsk, Russia
gimadi@math.nsc.ru
[2] Novosibirsk State University, Novosibirsk, Russia
shevash.97@gmail.com

**Abstract.** In this article we consider a three-level facility location problem on a tree-like network under the restriction that the transportation costs for a unit of production from one node to another is equal to the sum of the edges in the path connecting these nodes. As a result we construct an exact algorithm for this problem and prove his complexity equeled $O(nm^6)$, where n is the number of the production demand points and, m is an upper bound on the number of possible facility location sites of each level.

**Keywords:** Three-level facility location problem · Tree-like network · Polynomial-time algorithm

## 1 Introduction

The class of multistage facility location problems is characterized by existence of several stages of manufacturing where the raw materials are processed before the end-product arrives to the final consumer. Petroleum mining and processing, when the crude from a well first comes to an oil-processing plant and afterwards the petrol arrives to oil stations that are the final consumers in this case, is the classic example of a two-stage facility location problem. It is known that, in the general case, the facility location problem is NP-hard even in the classical one-stage variant [3,4]. A fairly modern overview of the multistage facility location problems is given in [9] (see also addition in [8]).

Therefore, it looks appropriate for the multistage facility location problem to carry out investigations in the following two ways:

1. Search for the special cases of the problem in which constructing exact polynomial time algorithms are possible. In [5,6] for the metric $k$-stage facility location problem on the path, an algorithm of time complexity $\mathcal{O}(n^3 \sum_{r=1}^{k} m_r)$

---

Supported by the program of fundamental scientific researches of the SB RAS, and by the Ministry of Science and Higher Education of the Russian Federation under the 5-100 Excellence Programme.

W. M. P. van der Aalst et al. (Eds.): AIST 2020, CCIS 1357, pp. 267–274, 2021.
https://doi.org/10.1007/978-3-030-71214-3_22

is developed, where $n$ is the number of the production demand points and $m_r$ is the number of possible places for the facilities opening at stage $r$. The algorithm is polynomial for the constant number of stages. For the metric three-stage problem on the path, in [5,6], algorithms of the time complexity $\mathcal{O}(nm_1m_2m_3)$, and $\mathcal{O}(nm^3)$, where $m$ is an upper bound on the number of possible facility location points at each stage. In [7], the exact algorithm constructed withtime complexity $\mathcal{O}(nm^3)$.

2. Development of approximate algorithms with guaranteed performance. For example, in [1] for the metric $k$-stage facility location problem, a combinatorial algorithm was constructed with the preciseness bound 3.27, which improves drastically the previous record equal to 6 [2]. For the cases $k = 2$ and $k = 3$, the algorithms were designed with bounds 2.4211 and 2.8446 respectively. Later, for the case $k = 2$ in [10] with the use of the greedy algorithms techniques, some polynomial algorithm was constructed with the preciseness bound 1.77.

In this article we would like to generelize the result showed in [7] for the three level problem. At the end we build an exact algorithm for this problem and prove his complexity equeled $O(nm^6)$, where n is the number of the production demand points and, m is an upper bound on the number of possible facility location sites of each leve

## 2   The General Formulation and Definitions

Let $N = 1, \ldots, n$ be the set of the end-product demands points and $M_r \subset N$ be the set of all possible facility location points at stage $r, r = 1, 2, 3$. The prices $g_i^r \geq 0$ are known for locating (opening) a facility of stage $r$ of point $i \in M_r$. For every demand point $j \in N$, the demand volumes $b_j \geq 0$ and the transportation fees $c_{ij} \geq 0$ related to the delivery of production unit from the point $i$ to $j$ are given. For satisfying the demand of $j \in N$, the necessary volume $b_j$ of the product must pass the following way: an open facility of the 3rd level $->$ an open facility of the 2nd level $->$ an open facility of level 1 $->$ the point $j$.

It is assumed that each end-product demand point and each facility point of every stage get the production only from one producer and, moreover, a facility of stage 1 gets the production from a facility of the stage 2 and this one gets the production from a facility of the stage 3. We will say below that $i \in M_r; r = 1, 2, 3$, participates in serving the demand point $j \in N$ if i is used in the path of the facilities via which $j$ gets the end-product. Note also that, at the same point, there can be the facilities of stages 1, 2 and 3 at the same time.

The goal is to choose subsets of the facility location points of each stage $I_r \subseteq M_r; r = 1, 2, 3$; and make an assignment of the chosen facilities to the demand points in such a way that the sum fees for opening all chosen facilities and transportation the production of all open facilities to the corresponding demand points would be minimized.

Let us define the general mathematical model of the three-stage location problem. It is necessary to minimize

$$\sum_{i\in M_1} g_i^1 x_i + \sum_{k\in M_2} g_k^2 y_k \sum_{l\in M_3} g_l^3 z_l + \sum_{j\in N} b_j \sum_{l\in M_3} \sum_{k\in M_2} \sum_{i\in M_1} (c_{lk}+c_{ki}+c_{ij})x_{lkij} \longrightarrow \min$$

by the variables $x_i, y_k, z_l, x_{lkij}$ under the following restrictions

$$\sum_{l\in M_3} \sum_{k\in M_2} \sum_{i\in M_1} x_{lkij} = 1, \; j \in N;$$

$$\sum_{k\in M_2} \sum_{i\in M_1} x_{lkij} \leq z_l, \; j \in N, \; l \in M_3;$$

$$\sum_{l\in M_3} \sum_{i\in M_1} x_{lkij} \leq y_k, \; j \in N, \; k \in M_2;$$

$$\sum_{l\in M_3} \sum_{k\in M_2} x_{lkij} \leq x_i, \; j \in N, \; i \in M_1;$$

$$x_i, y_k, z_l, x_{lkij} \in \{0,1\},$$

where $x_i, y_k$ and $z_l, i \in M_1, k \in M_2, l \in M_3$ are the choice variables for the facilities of stages 1, 2 and 3 respectively (if $x_i = 1$ then in the corresponding solution the facility $i \in M_1$ of stage 1 is open); $x_{lkij} \; l \in M_3, k \in M_2, i \in M_1$, and $j \in N$, are the transportation variables that define the facilities participating servicing the demand point $j$ (if $x_{lkij} = 1$ then in the corresponding solution the demand point $j$ gets the product from the facility $i \in M_1$ of stage 1 which, in its turn, gets the product from the facility $k \in M_2$ of stage 2, which, in its turn, gets the product from the facility $l \in M_3$ of stage 3).

Define the model in terms of the assignment variables. For this some additional definitions are necessary:

$\pi^r = (\pi_1^r, \ldots, \pi_n^r)$ is the assignment vector of the $r$-th stage facilities, where $\pi_j^r$ is the index of the point from $M_r$, where the facility of stage $r$ servicing the demand point $j$, $r = 1, 2, 3$; $1 \leq j \leq n$ is located (open);

$\pi = (\pi^1, \pi^2, \pi^3)$ is the triplet of the assignment vectors also referred to as the solution of the problem;

$I^r(\pi) = \cup_{j\in N}\{\pi_j^r\}$ is the set of facilities of stage $r$ which are used (open) in the solution $\pi$, $r = 1, 2, 3$.

The three-stage location problem in terms of the assignment variables can be written in conciser:

$$\sum_{i\in I^1(\pi)} g_i^1 + \sum_{k\in I^2(\pi)} g_k^2 + \sum_{l\in I^3(\pi)} g_l^3 + \sum_{j\in N} b_j(c_{\pi_j^3\pi_j^2} + c_{\pi_j^2\pi_j^1} + c_{\pi_j^1 j}) \longrightarrow \min_{\pi}$$

In the paper, we consider the three-stage location problem on a tree-like network. More specifically, we consider the problems where the transportation fee matrix $(c_{ij})$ corresponds to an acyclic network $G = (N, E)$, where $N = \{1, \ldots, n\}$ and $E = \{e_k \,|\, 1 \leq k < n\}$ are the sets of nodes and edges respectively. The nodes of the network correspond to the demand points. The transportation fee $c_{ij}$ from node i to node j for unit of the product is defined as the sum of the lengths of the edges in the path connecting these nodes. Note also that $c_{ij} = c_{ji}$ and $c_{ik} \leq c_{ij} + c_{jk}$ for every $i, j, k \in N$.

For each $j \in N$, denote by $N_j$ the set of descendants of node $j$ (the maximum subtree of the initial tree with the root vertex $j$), and by $I_j^r(\pi) = \cup\{\pi_k^r \mid k \in N_j\}$, the set of the $r$-th stage facilities servicing the clients of the set $N_j$ in the assignment $\pi$.

Introduce inductively the special notations $(\mu_j^2(\pi), \mu_j^3(\pi))$: for the root of the tree $j = 1$, let

$$(\mu_1^2(\pi), \mu_1^3(\pi)) = \arg\min\{c_{1k} + c_{kl} \mid k \in I^2(\pi), l \in I^3(\pi)\}$$

and, for the node $j$, $1 < j \leq n$, let

$$(\mu_j^2(\pi), \mu_j^3(\pi)) = \begin{cases} (\mu_i^2(\pi), \mu_i^3(\pi)), \text{where } i \text{ is the father of } j, \text{ if } c_{j\mu_i^2(\pi)} + c_{\mu_i^2(\pi)\mu_i^3(\pi)} \\ \qquad = \min\{c_{jk} + c_{kl} \mid k \in I^2(\pi), l \in I^3(\pi)\}; \\ \arg\min\{c_{jk} + c_{kl} \mid k \in I^2(\pi), l \in I^3(\pi)\}, \text{otherwise.} \end{cases}$$

Also introduce the special notations $\zeta_j(\pi)$, for the facility of stage 3 closest to node $j \in N$ used in the assignment $\pi$ (note that this facility must be open): for the root of the tree $j = 1$, let

$$\zeta_1(\pi) = \arg\min\{c_{1k} \mid k \in I^3(\pi)\}$$

and, for the node $j$, $1 < j \leq n$, let

$$\zeta_j(\pi) = \begin{cases} \zeta_i(\pi), \text{where } i \text{ is the father of } j, \text{ if } c_{j\zeta_i(\pi)} = \min\{c_{jk} \mid k \in I^3(\pi)\}; \\ \arg\min\{c_{jk} \mid k \in I^3(\pi)\}, \text{otherwise.} \end{cases}$$

Sometimes for the convenience we will omit $\pi$ in the notations of $\mu_j^r(\pi)$, $\zeta_j(\pi)$ and simply write $\mu_j^r, \zeta_j$, $r = 2, 3$ if it is clear what assignment is implied.

*Remark 1.* If node $j$ is a son of node $i$ then $\mu_j^2 \in N_i \cup \mu_i^2$

*Remark 2.* If $\mu_i^2 = \mu_j^2$, then $\mu_i^3 = \mu_j^3$. For each $i, j \in N$

*Remark 3.* If node $j$ is a son of node $i$ then $\zeta_j \in N_i \cup \zeta_i$

*Remark 4.* If node $j$ is a son of node $i$ then $\mu_j^3 \in N_i \cup \mu_i^3 \cup \zeta_i$

Introduce some additional definitions. Let some solution $\pi$ of the three-stage location problem on the network be given. For each node $j$, call the nodes $k \in N_j$ for which

$$\pi_k^1 \notin N_j \cup \{\pi_j^1\}$$

bad nodes of stage 1. By the bad nodes of stage 2 for $j$ we call the nodes $k \in N_j$ for which

$$\pi_k^2 \notin N_j \cup \{\pi_j^2\} \cup \mu_j^2$$

And by the bad nodes of stage 3 for $j$ we call the nodes $k \in N_j$ for which

$$\pi_k^3 \notin N_j \cup \{\pi_j^3\} \cup \mu_j^3 \cup \zeta_j$$

The nodes for which the opposite inclusions are true are called the good nodes of stage 1 for $j$, the good nodes of stage 2 for $j$ and the good nodes of stage 3 for $j$ respectively. Call node $k$ bad for $j$ if it is a bad node of stage 1, 2 or 3 for $j$. A node $k$ is called bad if it is bad for some vertex of the tree. A node $k$ is called good vertex for $j$ if it is a good node of stage 1, 2 and 3 for $j$. A node $k$ is called good if it is good for all vertices of the tree. Denote the number of bad nodes of stages 1, 2 and 3 for a vertex $j$ by $\nu_j^1(\pi), \nu_j^2(\pi)$ and $\nu_j^3(\pi)$ respectively. Put $\nu_j(\pi) = \nu_j^1(\pi) + \nu_j^2(\pi) + \nu_j^3(\pi)$. The parameter

$$\nu(\pi) = \sum_{j \in N} \nu_j(\pi)$$

is called the *index* of a solution $\pi$.

## 3   Main Statements

**Lemma 1.** *Given an optimal solution $\pi$, if a node $k$ is good for $j$ and $j$ is good for $i$ then $k$ is good for $i$.*

Now we prove the fundamental property of the optimal solutions of the problem. This property will be used for constructing some algorithm for searching optimal solution.

**Theorem 1.** *There is an optimal solution of the two-stage location problem on the network in which, for every node $t \in N$, Properties (I1) are satisfied:*

$$I_t^1(\pi) \subset N_t \cup \{\pi_t^1\},$$

$$I_t^2(\pi) \subset N_t \cup \{\pi_t^2\} \cup \mu_t^2(\pi).$$
$$I_t^3(\pi) \subset N_t \cup \{\pi_t^3\} \cup \mu_t^3(\pi) \cup \zeta_t(\pi).$$

*Using the definitions above, we can reformulate the statement of the theorem as follows: There is an optimal solution of the three-stage location problem on the network with zero index; i.e., there is an optimal solution $\pi$ for which $\nu(\pi) = 0$.*

## 4   Description of the Algorithm

Denote the initial three-stage location problem by $M = \langle M_1, M_2, M_3; N \rangle$. Consider the family of the subproblems

$$M_j(i, k, k', l, l', l'') = \{\langle M_1, M_2, M_3; N_j \mid \pi_j^1 = i, \ \pi_j^2 = k, \ \mu_j^2(\pi) = k', \ \pi_j^3 = l, \mu_j^3(\pi) = l', \ \zeta_j(\pi) = l'' \rangle,$$

$$i \in M_1, \ k, k' \in M_2, \ l, l', l'' \in M_3, \ 1 \le j \le n\}.$$

Let $F_j(i, k, k', l, l', l'')$ denote the optimal solution of Problem $M_j(i, k, k', l, l', l'')$ satisfying Property $(I1)$. Introduce the additional notations:

$$F_j(k, k', l, l', l'') = \min_{i \in M_1} F_j(i, k, k', l, l', l''); \quad F_j(k', l, l', l'') = \min_{k \in M_2} F_j(k, k', l, l', l'');$$

$$F_j(l, l', l'') = \min_{k' \in M_2} F_j(k', l, l', l''); \quad F_j(l', l'') = \min_{l \in M_3} F_j(l, l', l'');$$

$$F_j(l'') = \min_{l' \in M_3} F_j(l', l''); \quad F_j = \min_{l'' \in M_3} F_j(l'').$$

Note that, by Theorem 1, the optimal solution of the initial problem $M = \langle M_1, M_2, M_3; N \rangle$ (denote it by $F^*$) coincides with $F1$. Let

$$g^r_{kk'} = g^r_k + g^r_{k'}, k \neq k'; \quad g^r_{kk} = g^r_k; r = 2, 3;$$

$$g^3_{ll'l''} = g^3_l + g^3_{l'} + g^3_{l''}, l \neq l', l' \neq l'', l'' \neq l; \quad g^3_{lll'} = g^3_{ll'}; \quad g^3_{lll} = g^3_l;$$

$$G_j^{\beta_u, \ldots, \beta_v} = \min_{P \setminus \{\beta_u, \ldots, \beta_v\} \subset N_j} F_j(i, k, k', l, l', l''), \quad \{\beta_u, \ldots, \beta_v\} \subset \{i, k, k', l, l', l''\} = P;$$

$$\beta_u \neq \beta_v \text{ if } u \neq v.$$

Also introduce the set $\Gamma$, wich contains all subsets of $\{i, k, k', l, l', l''\}$ except all subsets, wich contains the element $i$, but doesn't contain the element $k$, as well as all subsets, wich contains the element $k$, but doesn't contain the element $i$. Obviously that $\Gamma$ contains 32 elements.

The recurrence for counting $F_j(i, k, k', l, l', l'')$ yields

**Theorem 2.** *For all $j \in N$, $i \in M_1$, $k, k' \in M_2$ and $l, l', l'' \in M_3$ we have*

$$F_j(i, k, k', l, l', l'') = g^1_i + g^2_{kk'} + g^3_{ll'l''} + b_j(c_{lk} + c_{ki} + c_{ij})$$

$$+ \sum_{t \in S_j} \min \left\{ G_t^{\beta_u, \ldots, \beta_v} - \sum_{\beta \in \beta_u, \ldots, \beta_v} g^{r_\beta}_\beta \mid (\beta_u, \ldots, \beta_v) \in \Gamma \right\},$$

*where*

$$r_\beta = \begin{cases} 1, & if \quad \beta = i, \\ 2, & if \quad \beta = k, k', \\ 3, & if \quad \beta = l, l', l''. \end{cases}$$

*(here and below, $S_j$ is the set of the descendants of node $j$).*

This theorem allows us to construct an exact algorithm solving initial problem.

# 5   The Algorithms

The recurrences of Theorem 2 imply Algorithm $\mathcal{A}$ for solving the problem. It consists of $h$ steps, where $h$ is the height of the initial tree $N$.

**Algorithm $\mathcal{A}$**

**Step** $s$, $1 \leq s < h$. For each node $j$, that is at distance $h - s + 1$ from the root of the tree, and for each vertices $i \in M_1, k, k' \in M_2$ and $l, l' \in M_3$ calculate $F_j(i, k, k', l, l', l'')$ and $\{G_j^{\beta_u, \dots, \beta_v} \mid (\beta_u, \dots, \beta_v) \in \Gamma\}$

**step** $h$ For each nodes $i \in M_1, k, k' \in M_2, l, l', l'' \in M_3$ we calculate $F_1(i, k, k', l, l', l'')$. After find the value $F_1$ coinciding with the optimal solution of the problem.

Using Algorithm $\mathcal{A}$, we can find the optimum value of the aim function of the initial problem. For finding the optimal permutation $\pi$ we need the inverse Algorithm $\widetilde{\mathcal{A}}$ that also follows from Theorem 2.

**Theorem 3.** *The time complexity of Algorithms $\mathcal{A}$ and $\widetilde{\mathcal{A}}$ is at most $\mathcal{O}(nm^6)$.*

# 6   Conclusion

The article considered the three-stage facility location problem on an acyclic network. The transportation fee from node i to node j for unit of the product is defined as the sum of the lengths of the edges in the path connecting these nodes.

Using the formulation the problem in terms of the assignment variables and the useful structure properties of optimal solutions, it was possible to construct a polynomial-time exact algorithm for solving the problem using the dynamic programming technique. The time complexity of Algorithm is at most $\mathcal{O}(nm^6)$. Thus, the proposed technique solves the problem in linear time.

However, for the further research it would be interesting to improve the achieved estimate of time complexity depending on the number of facilities. In addition, the intriguing question remains about the polynomial time solvability of the $k$-Stage Facility Location Problem on a tree-like networks with arbitrary number of stages.

# References

1. Ageev, A., Ye, Y., Zhan, J.: Improved combinatorial approximation algorithms for the k-Level facility location problem. SIAM J. Discrete Math. **18**(1), 207–217 (2004)
2. Bumb, A., Kern, W.: A simple dual ascent algorithm for the multilevel facility location problem. In: Goemans, M., Jansen, K., Rolim, J.D.P., Trevisan, L. (eds.) APPROX/RANDOM-2001. LNCS, vol. 2129, pp. 55–63. Springer, Heidelberg (2001). https://doi.org/10.1007/3-540-44666-4_10

3. Garey, M.R., Johnson, D.S.: Computers and Intractability. Freeman, San Francisco (1979)
4. Mirchandani, P., Francis, R. (eds.): Discrete Location Theory. Wiley, New York (1990)
5. Gimad, E.Kh.: Effective algorithms for multistage location problem on a path. Diskret. Anal. Issled. Oper. Ser. 1 **2**(4), 13–31 (1995). (in Russian)
6. Gimadi, E.Kh.: Exact algorithm for some multi-level location problems on a chain and a tree. In: Operation Research Proceedings of SOR 1996, Braunschweig, Germany, 1996, pp. 72–77. Springer, Heidelberg (1997). https://doi.org/10.1007/978-3-642-60744-8_14
7. Gimadi, E.Kh, Kurochkin, A.A.: An effective algorithm for the two-stage location problem on a tree-like network. J. Appl. Industr. Math. **7**(2), 177–186 (2013). https://doi.org/10.1134/S1990478913020063
8. Gimadi, E.Kh., Shamardin, Y.V.: On multi-level network facility location problem. In: CEUR Workshop Proceedings, 2018, vol. 2098, pp. 150–158 (2018). EID: 2-s2.0-85047982363
9. Ortiz-Astorquiza, C., Contreras, I., Laporte, G.: Multi-level facility location problems. Eur. J. Oper. Res. 1–15 (2017)
10. Zhang, J.: Approximating the two-level facility location problem via a quasi-greedy approach. In: Proceedings of 15th ACM-SIAM Symposium on Discrete Algorithms SODA, New Orleans, Louisiana, USA, 11–14 January 2004, pp. 808–817. SIAM, Philadelphia (2004)

# Process Mining

# Educational Data Mining for Prediction of Academically Risky Students Depending on Their Temperament

Marianna M. Korenkova[1,2], Elena V. Shadrina[3(✉)], and Olga E. Oshmarina[4]

[1] The Department of Literature and Cross-Cultural Communication, National Research University Higher School of Economics, Nizhniy Novgorod, Russia
[2] Center of Applied Psychological and Pedagogical Studies, Moscow State University of Psychology and Education, Moscow, Russia
[3] Department of Applied Mathematics and Informatics, National Research University Higher School of Economics Nizhny Novgorod, Nizhny Novgorod, Russia
eshadrina@hse.ru
[4] Department of Information Systems and Technologies, National Research University Higher School of Economics Nizhny Novgorod, Nizhny Novgorod, Russia

**Abstract.** The article discusses the influence of temperament on the academic performance of the first-year students at HSE-Nizhny Novgorod on the example of the Faculty of Informatics, Mathematics and Computer Science (IM&CS). The analyses were done with the help of statistics and educational data mining. The baseline data for the study is information about students, obtained by a survey: the information about temperament, degree of extraversion, stability, and other personality traits of students. The study involved students of the first and second years of the faculty of the IM&CS 2017–2018 academic year. Further, psychological factors affecting the average score and the probability of re-training for students with different temperaments were identified. A certain connection between temperament and academic success, which makes possible the prediction of "risky" students, was found. Various machine learning methods are used: the kNN-method and decision trees. The best results were shown by decision trees. As a result, first-year students are classified into three groups (Good, Medium, Bad) according to the degree of risk of getting academic debt. The practical result of the research was the recommendations to the educational office of the Faculty of IM&CS to pay attention to risky students and assist them in the educational process. After the end of the summer session, the classification results were checked. The article also presents an algorithm for finding risky students, taking temperament into account.

**Keywords:** Psychology of higher education · Academic achievements · Human temperament · Students · Data mining · Decision tree · On-line questionnaire

## 1 Introduction

Students' academic performance, successful or unsuccessful leading sometimes to expel from the university is a serious issue that should be carefully studied [1, 3, 12, 14,

W. M. P. van der Aalst et al. (Eds.): AIST 2020, CCIS 1357, pp. 277–290, 2021.
https://doi.org/10.1007/978-3-030-71214-3_23

24]. Students' performance is the main parameter, on the basis of which it is possible to evaluate how well knowledge has been transferred and understood. If there is an opportunity for early prediction of examination results, preventive measures can be taken: electives and additional course consultations may be arranged in order to lower the number of students with unsatisfactory results or academic failures and dropouts. In small groups inclinations of students can be determined by a teacher during classes due to personal contact. But if the number of students rises, it becomes more difficult to monitor those who will be likely to fail the examination.

The Faculty of Informatics, Mathematics, and Computer Science (HSE Nizhny Novgorod) grows and accepts more and more first-year students with different motivation level, natural aptitude, perseverance, and other personal features. These personal qualities and attitudes can give them an advantage or create obstacles on the way to the academic success [17].

Despite the fact that students who come enter the faculty have a high score on the Unified State Exam (school leaving tests), the percentage of those who are expelled after the first year is very high - about 30%. It is especially peculiar to the highly selective universities, where the percentage of dropouts is higher than in non-selective universities [18].

Since there is high competition among universities for students, higher education institutions have to take into account the needs of the students more than ever nowadays and do their best to meet these needs. IT faculty, for example, is often chosen by young people who tend to be introverts and are hardly outgoing or communicative (нужна ссылка). This fact requires from the teaching staff a special approach to such type of the audience. Universities that take into account the complex of students' needs (everyday, personal, academic) to a greater extent will have an undoubted competitive advantage. It is of equal importance to be able to understand who the successful students are, what they are interested in, and what their intentions and ambitions are. At the Faculty of Informatics, Mathematics, and Computer Science (HSE Nizhny Novgorod) students with a strong background in computer science have the opportunity to engage in sports (Olympiad) programming and successfully perform at international Olympiads. Consequently, some major goals for a university today are: to find efficient ways of attracting students; to hold their interest; to strike a balance between academic requirements and catering for students' needs and interests; how to retain students without reducing the quality of their studies.

Therefore, we are sure that the problem of predicting students' performance on the basis of their personal features becomes increasingly burning in the field of education.

During the last 20 years different authors conducted studies detecting factors that to a great extent affect the academic performance and students' dropouts [2, 9, 15, 23]. Sociological theories take into account the student's university and family environment, mechanisms of socialization, the influence of key people [13].

A large-scale study was conducted by Superby with co-authors at a Belgian university detecting the most influential factors [25]. A newer study was conducted by Lust and co-authors at Belgian and French universities with the help of advanced methods of data mining and machine learning [20]. Finally, a similar study was conducted at the Higher School of Economics in 2014 [7]. All these studies show that factors connected to academic behaviour such as attendance, keeping notes, confidence in selection of a university and profession, doing homework, attending electives, as well as personal history factors including parents' education, presence of both parents in the family, school grades are of the most importance. They also demonstrate that character features are also important, but not so significantly as factors specified. In his turn, Poldin with co-authors in his article explains that sociable students by means of communication develop personally, and therefore become more successful [22]. The work of Valeeva detects a relationship proving that in the course of time social isolation of students with academic failures takes place, creating additional risks of being expelled from higher education institutions for them [10].

In the works mentioned above the effect of friendship ties on student performance is noted, and the authors of the current study noticed that those who can make friends are also capable of academic work. On the other hand, the authors noticed that students who choose the IT field are more prone to isolation and it is difficult for them to open up in a new team, both with classmates and with teachers. Thus, based on personal experience and the experience of colleagues, the idea arose to study more closely the psychological aspects of the personality of students.

Methods used by the researchers are varied from Statistics to Complex Data and Process Mining. Last years the research focus shifts from static data classification and clustering [2, 11] to the control-flow discovery [5, 6]. Process discovery [4, 26], conformance checking [8], social network mining became more and more popular.

In our research we think not only about classification of students, but also about the educational process in which they are involved. We consider the dropout problem as a relevant outcome of the educational process (and use students score data for three modules) and personal characteristics that were obtained from the survey. So, the data and process mining techniques are working together for our purposes.

In this work we will base on the typology of four temperaments singled out by Jung [16]:

- The choleric temperament is characterized with intensity and power of emotional processes. Choleric people are quick-tempered, passionate and energetic;
- A sanguine individual is distinguished by comparatively weak intensity of psyche processes with a quick change of certain processes with other. Sanguine people are cheerful, hard-working, they easily cope with various tasks;
- A phlegmatic person is distinguished by slowness, sluggish movements, lack of energy. Feelings of a phlegmatic person are even and quiet. Phlegmatic people are devoted persons and it is difficult for them to switch into new activity types;
- The melancholic temperament is characterized with depth of emotional expressions, but slow flow of psyche processes. Feelings and emotions of a melancholic person are usually uniform, such people are sensitive to external circumstances and often prove themselves to be passive and sluggish.

This work is aimed at studying the influence of personal features (i.e. temperament as a personality basis) on academic performance and detection of students who will highly likely fail their examination.

Our main hypothesis is that there is a relationship between temperament and student's academic performance. The information about personal features of students received by means of questioning via Google online form was used as a material for the study. First- and second-year students of the Faculty of Informatics, Mathematics and Computer Science of academic year 2017–2018 were the study subject.

## 2   Obtaining and Processing Data

The famous scientist Boris Mirkin draws attention to the importance of classification in data analysis problems. Classification is the construction of a classification that structures the set of phenomena under consideration into a set of separate classes reflecting the important properties of these phenomena. Currently, this term is also applied to the problems of assigning individual objects to predefined classes [21].

The main task of our research is to classify students into three groups:

- Good students are "not risky" students, they will not have academic debts;
- Medium students are "with a medium likelihood of failing the exam", they can successfully pass the session, thanks to the support measures adopted at the university;
- Bad is "high risk students" fail the exam, drop out.

Thus, we need to create a dataset in which each student is described according to a set of personality characteristics.

To obtain data on the temperament of students and a number of other character traits, a survey was conducted in the internal system of the HSE-Nizhny Novgorod LMS (Learning Management System). 90 first-year and 50 second-year students of the Faculty of Informatics, Mathematics and Computer Science, NRU HSE-Nizhny Novgorod took part in it.

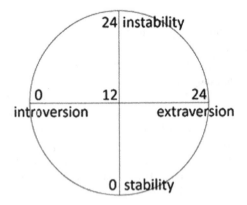

**Fig. 1.** Classification of temperaments by extraversion and rationality

The survey was compiled on the basis of a questionnaire test by G. Aysenck [16] to determine the types of temperament of each student in terms of extraversion-introversion and instability-stability. The numbers in Fig. 1 indicate the degree of introversion and extraversion from moderate 0–7 to significant 19–24. Questions about school activity, perseverance, the ability to prioritize that are all personal characteristics of the student were also included in the questionnaire. The block for identifying temperament consisted of 12 questions, each of which described one of the temperaments: choleric, sanguine, phlegmatic or melancholic. Using the students' answers, we calculated such parameters as extraversion and rationality (Fig. 1), then the temperament of each student was determined. In total, students had to answer 17 personal questions in the questionnaire. We took the data on midterm and final control (marks for examinations) after the first half of the academic year from the internal database of the NRU HSE student management system. Next, after combining all the information, we extracted 13 binary variables for each student (variables of the form 0/1). The main variable for making a decision used to validate our model is the average score across all disciplines (GPA) and pass/fail information from the overall student ranking. Based on this variable, the classification into the groups Good, Medium, Bad is made.

The distribution of temperaments in the sample is shown in Fig. 2. Melancholic (27%) and sanguine (26%) make up about one fourth of the respondents. The number of choleric people (34%) is three times higher than the number of phlegmatic people (13%). It turned out that there were more extroverts than introverts at the Faculty of Informatics, Mathematics, and Computer Science.

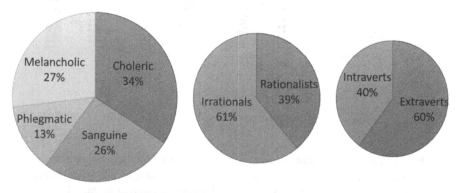

**Fig. 2.** Distribution of temperaments in the sample of students

As the number of surveyed first-year students was two times more than the number of second-year students, we decided to create a training sample of the total number of second-year students, and half of the first year, 100 people in total. The sample on which the model will be tested consists of 40 first-year students.

Students from the training sample, depending on the average score and the occurrence of retakes, were divided into categories Good, Medium and Bad, respectively, with a low, medium and high probability of not passing any of the exams (Fig. 3). The category of Good includes students of the first third of rating without retaken exams, the Medium category includes the second third of rating without retaken exams, and the Bad category includes all the rest students (see Fig. 3a, Fig. 3b).

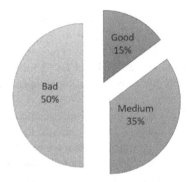

**Fig. 3.** Distribution of students by category.

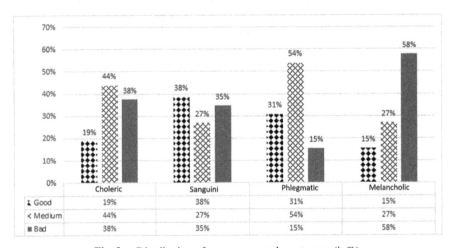

**Fig. 3a.** Distribution of temperaments by category (in%).

In Fig. 3 triangles represent students from the category Medium, squares represent students from the category Bad, both students marked by triangles and students marked by squares have to retake exams - that is why they are marked at point $-1$. Triangles have average grade more than 6. Circles are students from category Good -they do not have academic debts. All the circles are at point $+1$. Figure 3 shows that half of the students fell into the "Bad" category. Such breakdown of the data is explained by the fact that almost each second student of the interviewed had a retaken examination in a term. This is typical for the Faculty of Informatics, Mathematics and Computer Science since a number of disciplines studied at the beginning of the first and second years of education that are rather difficult. It is worth noting that the allocation to category Bad does not mean that the student will not pass three or more exams and is on the verge of expulsion. Many of the students assigned thereto have a high-grade average, but an academic debt in one of the disciplines. Since the aim of this study is to define all the students who are expected to have failures, therefore, even students with good progress and some academic failure may be categorized as Bad. From Fig. 3a it can be seen that

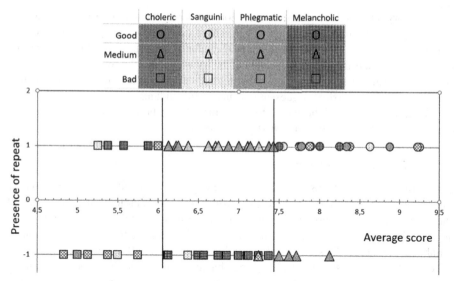

**Fig. 3b.** Distribution of temperaments by categories (Good/Medium/Bad) depending on the average score and the presence/absence of repeats

both general tendencies and differences among temperaments can be traced. The pattern of categorization in choleric and phlegmatic students is similar: the greatest number of students got into Medium category (44% of choleric, 54% of phlegmatic students). Among sanguine people, 38% are students from the Good category. For melancholic people, the largest number of students falls into category Bad.

Let us consider the factor of presence/absence of retake in more detail. Fig. 3b shows the distribution of temperaments by category (Good, Medium, Bad) depending on the average score and the presence/absence of retake. In Fig. 3b the following is clearly visible:

- straight A students (Good, marked with a circle figure) have a high average score and no retakes;
- Students category Medium (marked with a triangle figure) have a grade point average of 6.3 to 7.4 and have no retakes;
- students from category Bad (marked with a square figure) have a low average score (below 6) and/or retakes; most Bad students have retakes.

## 3    Methods

The task of dividing students into the categories of high, medium and low probability of having academic failures is a task of classification based on supervised learning [19].

At the first stage of work, we have to select these characteristics that significantly influence the average grade and probable retakes of examinations.

At the second stage of work, we were looking for the algorithm of the optimal classification of students from the training sample (with the greatest number of guessed

Bad category students). It was important for our work that the algorithm can assign each student to the necessary category (Bad/Medium/Good) based on each student's formalized characteristics. Machine learning methods were used for this such as the kNN algorithm and the decision tree [11, 25]. As we did not know beforehand which method would give the most exact result, we tested both of them and found the parameters for the maximum precision of guessing unsuccessful students.

At the third stage of the work, the best model was be determined and used for predictions for the tested group of first-year students.

## 3.1  Finding Meaningful Parameters

The correlation factor helps to understand the degree of dependence between two or among more parameters and is quite successfully used in Sociology of Education and Data Mining [16]. For subsequent analysis and interpretation of the data we represented the data obtained in a convenient form: students' answers were translated into Boolean variables: 1 was used if a student agreed with a statement and 0 if otherwise. Using the correlation factor, we selected from all the characteristics those that produce the most impact on the grade average and retaken examinations. The calculation results are shown in Table 1. For the purpose of our investigation the value of the correlation coefficient is significant if its absolute value is more than 0.2 (in the table in italics) and insignificant, if it is between 0 and 0.2.

**Table 1.** Dependence between academic performance and temperament

| Characteristic | Grade average correlation | Retaken examination correlation |
|---|---|---|
| Activeness at school | 0.06 | 0.09 |
| Perseverance | 0.19 | −0.24 |
| Setting of priorities | 0.29 | −0.18 |
| Living in a dormitory | 0.06 | −0.05 |
| Living with mates | −0.01 | 0.01 |
| Living with parents | −0.01 | 0.02 |
| Living alone | −0.04 | 0 |
| Extroversion | 0.07 | −0.18 |
| Rationality | −0.004 | −0.08 |
| *Choleric* | −0.1 | 0.044 |
| *Sanguine* | 0.05 | −0.16 |
| *Phlegmatic* | 0.13 | 0.017 |
| *Melancholic* | −0.04 | 0.2 |

As it is possible to see from the Table 1, the characteristics of "perseverance" and "setting priorities" depend significantly on academic performance, which seems obvious and explained, while temperament and the resulting psychological characteristics are less correlated with the studying success. But since we study the influence of temperaments (Choleric/Sanguine/Phlegmatic/Melancholic) on academic success, they were highlighted in italics, because their absolute value of correlation with at least one of the signs is greater than 0.1.

So, the dependence between academic performance and temperament does exist, and it is rather significant in some cases. For example, cholerics have a lower grade average (cor $= -0.1$) while perseverance of phlegmatics provides for higher marks (cor $= 0.13$). Sanguine re-take examinations rarely (cor $= -0.16$), and for melancholics the probability of retaking exams is considerable (cor $= 0.2$).

### 3.2 Comparison of Classification Using the kNN Method and the Decision Tree

The first method to test was kNN-method. Three series of calculations were carried out:

- using all characteristics with high correlation and a grade average. The best result was achieved with 8 nearest neighbors (73%);
- all characteristics without grade average. The best result was shown with 5 neighbors (60%). This is a good indication as probability of allocation to a required category is twice as higher as random guessing of category (which is 33%);
- using only grade average. The best result was obtained with 4 neighbors and was 77% of guessing. This is well-reasoned because a student with a low-grade average is more likely to be assigned to the bad students' category and, accordingly, more neighbors of such student with a low-grade average will be assigned to Bad.

It is important to mention that when applying kNN method psychological characteristics decrease the accuracy of calculations, therefore, the resulting classification is not suitable for our purposes.

The second method to test was the decision tree method:

- when all the characteristics were used for construction, the accuracy reached 74%;
- without taking into account the average score, the tree was built on the characteristics "Extraversion" and "Rationality" and the accuracy reached 62%;
- based on the average score only, the tree gave an accuracy of 76%;
- the best result was obtained when building a tree on the traits "Perseverance", "Ability to prioritize", "Extraversion", "Rationality" and type of temperament, and the accuracy reached 84%.

As it is possible to see, a different combination of parameters can give an increase in guessing accuracy compared to using all features.

The accuracy of answers using both methods fluctuated around 75%, but the decision tree showed the best result. Therefore, this method was chosen for the analysis. The final decision tree with the parameters is shown in Fig. 4. The final distribution included only choleric people, probably because their psychological state can be unstable, while other the temperaments do not significantly affect the results of studies. On the other hand, we conducted a study on two streams of students, the next stage is to analyze more data in order to confirm or deny the influence of temperament on student dropout.

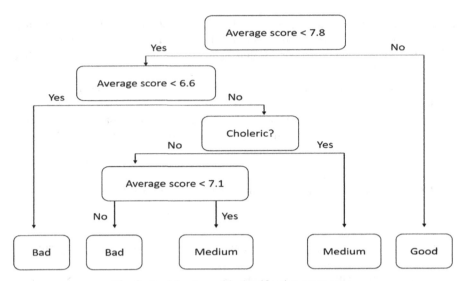

**Fig. 4.** Decision tree with classification parameters

## 4   Research Results

With the help of the decision tree generated at the previous step the students of the first-year were divided into three groups: Bad, Medium and Good. In order not to reveal personal information, the last name of each student was replaced with the symbol Student 1, Student 2, ..., Student N. We found that 22 students were in the Bad category (the category of risky students), in the Medium category - 12 students, in the Good category - 6 students. After the final session in module 4 and all the retakes in the fall, it became possible to check the research results.

Table 2 presents data on the real rating and the occurrence of student retakes at the end of the first year of study.

**Table 2.** Data summary

| Predicted category | Total students | Students with retakes | | Dropouts or Individual plans | | A high score (strictly over 7, 5) | |
|---|---|---|---|---|---|---|---|
| | | num | % | num | % | num | % |
| Bad | 22 | 14 | 64% | 6 | 27% | 0 | 0% |
| Medium | 12 | 3 | 25% | 1 | 8% | 4 | 33% |
| Good | 6 | 0 | 0% | 0 | 0% | 6 | 100% |

Twenty-two students fell into the Bad category, 14 students (64%) had retakes, according to the results of all autumn retakes 6 students (27%) remained with academic debts or were expelled. It should be noted that 4 out of 6 students still had an academic debt in one discipline either Mathematical Analysis or Linear Algebra. In such a case they were offered an Individual Learning Plan (ILP), according to which that study these disciplines again and continue to study on a commercial basis.

Twelve students fell into the Medium category, 3 students (25%) had retakes, but successfully retook the exams and were transferred to the 2nd year without academic debts. One student from the Medium category was expelled of her own free will, the reason was not any academic debts, but she decided to radically change her field of studies and career.

Among the students who fell into the Good category (6 people) there were no retakes and none of the students in this category were expelled.

Thus, we can conclude that the result of the distribution of students by categories is in good agreement with the real situation. It is also important to pay attention to the average score of students: students with a high GPA (strictly over 7.5) are more in the Good category, fewer in the Medium category and none in the Bad category.

- As a result of the study, the following recommendations were given to the educational office of the Faculty of Informatics, Mathematics and Computer Science HSE - Nizhny Novgorod:
- pay close attention to the students from category Bad (e.g. talk with students or their legal representatives);
- provide availability of additional classes for the students of categories Bad and Medium;
- engage training assistants into helping students with understanding and fulfilling home assignments;
- pay additional attention to students from the category Medium with a low average score after the first half of the academic year.

In conclusion, we will describe the algorithm for finding risky students based on temperament:

1. Conducting a survey of 1st and 2nd year students (September-October of the new academic year) in order to obtain data on psychological characteristics.
2. Converting survey results to Boolean variables (in accordance with the parameters in Table 1).
3. Formation of a training sample from 2nd year students.
4. Obtaining information about the rating and the occurrence of retakes (for students from the training sample).
5. Calculation of the correlation of psychological characteristics with the average score and the occurrence of retakes.
6. The choice of parameters that are significant for our study.
7. Classification of students from the training sample into categories Good, Medium, Bad, depending on the rating and the occurrence of retakes.
8. Building a decision tree for significant parameters in accordance with the resulting classification.
9. Formation of a test sample of the 1st year students.

10. Obtaining information about the rating and the occurrence of retakes after 1 module (November of the current academic year).
11. Classification of the 1st year students using the constructed decision tree.
12. Recommendations for the study office to pay attention to the performance of students who fall into the Bad category, as well as in the Medium category with a low GPA (no later than the beginning of December of the current academic year).

## 5   Conclusion

Studying the influence of temperament on student performance at the NRU HSE-Nizhny Novgorod at the Faculty of the IT Sciences we identified the most important psychological factors. They turned out to be "Perseverance" and "Ability to prioritize" - the presence of these factors sharply increased the average score and reduced the likelihood of retakes. It has been observed that hot-tempered choleric people have a lower GPA, while the calmness and measuredness of phlegmatic people help them study better. Sanguine people are less likely to have retakes, while the chance of a melancholic to retake exams is much higher. The more extraversion a student demonstrates, the higher his GPA.

We believe that our research might be useful to other universities for:

1. identifying academically unsuccessful students and focusing on "risky" students. One of the most important goals of the Faculty of Informatics, Mathematics, and Computer Science (HSE Nizhny Novgorod) is to train and transfer without academic debts to the 2nd as many first-year students as were accepted for the program. Otherwise, the resources of the state (in the case of government-subsidized education) or a student personal funds (in the case of paid education) spent on education will not be used rationally.
2. forming an individual educational trajectory. At HSE-Nizhny Novgorod today, there are flexible opportunities for switching from one educational program to another using Individual Learning Plan (ILP). We assume that forming studying groups, taking into account students personal characteristics will increase the performance of each student.

We understand that there are certain limitations to our study. In this paper, a small pool of baseline data is presented (140 student responses). And the Faculty of Informatics, Mathematics, and Computer Science (HSE Nizhny Novgorod) is not large enough to talk about "real" Big Data and use all the opportunities of machine learning methods. In the future, we plan to use the longitude data for results verification over several years. It would also be interesting to try other data mining methods to predict academic failures and dropout.

**Acknowledgements.** We would like to thank the education office specialists from HSE University in Nizhny Novgorod for providing data for conducting the research.

# References

1. Anuwatvisit, S., Tungkasthan, A., Premchaiswadi, W.: Bottleneck mining and Petri net simulation in education situations. In: ICT and Knowledge Engineering (ICT & Knowledge Engineering), 10th International Conference, Bangkok, Thailand, pp. 244–251. IEEE (2012)
2. Arulselvan, A., Boginski, V., Mendoza, P., Pardalos, P.: Predicting the nexus between postsecondary education affordability and student success: an application of network-based approaches. In: ASONAM 2009 International Conference on Advances in Social Networks Analysis and Mining, pp. 149–154 (2009)
3. Bannert, M., Reimann, P., Sonnenberg, C.: Process mining techniques for analysing patterns and strategies in students' self-regulated learning. Metacognition Learn. **9**(2), 161–185 (2013). https://doi.org/10.1007/s11409-013-9107-6
4. Barreiros, B.V., Lama, M., Mucientes, M., Vidal, J.C.: Softlearn: a process mining platform for the discovery of learning paths. In: 14th International Conference on Advanced Learning Technologies, pp. 373–375. IEEE, Athens, Greece (2014)
5. Bogarin, A., Cereso, R., Romero, C.: A survey on educational process mining. WIREs Data Min. Knowl. Disc. **9**(1) (2018). https://doi.org/10.1002/widm.1230
6. Bogarín, A., Romero, C., Cerezo, R., Sánchez-Santillán, M.: Clustering for improving educational process mining. In: Proceedings of the Fourth International Conference on Learning Analytics and Knowledge, pp. 11–15. ACM, Indianapolis (2014)
7. Bulycheva, P., Oshmarina, O., Shadrina, E.: Identifying academically "unsuccessful" first-year students: a case study of Higher School of Economics – Nizhny Novgorod. In: Vestnik of Lobachevsky State University of Nizhny Novgorod. Series: Social Sciences, vol. 2, no. 42, pp. 136–143 (2016)
8. Cairns, A.H., Gueni, B., Assu, J., Joubert, C., Khelifa, N.: Analyzing and improving educational process models using process mining techniques. In: The Fifth International Conference on Advances in Information Mining and Management, pp. 17–22, Brussels, Belgium (2015)
9. Campbell, J.P.: Utilizing student data within the course management system to determine undergraduate student academic success: An exploratory study. Purdue University (2007)
10. Dokuka, S.V., Valeeva, D.R., Yudkevich, M.M.: How academic failures break up friendship ties: social networks and retakes. Educ. Stud. Moscow **1**, 8–21 (2017)
11. Friedl, M.A., Brodley, C.E.: Decision tree classification of land cover from remotely sensed data. Remote Sens. Environ. **61**(3), 399–409, (1997)
12. Gorbunova, E.: Vybytiya studentov iz vuzov: Issledovaniya v Rossii i SShA [Elaboration of Research on Student Withdrawal from Universities in Russia and the United States]. Voprosy obrazovaniya / Educ. Stud. Moscow **1**, 110–131 (2018)
13. Gorbunova, E.V., Gruzdev, I.A., Frumin, I.D.: Studencheskii otsev v rossiiskikh vuzakh: K postanovke problemy [Student Dropout in Russian Higher Education Institutions: The Problem Statement]. Voprosy obrazovaniya [Educ. Stud.] **2**, 67–81 (2013)
14. Howard, L., Johnson, J., Neitzel, C.: Examining learner control in a structured inquiry cycle using process mining. In: Proceedings of the 3th International Conference on Educational Data Mining, pp. 71–80, Pittsburgh, USA (2010)
15. Ignatov, D., Mamedova, S., Romashkin, N., Shamshurin, I.: What can closed sets of students and their marks say? In: Pechenizkiy, M., et al. (ed.) Proceedings of 4th International Conference on Educational Data Mining, EDM-2011, TU/e Eindhoven, pp. 223–228 (2011)
16. Izenk, G., Vilson, G.: Kak izmerit lichnost [How to measure a personality]. M.: Kogito tsentr, pp. 156–159 (2000)
17. Keyek-Franssen, D.: Praktiki uspeshnosti studentov: Ot ochnogo obucheniya k masshtabnomu i obratno [Practices for Student Success: From Face-to-Face to At-Scale and Back]. Voprosy obrazovaniya / Educ. Stud. Moscow **4**, 116–138 (2018)

18. Kochergina, E., Prakhov, I.: Vzaimosvyaz' mezhdu otnosheniem k risku, uspevaemost'yu studentov i veroyatnost'yu otchisleniya iz vuza [Relationships between Risk Attitude, Academic Performance, and the Likelihood of Drop-outs]. /Voprosy obrazovaniya [Educational Studies] Moscow, vol. 4, pp. 206–228 (2016)

19. Luan, J.: Data mining and its applications in higher education. New Dir. Inst. Res. **113**, 17–36 (2002)

20. Lust, T., Meskens, N., Ahues, M.: Predicting academic success in Belgium and France Comparison and integration of variables related to student behavior. arXiv preprint arXiv:1408.4955 (2014)

21. Mirkin, B.G.: Vvedeniye v analiz dannyh: uchebnik I praktikum dlya bakalavriata I magistratury [Introduction to data analyses: theory and practice for bachelor and master courses], M.: Izdatelstvo Urite, p. 174 (2014)

22. Poldin, O., Valeeva, D., Yudkevich, M.: How social ties affect peer group effects: Case of university students. SSRN Electron. J. (2013). https://doi.org/10.2139/ssrn.2207666

23. Pal, S.: Mining educational data using classification to decrease dropout rate of students. arXiv preprint arXiv:1206.3078 (2012)

24. Southavilay, V., Yacef, K., Callvo, R.A.: Process mining to support students' collaborative writing. In: Proceedings of the 3th International Conference on Educational Data Mining, pp. 257–266, Pittsburgh, USA (2010)

25. Superby, J.-F., Vandamme, J., Meskens, N.: Determination of factors influencing the achievement of the first-year university students using data mining methods. In: Workshop on Educational Data Mining, vol. 32, p. 234 (2006)

26. Vidal, J.C., Lama, M., Bugarín, A.: Petri net-based engine for adaptive learning. Expert Syst. Appl. **39**, 12799–12813 (2012)

# Posters

# The Information-Analytical Bot Detection System Based on the Assembly of Classifiers

Vladimir N. Kuzmin$^{(\boxtimes)}$ ⓘ, Artem B. Menisov ⓘ,
and Ivan A. Shastun ⓘ

Space Military Academia Named By A.F.Mozhaysky, Zhdanovskay Street, 13,
Saint-Petersburg 197198, Russia
vka@mil.ru

**Abstract.** Currently, the use of bots, disguised as ordinary users of social networks and guidance with special programs, has serious consequences. For example, bots were used to influence political elections, distort information on the Internet, and manipulate stock prices on the stock exchange. The detection of bots in social networks is carried out by many research teams, the areas of research of which include the use of machine learning methods. However, the practical results of detecting bots on social networks indicate significant limitations, since the methodological tools used have language limitations and ineffective criteria for determining bots.

The report provides a description of the information-analytical system (client-server application) that allows for collection and analysis data of social networks in order to identify bots. The application is based on a bot detection module based on the assembly of classifiers of social network accounts, the capabilities of which allowed to minimize the risk of bot detection errors. The practical utilizing of the application allows increasing the operatively and effectiveness of detecting bots in comparison with other approaches.

**Keywords:** Bot detection · Social networks · Machine learning · Ensemble of models · Association of classifiers

## 1 Summary

The development of new approaches to improving the security of state organizations and users of information web-systems is a constant and urgent task.

The information-analytical system for detecting bots on social networks based on a special association of classifiers is a client–server application that allows to:

1) provide a decentralized management system, access to resources from all devices;
2) evaluate informational occasions;
3) identify bots among subscribers and in individual communities of social networks;
4) interactively display on maps and graphs the results, generate reporting and information documents;
5) provide API to third-party services.

© Springer Nature Switzerland AG 2021
W. M. P. van der Aalst et al. (Eds.): AIST 2020, CCIS 1357, pp. 293–296, 2021.
https://doi.org/10.1007/978-3-030-71214-3

The application is designed to improve the existing decision support system for identifying bots and developing organizational and technical measures to neutralize the consequences of using bots.

The main consumers are state and commercial organizations, as well as well-known personalities, against which negative information can be disseminated.

The connection to the server is established using the HTTPS protocol using encryption in order to increase security.

The application can be installed on any computer on the local network with any operating environment, including the Astra Linux Special Edition.

Due to the large amount of data transferred, it is necessary to have a high-speed connection between the server and the client.

To perform remote administration of the application via the web interface, the Apache webserver is required. The architecture of the application is shown in see Fig. 1.

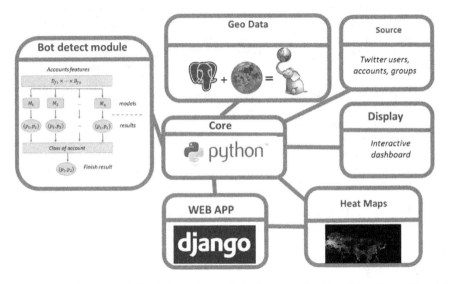

**Fig. 1** The architecture of the information-analytical system for identifying bots of social networks

An element of the scientific novelty of the developed approach for identifying bots of social networks is the recommended combination of the following features: thematic relationship of accounts, activity, anonymity, and data inconsistency. A feature of this approach is to take into account the growing tendency to use one set of bots to solve different informational goals.

The developed approach to detecting bots on the Twitter social network based on a special association of classifiers has an advantage in terms of performance over modern machine-learning algorithms and can reduce the error of detecting bots. Since the activity of social network bots includes categorical attributes, adaptation to the source, data is necessary to use ensembles of models.

However, despite the advantages of machine learning, one of the main drawbacks of the developed approach maybe its impracticality when there are too many unique records, for example, because string representations of categorical features display typos or combinations of several data in the same records.

The practical application of the application allows for increasing the efficiency and effectiveness of detecting bots in comparison with other approaches.

As a further development of research, we can distinguish:

– study of the issues of collecting additional data about social network accounts;
– analysis of the effect of data imbalance on model training;
– study of the possibilities of increasing the productivity of detecting bots of social networks.

# References

1. Williamson, W., Scrofani, J.: Trends in detection and characterization of propaganda bots. In: 52nd Hawaii International Conference on System Sciences, Maui, pp. 7118–7123 (2019)
2. Lukyanov, R.V.: Methodology for monitoring the state of information security of automated systems in the context of heterogeneous mass incidents. Transactions of the Military Space Academy named by A.F.Mozhaysky, 660, pp. 111–115 (2018)
3. As many as 48 million Twitter accounts aren't people, says study. https://www.cnbc.com/2017/03/10/nearly-48-million-twitter-accounts-could-be-bots-says-study.html. Accessed 11 Apr 2020
4. Massive networks of fake accounts found on Twitter. https://www.bbc.co.uk/news/technology-38724082. Accessed 11 Apr 2020
5. Terdima, D.: Here's how facebook uses AI to detect many kinds of bad content. Fast Company. https://www.fastcompany.com/40566786/heres-how-facebook-uses-ai-to-detect-many-kinds-of-bad-content. Accessed 11 Apr 2020
6. Fighting disinformation online. RAND. https://www.rand.org/research/projects/truth-decay/fighting-disinformation.html. Accessed 11 Apr 2020
7. Bacciu, A., La, M.M., Nemmi, E., Neri, V., Mei, A., Stefa, J.: Bot and gender detection of Twitter accounts using distortion and LSA. In: PAN at CLEF 2019 (2019)
8. Gamallo, P., Almatarneh, S.: Naïve-bayesian classification for bot detection in Twitter. In: at CLEF 2019 (2019)
9. Vogel, I., Jiang, P.: Bot and gender identification in Twitter using word and character N-grams. In: PAN at CLEF 2019 (2019)
10. Mahmood, A., Srinivasan, P.: Twitter bots and gender detection using tf-idf. In: PAN at CLEF 2019 (2019)
11. Farber, M., Qurdina, A., Ahmedi, L.: Identifying Twitter bots using a convolutional neural network. In: PAN at CLEF 2019 (2019)
12. Lundberg, J., Nordqvist, J., Laitinen, M.: Towards a language independent bot detection. In: DHN 2019, Copenhagen, pp. 308–319 (2019)
13. Sahoo, S.R., Gupta, B.B.: Hybrid approach for detection of malicious profiles in Twitter. Comput. Electr. Eng. **76**, 65–81 (2019)

14. Novotny, J.: Twitter Bot Detection & Categorization - A Comparative Study of Machine Learning Methods. Lund University, Lund (2019)
15. Davoudi, A., Klein, A.Z., Sarker, A., Gonzalez-Hernandez, A.: Towards automatic bot detection in Twitter for health-related tasks. Working paper ArXiv: 1909.13184 (2019)
16. Mazza, M., Cresci, S., Avvenuti, M.,Quattrociocchi, W., Tesconi, M.: RTbust: exploiting temporal patterns for dotnet detection on Twitter. In: 10th ACM Conference on Web Science, Amsterdam, pp. 183–192 (2018)
17. Beskow, D.M., Carley, K.M.: Its all in a name: detecting and labeling bots by their name. Computational and Mathematical Organization Theory 25(1), 24–35 (2018). https://doi.org/ 10.1007/s10588-018-09290-1
18. Varol, O., Ferrara, E., Davis, C.A., Menczer, F., Flammini, A.: Online humanbot interactions: detection, estimation, and characterization. In: Eleventh International AAAI Conference on Web and Social Media. https://arxiv.org/pdf/1703.03107.pdf. Accessed 16 May 2020
19. Minnich, A., Chavoshi, N., Koutra, D., Mueen, A.: BotWalk: efficient adaptive exploration of Twitter bot networks. In: IEEE/ACM International Conference on Advances in Social Networks Analysis and Mining 2017, pp. 467–474 (2017)
20. Chavoshi, N., Hamooni, H., Mueen, A.: DeBot: Twitter bot detection via warped correlation. In: ICDM, pp. 817–822 (2016)
21. Ferrara, E., Varol, O., Davis, C., Menczer, F., Flammini, A.: The rise of social bots. Commun. ACM 59(7), 96–104 (2016)
22. Mazza, M., Cresci, S., Avvenuti, M., Quattrociocchi, W., Tesconi, M.: Rtbust: Exploiting temporal patterns for botnet detection on twitter. https://arxiv.org/pdf/1902.04506.pdf. Accessed 19 May 2020
23. Twitter API Documentation. https://www.developer.twitter.com/docs. Accessed 21 May 2020
24. Vorontsov, K.V.: Mathematical methods of teaching by procedures (theory of machine learning). https://www.machinelearning.ru. Accessed 21 May 2020
25. Zhang, W., Du, T., Wang, J.: Deep learning over multi-field categorical data. In: European Conference on Information Retrieval. ArXiv, abs/1601.02376. Italia, Padua (2016)
26. Menisov, A.B., Shastun, I.A., Kapitsyn, S.U.: An approach to the identification of malicious Internet sites based on the processing of lexical signs of addresses (URLs) and an average ensemble of models. Inf. Technol. 25(11), 691–697 (2019)
27. Vorontsov, K.V.: Lectures on methods for evaluating and selecting models. https://www. machinelearning.ru. Accessed 21 May 2020
28. Gorodetsky, V.I., Serebryakov, S.V.: Collective recognition methods and algorithms: a review. Trans. SPIIRAS 1(3), 139–171 (2006)
29. Niyogi, P., Pierrot, J.-B., Siohan, O.: Multiple classifiers by constrained minimization. In: International Conference on Acoustics, Speech, and Signal Processing, pp. 3462–3465. Istanbul, Turkey (2000)
30. Prodromidis, A., Chan, P., Stolfo, S.: Meta-learning in distributed data mining systems: Issues and approaches. Advances in Distributed Data Mining 3, 81–114 (1999)
31. Gnidko, K.O., Makarov, S.A., Sergeev, A.S.: A model of an intellectual decision support system in order to identify the negative informational and psychological impact on students of educational organizations of the Ministry of Defense of Russia and to protect against it. Transactions of the Military Space Academy named by A.F.Mozhaysky 666, 142–147 (2019)
32. Kachura, Ya.O., Saprykin, D.I., Faleev, P.A.: Modeling of the military-political activity of states by the methods of associative analysis in decision support systems. Transactions of the Military Space Academy named by A.F.Mozhaysky 660, 19–29 (2018)

# Machine Learning Methods for Demographic Data Analysis

Anna Muratova$^{(\boxtimes)}$ , Dmitry Ignatov , and Ekaterina Mitrofanova

National Research University Higher School of Economics, Moscow, Russia
{amuratova,dignatov,emitrofanova}@hse.ru

**Abstract.** This is the extended abstract of a case study on demographic sequences analysis by machine learning and data mining methods.

**Keywords:** Data mining · Demographics · Neural network · Decision trees · Classification · Interpretation

## 1 Summary

The analysis of demographic data is very important, it helps to understand how people behavior changes depending on gender, generation and other factors. The data for the work was obtained from the scientific laboratory of socio-demographic policy at the HSE University. It contains results of a survey of 6626 respondents (3314 men and 3312 women). In the database, the dates of first significant events in respondents' lives are indicated, such as partner, marriage, break up, divorce, education, work, separation from parents and birth of a child. Also, for each person indicated different features: type of education (general, higher, professional), location (city, town, country), religion (religious or not), frequency of church attendance (once a week, several times in a week, minimum once a month, several times in a year or never), generation (1930–1969 or 1970–1986) and gender (male or female).

In this work we considered two main demographer's tasks. The first one is prediction of the next event in person's life, based on the previous events in their life and features. The second task is to find the dependece of events for the gender feature, does the behaviour of men differ from the women behavior in terms of events and other features, like education type, location and other. Let us call this task gender prediction.

The main goal of this work is to compare different methods, both interpretable and not-interpretable. Interpretable methods are good for demographers for the further interpretaion and working with results. Non-interpretable methods allow us to find how accurate the prediction is and to find the best method for these types of data.

Among interpretable methods we considered decision trees with different event encodings: binary (1 if event happened and 0 if not), pairwise, time encoding (with the age of when event happened in months) and different combinations

W. M. P. van der Aalst et al. (Eds.): AIST 2020, CCIS 1357, pp. 297–299, 2021.
https://doi.org/10.1007/978-3-030-71214-3

of these encodings together. The best accuracy result for the task of next event prediction is 0.88 with binary encoding and for the task of gender prediction is 0.69 with all of the encodings together (binary, pairwise and time).

Among non-interpretable methods we considered SVM with custom kernels and neural networks. The formulas for similarity measures without discontinuities (or gaps) were derived and used as custom kernels: ACS (all common sequences), LCS (longest common sequence) and CP (common prefix). The best accuracy for the next event prediction is 0.91 with similarity measure ACS and for the task of gender prediction the best accuracy is 0.68 with similarity measure CP.

Among neural networks we considered Simple RNN and Convolutional Neural network. The best accuracy result for the next event prediction, as well as for the gender prediction was obtained with Convolutional Neural network. The accuracy is 0.94 for the next event prediction and 0.78 for the gender prediction.

Among all of the methods, the best method in terms of accuracy is Convolutional Neural network. Decision tree method is good for interpretation of demographers, which is also important. So, there are different best methods for different purposes.

**Acknowledgments.** The study was implemented in the framework of the Basic Research Program at the HSE University and funded by the Russian Academic Excellence Project '5-100'. This research is supported by the Faculty of Social Sciences, National Research University Higher School of Economics.

# References

1. Egho, E., Raïssi, C., Calders, T., Jay, N., Napoli, A.: On measuring similarity for sequences of itemsets. Data Min. Knowl. Discov. **29**(3), 732–764 (2014). https://doi.org/10.1007/s10618-014-0362-1
2. Elzinga, C.H., Rahmann, S., Wang, H.: Algorithms for subsequence combinatorics. Theor. Comput. Sci. **409**(3), 394–404 (2008)
3. Elzinga, C.H., Liefbroer, A.C.: De-standardization of family-life trajectories of young adults. A cross-national comparison using sequence analysis. Eur. J. Popul. **23**(3), 225–250 (2007)
4. Gizdatullin, D., Baixeries, J., Ignatov, D.I., Mitrofanova, E., Muratova, A., Espy, T.H.: Learning interpretable prefix-based patterns from demographic sequences. In: Strijov, V.V., Ignatov, D.I., Vorontsov, K.V. (eds.) IDP 2016. CCIS, vol. 794, pp. 74–91. Springer, Cham (2019). https://doi.org/10.1007/978-3-030-35400-8_6
5. Gizdatullin, D., Ignatov, D., Mitrofanova, E., Muratova, A.: Classification of demographic sequences based on pattern structures and emerging patterns. In: 14th International Conference on Formal Concept Analysis, Supplementary proceedings, ICFCA, Rennes, France (2017)
6. Ignatov, D.I., Mitrofanova, E., Muratova, A., Gizdatullin, D.: Pattern mining and machine learning for demographic sequences. In: Klinov, P., Mouromtsev, D. (eds.) KESW 2015. CCIS, vol. 518, pp. 225–239. Springer, Cham (2015). https://doi.org/10.1007/978-3-319-24543-0_17

7. Muratova, A., Sushko, P., Espy, T.: Black-box classification techniques for demographic sequences: from customised SVM to RNN. In: Proceedings of the Fourth Workshop on Experimental Economics and Machine Learning, EEML 2017, vol. 1968, pp. 31–40. Aachen : CEUR Workshop Proceedings, Dresden (2017)

8. Lodhi, H., Saunders, C., Shawe-Taylor, J., Cristianini, N., Watkins, C.: Text classification using string kernels. J. Mach. Learn. Res. **2**, 419–444 (2002)

9. Cule, B., Feremans, L., Goethals, B.: Efficiently mining cohesion-based patterns and rules in event sequences. Data Min. Knowl. Discov. **33**(4), 1125–1182 (2019). https://doi.org/10.1007/s10618-019-00628-0

10. Blockeel, H., Fürnkranz, J., Prskawetz, A., Billari, F.C.: Detecting temporal change in event sequences: an application to demographic data. In: De Raedt, Luc, Siebes, Arno (eds.) PKDD 2001. LNCS (LNAI), vol. 2168, pp. 29–41. Springer, Heidelberg (2001). https://doi.org/10.1007/3-540-44794-6_3

11. Lee, V.E., Jin, R., Agrawal, G.: Frequent Pattern Mining in Data Streams. In: Aggarwal, C.C., Han, J. (eds.) Frequent Pattern Mining, pp. 199–224. Springer, Cham (2014). https://doi.org/10.1007/978-3-319-07821-2_9

# International Trade of Wood: Stability Issues Within Network Paradigm

Andrey Varkentin[✉]🆔

Moscow State University, Moscow, Russia
andrei.varkentin@yandex.ru

**Abstract.** Network approach in Economics makes it possible to model economic relationships not only taking into account their scale and dynamics, but also with regarding the underlying topology. Thus, the main aim of this research is to define the characteristics of the wood trade network.

To achieve these goals criteria of average shortest path, density, number of strongly connected components and reciprocity has been evaluated on the international trade data of raw, processed wood and finished wooden products.

The results incline that a)the network becomes more complex and dense during the analized period, b)the network resistance towards crisises increases with the growth of goods technological complexity, c) the more sophisticated products are traded, the less fragmented network becomes.

**Keywords:** International trade · Networks contingence · Wood exports · Technological sophistication

## 1 Introduction

Network approach in Economics makes it possible to model economic relationships not only taking into account their scale and dynamics, but also with regarding the underlying topology of a graph. Thus, the main aim of this research is to clarify the characteristics of the wood trade network and identify communities of countries, involved in trade relationships.

At first, we are going to present some papers on the similar topic, then describe the data and applied methodology, and finish with discussing the results.

## 2 Theoretical Overview

There are several works, which incorporate network analysis into their methodology. For example, Daisuke Fujii [4] models international trade with a domestic interfirm production network, which enables him to study indirect exporters in Japan and the effect of trade liberalizations on firm size distributions or industry dynamics.

© Springer Nature Switzerland AG 2021
W. M. P. van der Aalst et al. (Eds.): AIST 2020, CCIS 1357, pp. 300–302, 2021.
https://doi.org/10.1007/978-3-030-71214-3

Furthermore, one should outline "Networks, crowds, and markets: reasoning about a highly connected world" [3] - a comprehensive review of graph theory with real-life applications, including market analysis.

Finally, an article by Du et al. [2] can be mentioned. Researchers identified main players on the oil market and stated, that major oil importers could influence other countries, as well as major exporters.

# 3   Data and Methodology

Data were inspired by The Observatory of Economic Complexityof Cesar Hidalgo [1], while that project was based upon data from the BACI International Trade Database [5]. The period covered by data is 2003-2017, the export scales were taken for 4401-4421 HS6 codes on a country level, which were aggregated into 3 groups: 4401-4405 - for raw wood, 4406 – 4413 – for processed wood and 4414-4421 – for wooden goods.

This data was used to build a directed graphs of international trade of wood. Thus, for each year in our period and for each group a separate graph was built and the following criteria were evaluated:

- Average shortest path – the higher number of hops need to be made, the less trade partners have an average country.
- Density – the degree to which number of edges is close to possible maximum. The higher density, the more complex is network.
- Number of strongly connected components – graph is strongly connected if any node is reachable from any other node. The higher the number strong components, the higher amount communities are likely to appear, the more fragmented the network is.
- Reciprocity – the likelihood that nodes in a directed graph are mutually linked. The higher reciprocity, the fairer and more competitive are economic relationships.

For determining the number of communities was used asynchronous label propagation algorithm, described by Usha Nandini et al. [6] and realized in Networkx Python library.

Our *hypotheses* can be formulated as following:

H1. As time went on, trade developed and became more active.
H2. The higher goods complexity, the more stable the trade is.

# 4   Results and Conclusions

This work has analyzed three segments with increasing technological complexity in the international market of wood. The results are supposed to be visualized. Their dynamics inclines that:

1. As time goes by, the network becomes more complex, developed and dense.
2. The higher technological complexity of exporting goods, the wider is the portfolio of partners, and the less trade is affected by a crisis.
3. The network of more technological advanced goods is less fragmented.

Thus, we can state that two hypotheses, set in the beginning of this work, has been proven.

# References

1. Simoes, A.J.G., Hidalgo, C.A.: The economic complexity observatory: an analytical tool for understanding the dynamics of economic development. In: Workshops at the Twenty-Fifth AAAI Conference on Artificial Intelligence (2011)
2. Du, R., et al.: A complex network perspective on interrelations and evolution features of international oil trade, 2002–2013. Appl. Energy **196**, 142–151 (2017)
3. Easley, D., Kleinberg, J.: Networks, Crowds, and Markets, vol. 8. Cambridge University Press, Cambridge (2010)
4. Fujii, D.: International Trade and Domestic Production Networks. No. 17116 (2017)
5. Gaulier, G., Zignago, S.: BACI: international trade database at the product-level. In: The 1994–2007 Version CEPII Working Paper, N°2010-23, October 2010
6. Raghavan, U.N., Albert, R., Kumara, S.: Near linear time algorithm to detect community structures in large-scale networks. Phys. Rev. E **76**(3), 036106 (2007)

# Virus Model Evaluation Framework

Nikolay Veld[(✉)]

Higher School of Economics, Moscow, Russia
`novel8mail@gmail.com`

**Abstract.** During COVID-19 outbreak more interest in forecasting models has appeared. However, one of the popular sources of selecting the best models relies on specific leaderboards. The leaderboard maintainers utilize their own test configurations but it is unhelpful when they are trying to fit a model to not listed configuration or a model cannot be shown to anyone (if the leaderboard evaluates the model). In this case, a downloadable evaluation framework with various configuration parameters is needed. This work presents such a framework and the link to its repository.

**Keywords:** Evaluation framework · Forecasting model · Viruses · Epidemics

## 1 Background

Nowadays many scientific and business areas with time series occurrences can be researched in terms of forecasting algorithms. One of such areas is epidemics. Only for epidemics many different forecasting models have been discovered. [1] Some people want to compare all existent prediction methods in research or utilization purposes. A popular way to do it is competitions' leader boards checking [3] or specialized hubs [6]. However, it is quite efficient brief acquaintance, just go through an online resource. But it can be insufficient for testing particular countries/regions (maybe even not considered by the popular data), metrics and especially only just implemented model versus other models. Such resources are mostly inflexible and consider only selected by a organizer or developer entities.

Thus the moment of evaluation frameworks comes. The example of the implemented framework can be seen here [5]. The described frameworks provides many features, many errors measures and ranking methods on paper. But in fact, no information about actual application has been provided, even no link to it has been also presented, only a promise to publish the software. Thus the article can be viewed as a guide for advanced evaluation tool implementing.

## 2 Framework Description

On the contrary to viewed works, this paper focuses on actual application implementation. Let us describe the framework in the way that a user walks through the application.

First of all, the input needs to be prepared according to the following rules:

© Springer Nature Switzerland AG 2021
W. M. P. van der Aalst et al. (Eds.): AIST 2020, CCIS 1357, pp. 303–306, 2021.
https://doi.org/10.1007/978-3-030-71214-3

1. An input files are CSV-files (a comma is a separator) prediction of model for one or several horizons ("the length of time into the future for which forecasts are to be prepared" [2]).
2. The columns of the prediction have such names in the same order: Region, Country, TrialDate, Confirmed_X, Death_X, Recovered_X, etc. The last three can be repeated many times with different values of $X$. Here, $X$ is a horizon value.
3. If the algorithm does not produce some type of prediction (Recovered, for example) but produces for at least one type, all the columns must be defined. Prediction's absence would be marked with empty string.
4. The date format should be standard (for example, 1/22/20), time part is not considered but can be included.
5. The index column would be omitted in the file.

Uncovering input it is crucial to describe the usage options and user interface. The application is designed in the way that allow to call it as a CLI application (`python3 model_evaluation.py args`) and as executable Python-module (`import model_evaluation; model_evaluation.main(args_list)`) for incorporating to another program.

The following arguments are supported:

- `--predictions-list` (`-p`): prediction file paths list (limited support of unix patterns);
- `--predictions-filter` (`-f`): helps to select one model from the family, type in format `family_starting_tag:selected_model`;
- `--output-filename` (`-o`): output file name without extension, it is used for saving reporting results;
- `--metrics-list` (`-m`): metric names list (support sklearn metrics);
- `--whole-country` (`-c`): set this flag in order to sum data over regions and set country as minimal location entity;
- `--horizons-list` (`-n`): selection of specific horizons from predictions;
- `--date-selector` (`-d`): anchor date(s) for forecasting plots.

Some additional details: some metrics are implemented but the user can implement own metrics and easily use them using the same interface; prediction and output file path can be defined by short way to special directory inside project or by absolute/relative path.

In turn, output is more complex than just a plot/table. Three types output has been extracted: metrics values, forecast plot and comparison plot. Further, the details about each type can be found. Metrics values are saved not only in CSV format but also in TeX format with value rounding for fast pasting into a TeX-based paper. Two plots are built: a forecast plot allows to compare actual data with the model forecast for different future time by plotting prediction for different horizons for model trained on one piece of data. Using `--date-selector` argument, the particular date(s) can be chosen as a end of training series. In contrary, a comparison plot demonstrates how to change model

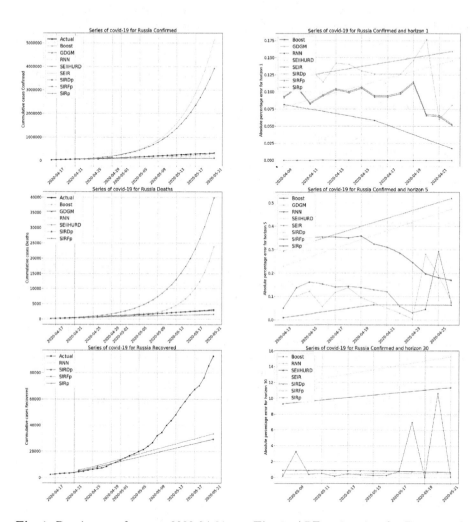

**Fig. 1.** Russia cases forecast, 2020-04-21 is a end of training series

**Fig. 2.** APE estimation for Russia and 1, 5, 30 horizons

prediction over training end date shifting for a fixed horizon. Additionally, the plots of selected metrics between estimation and actual datum are created.

Since the framework is posed as an open source project, the code has been written according to common practices. The code is cleaned using "pylint" advises and tested using pytest framework with self-developed tests. The docstrings have been provided for key elements. The repository is open for new contributions. Its "Issue" section is open for discussions.

The developed framework is placed on [4]. The example prediction files can be found in the "`prediction`" directory. "`run.py`" is an example of python-module-like run with the defined arguments. Finally, the computed output files are placed in the "`output`" directory. Example forecast and comparison plots can be observed on Fig. 1 and Fig. 2 accordingly.

**Acknowledgements.** Thanks to Alexander Chernyavskiy, Anton Chernyavskiy, Sergey Bazhmin, Nikita Aleksandrov, Maxim Shalankin and Anastasia Borneva for providing valuable feedback and prediction files that has become a base for the framework output.

# References

1. Chretien, J.P., George, D., Shaman, J., Chitale, R.A., McKenzie, F.E.: Influenza forecasting in human populations: a scoping review. PLOS ONE **9**(4), 1–8 (2014). https://doi.org/10.1371/journal.pone.0094130
2. Eurostat: Glossary:forecast horizon - statistics explained (2014). https://ec.europa.eu/eurostat/statistics-explained/index.php/Glossary:Forecast_horizon
3. Kaggle: Covid19 global forecasting (week 5) leaderboard (2020). https://www.kaggle.com/c/covid19-global-forecasting-week-5/leaderboard
4. NickVeld (2014). https://github.com/NickVeld/model_evaluation
5. Tabataba, F., et al.: A framework for evaluating epidemic forecasts. BMC Infect. Dis. **17**, (2017). https://doi.org/10.1186/s12879-017-2365-1
6. at UMass-Amherst, T.R.L.: Viz - covid-19 forecast hub (2020). https://viz.covid19forecasthub.org/

# Author Index

Printed in the United States
by Baker & Taylor Publisher Services